Women in Engineering and Science

Series Editor
Jill S. Tietjen, Greenwood Village, CO, USA

The Springer Women in Engineering and Science series highlights women's accomplishments in these critical fields. The foundational volume in the series provides a broad overview of women's multi-faceted contributions to engineering over the last century. Each subsequent volume is dedicated to illuminating women's research and achievements in key, targeted areas of contemporary engineering and science endeavors. The goal for the series is to raise awareness of the pivotal work women are undertaking in areas of keen importance to our global community.

More information about this series at http://link.springer.com/series/15424

Peggy Layne • Jill S. Tietjen

Editors

Women in Infrastructure

 Springer

Editors
Peggy Layne
Virginia Tech (retired)
Blacksburg, VA, USA

Jill S. Tietjen
Technically Speaking Inc
Greenwood Village, CO, USA

ISSN 2509-6427   ISSN 2509-6435   (electronic)
Women in Engineering and Science
ISBN 978-3-030-92823-0   ISBN 978-3-030-92821-6   (eBook)
https://doi.org/10.1007/978-3-030-92821-6

This Springer imprint is published by the registered company Springer Nature Switzerland AG
The registered company address is: Gewerbestrasse 11, 6330 Cham, Switzerland

Foreword I

Infrastructure is critical to the public health, safety, and welfare and is a critical index of a nation's economic vitality. Infrastructure supports nearly every aspect of life comprising more than the roads and bridges that are traditionally thought of as "infrastructure." Infrastructure encompasses not only transportation but water resources, water treatment, power, transmission and distribution, and most of what we as people desire and, in most cases, need to live and have a quality of life. To build a quality of life takes multiple engineering disciplines employed by governments, engineering companies, construction companies, financial institutions, universities, and consulting engineers advising those entities. I have witnessed firsthand the desire of these entities in not only educating our politicians but the public as to the importance of infrastructure to maintain a quality of life as we know it today. From becoming the first woman president of the American Society of Civil Engineers (ASCE) in 2004 to a member of the National Science Board from 2006 to 2012, to then becoming a member of the board of directors of both "America's Infrastructure Company®"– Granite Construction, Inc., and one of the world's largest engineering companies advancing the quality of life across the globe – Stantec, to also having served as President and CEO of my own companies and now Chairman of Pegasus Global Holdings, Inc., I recognize the time and effort it takes to maintain core values, and the integrity in our messaging, our actions, and our work we undertake as engineers and constructors to make this world a better place to live. As engineers, it is imperative that we strive to ensure the world's and our nation's infrastructure is repaired, replaced, and built with the public interest as our top priority.

To that end, ASCE is a leader in assessing our nation's infrastructure. In 1988, the U.S. Government issued the first "report card" giving our nation's infrastructure a "C" grade, barely adequate to support the then current demands. When the Federal Government indicated it would no longer be updating *"Fragile Foundations: A Report on America's Public Works,"* ASCE took over that report card in 1998. A decade later, however, our nation suffered an average grade of "D." ASCE has continued to release a report card every four years taking into consideration all the changing elements that affect America's infrastructure. It provides the necessary information and guidance to our nation's leaders and policy makers as to not only the importance of replacing, repairing, and building our nation's infrastructure, but what needs attention in both federal and state jurisdictions.

When I became President of ASCE in 2004, the 2003 Report Card Progress Report had only bettered the nation's average grade to a "D+" with an investment need of $1.6 trillion. Today, ASCE's 2021 Report Card reveals that the nation has made progress in restoring our nation's infrastructure as for the first time in 20 years, the average grade has raised to a C+. However, there is still a lot of work to be done as 11 of the 17 categories assessed still remain a D grade. Significant challenges remain including the COVID-19 pandemic's impacts on infrastructure revenue streams that have threatened the progress made to date. However, the long-term investment gap needed to improve our infrastructure from a C+ to a grade of B continues to grow. ASCE's 2021 estimate has gone from $2.1 trillion over 10 years to nearly $2.59 trillion over 10 years. As noted in the 2021 ASCE Report Card, by 2039, a continued underestimate in our infrastructure at current rates will cost $10 trillion in GDP, more than 3 million jobs and $2.4 trillion in exports over the next 20 years. By 2039, America's overdue infrastructure bill will cost the American household $3,300 a year. At the time of this writing, the United States Congress is discussing various bi-partisan bills that would finally infuse dollars into our nation's infrastructure and reduce the future economic burden. But to do so will take action from our policy makers. Thus, we as engineers have an obligation to step up and educate our policy makers and the public as to why not acting now will jeopardize not only our nation's economic viability but our quality of life as we know it today.

In 2006, I was honored to write the foreword to Sybil Hatch's book on women engineers entitled *Changing Our World: True Stories of Women Engineers*, which told the inspirational stories of great women engineers, past and present, who truly changed our world. Now, I am equally honored to prepare the foreword to this book which is authored by women leaders from all industry sectors and engineering disciplines that speak to the importance of infrastructure and continue to change our world. These inspirational women leaders as well as future women leaders are and will continue to be role models for young women as they consider what career they may aspire to one day. As engineers, we pledge to protect the public health, safety, and welfare – the very cornerstone of what our infrastructure does and what the public takes for granted will not fail and will improve our quality of life.

In an era where the environment, sustainability, and governance (ESG) has become a world focus and our infrastructure continues to crumble, we need more young women to enter the engineering profession and women who are in the profession to stay there. Given that ESG and infrastructure have taken center stage, young women can aspire to have a career where they can contribute to truly making a difference to making the world a better place to live. The women engineers who are already making a difference to our world and nation's infrastructure are the heroes in helping to clean up our environment, ensuring that sustainability has a place in all our decision-making, and serving as leaders governing the actions of governments, companies, and our education regarding infrastructure.

I hope you will enjoy reading how these inspiring women continue to make a difference and I hope you will share their messages with young women, educators, companies, policy makers, and your fellow neighbors as to why we must urgently prioritize infrastructure so as to ensure the quality of life not only for the population of today but for the future generations of tomorrow.

Dr. Patricia D. Galloway, P.E., Chartered Engineer, F. ASCE,
Past ASCE President
Chairman
Pegasus Global Holdings, Inc

Foreword II

Every day civil engineers are tasked with overcoming obstacles and finding solutions to complex problems to provide a better quality of life for all citizens. The health and well-being of the nation is in our hands. For example, when the U.S. Centers for Disease Control and Prevention stressed the importance of washing hands thoroughly during the COVID-19 pandemic, civil engineers needed to ensure reliable drinking water systems would get clean water to every American's doorstep. When major storms roll through a region, civil engineers are responsible for ensuring systems are in place to manage stormwater and keep the electric grid running. Our health, quality of life, and economic well-being are directly tied to the transportation network that facilitates passage of goods and services throughout the country and to the world.

A rapidly changing climate and constantly evolving variables – such as emerging contaminants in water systems and the transition to electric, connected, and autonomous vehicles – add to the challenges engineers face. Fortunately, we are taught to keep moving forward until we have found an answer. The opportunity to work in a challenging and evolving field that has a direct impact on my community is what brought me to civil engineering.

Structural concerns and external threats to the built environment are typically the most thought-of issues faced by engineers, but we also have internal challenges to address, particularly regarding the disparity between men and women in the workforce. According to the U.S. Bureau of Labor Statistics, in 2020, women only represented 16% of all civil engineers. On average, we continue to earn lower incomes than our male colleagues. Whether driven by a lack of awareness of the issue or a lack of confidence joining a male-dominated field and negotiating a salary, the female representation and wage disparity is a pressing issue our community must solve. While we have made modest progress considering only 9.7% of the industry was represented by women in 2010, we need to continue encouraging women to confidently step into this rewarding field to add balance to the workforce and diversify and improve solutions.

Throughout my career I have made it a habit to routinely ask questions and reach out to other engineers to further my understanding of practices in the trade. Getting out of my comfort zone, I moved from Las Vegas to Washington, D.C. in 2009 to serve as a Science and Technology Policy Fellow through the American Society of Civil Engineers (ASCE) and the American Association for the Advancement of Science. Through my fellowship and getting involved in ASCE's Infrastructure Report Card, which serves as an important advocacy tool highlighting infrastructure needs nationwide, I came to understand that engineers need to do more than solve technical problems. Engineers need a seat at the table during critical discussions and decisions to ensure needs are met. Further, the engineers at the table need to represent diverse backgrounds to ensure the best decisions for our communities are advanced.

The stories of extraordinary female accomplishments in this book clearly demonstrate that, when given an opportunity, we have the ability to revolutionize any field, even those that have been led predominantly by men for generations. Serving as the 2018 president of ASCE was truly an honor. But perhaps what was most encouraging was being the second in a line of three consecutive female presidents, a first for the Society. When my time as president concluded, I became chair of ASCE's Committee on America's Infrastructure, which is responsible for creating the nationally recognized Infrastructure Report Card. The 2021 Report Card for America's Infrastructure, which assessed our nation's infrastructure a cumulative grade of C-minus, served as a rallying cry for legislative action and a key reference point for mainstream media outlets, which use the report to shape the national dialogue behind America's infrastructure crisis. Less than a month after the Report Card was released, President Joe Biden unveiled the American Jobs Plan, calling for once-in-a-generation infrastructure investment. Featured prominently in the text of the plan: findings from the ASCE Report Card. We had our seat at the table.

However, overrepresentation in leadership roles can provide a false sense of having solved a problem when perhaps that same representation is not felt throughout the industry at other levels. In my role as Director of the Nevada Department of Transportation, I have the privilege of working with remarkable people who dispel the common stereotypes of traditional engineers but there is still room to create space for more diverse perspectives. We must use our problem-solving abilities throughout each of our organizations to determine how we can retain more of our female students and engineers. Nearly half of civil engineering students are women and women tend to be overrepresented in leadership at the student level and younger member groups, but somewhere along the way we lose many of them to other paths.

To the women reading the stories in this book, just remember – anything is possible. Think bigger, think bolder, and together we will solve any problems we set our minds to. Engineering is done better when women have their voices heard. Now, we just need more of them.

Kristina Swallow, P.E., F.ASCE
Director, Nevada Department of Transportation

Preface

The Past Presidents' Reception at the 2016 Society of Women Engineers' National Conference was in full swing when Peggy Layne suggested to Shelley Wolff and Jill Tietjen that the three of us should be co-editors of a volume in the Springer Women in Engineering and Science series on Infrastructure. As we batted the idea around, we realized that the framework already existed – the American Society of Civil Engineers' (ASCE) Infrastructure Report Card. And so it began.

The country's infrastructure has been underfunded for years but nevertheless underpins the economy and lifestyle of our nation. Peggy and Jill embarked on the hunt for authors and Shelley assisted – although she did not want to take on the responsibility of co-volume editor.

As an environmental engineer and ASCE member, Peggy had served on various society committees and knew several women leaders in ASCE who have served as national president and as members of the committee that prepares the Report Card on America's Infrastructure. Initial contacts led to introductions to other women working in many different roles across the infrastructure spectrum. Jill's many connections opened doors in other engineering disciplines that contribute to infrastructure in different ways. Many emails and phone calls later, a list of possible authors began to emerge.

As engineers, we each practice in our area of expertise – but also within our industry. Infrastructure spans so many different areas and engineering disciplines. Our cross-cutting volume brings insight into each corner of our country's backbone. And not a moment too soon.

Blacksburg, VA, USA   Peggy Layne
Greenwood Village, CO, USA   Jill S. Tietjen

Acknowledgments

No one writes a book alone. Many, many people provide support, guidance, direction, and input. We apologize in advance for any omissions in this list, however, unintended.

To Shelley Wolff, like us, a Society of Women Engineers' Past President, who helped us find authors and provided assistance whenever we called upon her.

To the American Society of Civil Engineers (ASCE) communications and government relations staff who helped with background on the Report Card project and contacts with potential contributors.

To Pat Galloway, Kristina Swallow, and Maria Lehman, national leaders in ASCE for their support, encouragement, and contributions to this volume.

To Mary James, our editor at Springer, who continues her amazing support for this book and the entire Springer Women in Engineering and Science series.

To Jill's husband, David Tietjen, who always supports her many endeavors.

Co-Volume Editor Biography

Peggy Layne, P.E., decided to study environmental engineering after attending a nature study summer camp where her love of the natural world combined with an awareness of the damage caused by human activity. In the 1970s, environmental issues were in the news, and, in response, Congress was passing laws that required cleanup of air, water, and solid and hazardous waste disposal practices. An engineering approach to pollution control appealed to her scientific mindset, and jobs were plentiful. After completing a bachelor's degree from Vanderbilt University and a master's degree in environmental and water resources engineering from the University of North Carolina School of Public Health, Peggy worked for several consulting firms on the design of water and wastewater treatment and solid and hazardous waste management systems for industries and municipalities. Clients ranged from local municipal water systems to Fortune 500 companies to the U.S. Department of Defense. Environmental cleanup work was then and is still driven by government regulations and requirements, so keeping up with the latest public policy is important. This interest in public policy led Peggy to spend a year as an American Association for the Advancement of Science (AAAS) Science and Technology Policy Fellow in the office of Senator Bob Graham (D-FL), where she was responsible for water, wastewater, and hazardous waste policy issues, sponsored by the American Society of Civil Engineers. While working as a consulting engineer, Peggy became involved in the Society of Women Engineers to encourage more young women to pursue engineering and advocate for women in the profession. After her year working in Congress, she made a career change to work full time as an advocate for broadening participation in engineering. She spent two years as director of the program on diversity in the engineering workforce at the National Academy of Engineering and a year as a diversity consultant for the American Association of Engineering Societies before joining Virginia Tech, where she served as Assistant Provost for Faculty Development and director of *AdvanceVT*, a National Science Foundation sponsored program to increase the number and success of women faculty in science and engineering.Now retired, Peggy currently serves on the boards of the League of Women Voters of Montgomery County VA, the Blacksburg Master Chorale, the Village at Tom's Creek Homeowners' Association, and the Canadian Institute for Women in Science and Engineering. She is a member of the American Society of Civil Engineers, where she has served on the diversity, communications, and policy committees, and the Society of Women Engineers, where she has served in numerous local, regional, and national committee and board leadership positions, including national president in 1996–97.Peggy is the editor of *Women in Engineering: Pioneers and Trailblazers* and *Women in Engineering: Professional Life* published by ASCE Press in 2009. She is a regular contributor to the *Society of Women Engineers Magazine*, and has given numerous presentations on engineering and public policy and diversity at national and international conferences.

Contents

List of Figures

List of Tables

Chapter 1
Introduction

Jill S. Tietjen and Peggy Layne

Abstract Infrastructure is defined as the physical and organizational structures needed for the functioning of a society or enterprise. Most of the public just takes infrastructure for granted – until they don't have it. It isn't just weather and cyberattacks that focus our attention on infrastructure and help us understand why it is important to our quality of life and standard of living. The lack of funds to maintain, upgrade and expand US infrastructure costs businesses – which in turn affects the cost of products. The American Society of Civil Engineers has been grading the infrastructure in the U.S. for a number of years now. The bad news is that the 2021 grades were not the kind of grades we would have wanted to bring home from school when we were kids to have to show our parents. As you read this volume, we hope that you will begin to understand not only the important role that infrastructure plays in our lives but why spending money to build, maintain, upgrade, and expand infrastructure is in our country's best interest. We also hope that you will come to appreciate the many and varied contributions of women engineers to infrastructure in the US.

Keywords Infrastructure · American Society of Civil Engineers · Infrastructure Report Card · Infrastructure funding gap

1.1 What Is Infrastructure?

Infrastructure is defined as the physical and organizational structures needed for the functioning of a society or enterprise [1]. Think roads, bridges, buildings, pipelines, drinking water, and so forth. Most of the public just takes infrastructure for granted – until they don't have it. In 2021, infrastructure has been at the forefront of

J. S. Tietjen (✉)
Technically Speaking Inc, Greenwood Village, CO, USA

P. Layne
Virginia Tech (Retired), Blacksburg, VA, USA

© The Author(s), under exclusive license to Springer Nature Switzerland AG 2022
P. Layne, J. S. Tietjen (eds.), *Women in Infrastructure*, Women in Engineering and Science, https://doi.org/10.1007/978-3-030-92821-6_1

discussion in the USA because of Presidential support and Congressional action as well as because of the Texas power crisis in February 2021, the Colonial Pipeline Cyber Attack in May 2021, and Hurricane Ida and others during the 2021 hurricane season.

1.2 Why Infrastructure Matters

It isn't just weather and cyberattacks that focus our attention on infrastructure and help us understand why it is important to our quality of life and standard of living. A *Business Week* article titled "Expect Delays: One factory's logistics show the cost to businesses of strained U.S. infrastructure" describes how lack of funds to maintain, upgrade, and expand US infrastructure costs businesses – which in turn affects the cost of products [2].

The Volvo manufacturing facility profiled in the article makes wheel loaders, soil compactors, and other industrial vehicles in its manufacturing facility in Shippensburg, Pennsylvania. Infrastructure issues including bad traffic on Interstate 81 and congestion at East Coast ports mean the factory can no longer rely on just-in-time deliveries and instead needs to stockpile many of its parts. It takes 1574 parts to assemble a wheel loader from 226 suppliers in Germany, Sweden, China, Arizona, Iowa, Georgia, and elsewhere. Infrastructure deficiencies cause the plant to be shut down on multiple occasions each year with the workforce idled – costing 5% to 10% of productivity. In the end, the consumer pays for the infrastructure deficiencies [2].

1.3 The American Society of Civil Engineers' Infrastructure Report Card

The American Society of Civil Engineers (ASCE) decided to start grading America's infrastructure, providing a cost estimate of what it would take to remedy the poor grades and showing the level of the gap in projected funding (see more information in Chap. 2). When we started working on recruiting the chapter authors for this volume, the 2017 Report Card had been issued. As we are nearing completion of the volume, the 2021 Report Card has now been issued. Although there is some good news, the grades from both of those Report Cards (Table 1.1) are not the kind of grades we would have wanted to bring home from school when we were kids to have to show our parents. We as a country can – and have to – do better.

Table 1.1 2017 and 2021 ASCE Infrastructure Report Card Grades [3, 4]

Category	2017 Grade	2021 Grade
Aviation	D	D+
Bridges	C+	C
Broadband (introduced into 2021)	–	(Not graded)
Dams	D	D
Drinking water	D	C–
Energy	D+	C–
Hazardous waste	D+	D+
Inland waterways	D	D+
Levees	D	D
Parks & recreation (renamed public parks in 2021)	D+	D+
Ports	C+	B–
Rail	B	B
Roads	D	D
Schools	D+	D+
Solid waste	C+	C+
Stormwater (added in 2021)	–	D
Transit	D–	D–
Wastewater	D+	D+
Cumulative grade	D+	C–
Funding gap in total 10 year needs (billion $)	$2064	$2590

Note: Grades are for the entire USA

1.4 The Organization of this Book

Using the framework of the ASCE Infrastructure Report card, we organized like infrastructure categories together. Under "Moving People and Things," you will find chapters on aviation, roads, rail, and transit. Under "Making Connections," you will find chapters on bridges, inland waterways, seaports, and tunnels. Under "Controlling Water," you will find chapters on dams and levees. Under "Cleaning Up," you will find chapters on hazardous waste and water and wastewater. And, under "Improving the Quality of Life," you will find chapters on energy and public parks.

As you read, we hope that you will no longer take our infrastructure for granted and that you will understand not only the important role that it plays in our lives but why spending money to build, maintain, upgrade, and expand infrastructure is in our country's best interest. We also hope that you will come to appreciate the many and varied contributions of women engineers to infrastructure in the USA.

References

1. Oxford Language Dictionary, Google Dictionary, accessed September 18, 2021.
2. Hurley, Amanda Kolson and David Rocks, editors. "Expect Delays: One factory's logistics show the cost to businesses of strained U.S. infrastructure, *Business Week*, August 9, 2021, pp. 34–37.
3. American Society of Civil Engineers, 2017 Infrastructure Report Card: A Comprehensive Assessment of America's Infrastructure, https://www.infrastructurereportcard.org/wp-content/uploads/2017/04/2017-IRC-Executive-Summary-FINAL-FINAL.pdf, accessed September 18, 2021.
4. American Society of Civil Engineers, America's infrastruture scores a C-, https://infrastructurereportcard.org/, accessed September 18, 2021.

Chapter 2
The American Society of Civil Engineers' Report Card on America's Infrastructure

Maria Lehman

Abstract Civil engineering has been a part of human life since the beginning of human history, whether it be shelter, drinking water, the development of the wheel and sailing for transportation, or using the sun to heat and dry various food products. Founded in 1852, the American Society of Civil Engineers (ASCE) is the nation's oldest engineering society. ASCE stands at the forefront of a profession that plans, designs, constructs, and operates society's economic and social engine – the built environment – while protecting and restoring the natural environment. Since 1998, the ASCE has issued a Report Card grading America's Infrastructure. The 2021 Report Card was a public relations success. The overall infrastructure grade was a C-. Investment needs for the next 10 years total $5.94 trillion with a $2.59 trillion funding gap.

Keywords Infrastructure Report Card · ASCE · Civil engineering · Engineering public policy · Engineering legislative engagement · Infrastructure grades · Failure to act · Infrastructure investment

In order to provide context for the American Society of Civil Engineers (ASCE) Report Card for America's Infrastructure, it is important to start with the history of civil engineering and ASCE and then add the layer of my entry into the profession and my personal journey.

2.1 History of Civil Engineering

Civil engineering has been a part of human life since the beginning of human history, whether it be shelter, drinking water, the development of the wheel and sailing for transportation, or using the sun to heat and dry various food products. There are

M. Lehman (✉)
GHD, Buffalo, NY, USA

examples of what is now considered spectacular civil engineering back in Greek and Roman times. Consider the Parthenon, or the Roman road network and aqueducts. We are still amazed by the pyramids of Egypt. And when I recall my trip to Machu Picchu, as I turned the corner to see the entire site open up in view, it was a sacred experience. It was a place that even though you were outside, you spoke in hushed tones as the experience engulfed you. And then as you walk through the site, you are in utter amazement at the advanced engineering principles that were used, long before we understood the mathematics to make it safe and repeatable.

These are examples of the beginning of what we still call "civil engineering."

Prior to the eighteenth century, engineering was primarily done by the military. As we evolved and needed more human built environment, it became important to have more people involved in engineering, and thus the term "civil" engineering was coined. Thus, all current engineering disciplines were specialties that became their own disciplines as time progressed such as mechanical engineering, electrical engineering, industrial engineering, and so forth. That explains why to this current day, civil engineering is the most broad discipline, encompassing the entire project lifecycle and acting as the facilitator of all specialty disciplines as the orchestra conductor of the system of systems.

The National School of Bridges and Highways, France, was opened in 1747 as the first civil engineering school in the world. John Smeaton, who constructed the Eddystone Lighthouse, proclaimed himself as the first civil engineer. The first organization was formed in 1771, when Smeaton and some of his colleagues formed the Smeatonian Society of Civil Engineers, which was considered more or less a social club [1–3].

The Institution of Civil Engineers was founded in London, in 1818, and is considered to be the world's first engineering society. Thomas Telford became its first president in 1820 [4]. In 1828, the institution received a Royal Charter, formally recognizing civil engineering as a profession. Its charter defined civil engineering as [5]:

> The art of directing the great sources of power in nature for the use and convenience of man, as the means of production and of traffic in states, both for external and internal trade, as applied in the construction of roads, bridges, aqueducts, canals, river navigation and docks for internal intercourse and exchange, and in the construction of ports, harbours, moles, breakwaters and lighthouses, and in the art of navigation by artificial power for the purposes of commerce, and in the construction and application of machinery, and in the drainage of cities and towns.

Concurrently, the United States' first private college to teach civil engineering, Norwich University, was founded in 1819 by Captain Alden Partridge. Rensselaer Polytechnic Institute awarded the first degree in US civil engineering in 1835. Elizabeth Bragg became the first woman to receive a civil engineering degree in the United States when she graduated from the University of California Berkeley in 1876 [1, 6].

2.2   ASCE History and Role and Profession Definition

ASCE's history is articulated on its website as follows [7]:

On Nov. 5, 1852, a dozen eminent civil engineers gathered at the Croton Aqueduct in New York City in the office of Chief Engineer Alfred W. Craven to establish the American Society of Civil Engineers and Architects. In 1868, a few years after architects had formed a professional society of their own, ASCE adopted its current name. For the first 144 years of its existence, ASCE maintained its headquarters in New York City, relocating six times to progressively larger facilities. In 1996, ASCE moved to its current global headquarters in Reston, Virginia, just outside Washington, D.C.

Further ASCE defines the profession of civil engineering and the role of civil engineers as:

Civil engineers design, build, and maintain the foundation for our modern society – our roads and bridges, drinking water and energy systems, sea ports and airports, and the infrastructure for a cleaner environment, to name just a few.

Civil engineering touches us throughout our day. Think of a civil engineer when you:

- Turn on your tap to take a shower or drink clean water
- Flick on your lights and open your refrigerator
- Drive to work on roads and bridges through synchronized traffic lights
- Take mass transit or take a flight for a vacation
- Toss your empty coffee cup in the recycling bin

The ASCE represents more than 150,000 members of the civil engineering profession in 177 countries. Founded in 1852, ASCE is the nation's oldest engineering society.

ASCE stands at the forefront of a profession that plans, designs, constructs, and operates society's economic and social engine – the built environment – while protecting and restoring the natural environment.

Through the expertise of its active membership, ASCE is a leading provider of technical and professional conferences and continuing education, the world's largest publisher of civil engineering content, and an authoritative source for codes and standards that protect the public.

The Society advances civil engineering technical specialties through nine dynamic Institutes and leads with its many professional- and public-focused programs.

2.3   Personal Journey to Become a Civil Engineer

I am a first-generation American born to Polish parents who were World War II refugees. After several years in refugee camps in England, my parents decided that coming to the United States was for the future benefit of my brother and sister. They, along with my older sister Anne and older brother George, were sponsored to emigrate to the United States by my dad's cousin Leon Nowakowski, who was a doctor in Buffalo, New York. So through a long application process and saving everything they could, they bought passage on the Queen Mary and took the difficult journey. Upon arrival in New York, they were processed at Ellis Island and traveled by train to Buffalo, New York, with $50 in their pocket. They had to learn the language and find a way to earn a living and raise a family.

My mother came from a wealthy family in Poland and was a graduate of a Swiss finishing school. She was a buyer for her father's department stores, so she would regularly travel to Paris, Rome, London, and Milan, among other large European cities. While she did not know how to cook or maintain a household, she did know how to drive a motor car, and her favorite cars were a Bugatti and a Pierce-Arrow.

My dad was born in a very small and poor farming village in western Poland and went to seminary in order to get a high school degree. He then went on to get a Juris Doctor at the University of Poznan. He went on to build his own empire as a corporate attorney and a judge.

They were married on December 26, 1938. They had been married for 8 months, when the storm clouds of war were overhead and my father left Poland on foot, prior to the German invasion. My mother stayed with her dad until my grandfather was able to get her a visa to go to Malta as she was distraught with the disappearance of her husband and the war. They had a plan and met in Cyprus. Overnight they were penniless.

That experience forced their life view that there are only two things important in life: your family and what's in your head. So education was the primary driver in our family. And from the time I was a little girl, I was always encouraged to lead, as I had the grit to stand up for what was right. I remember my dad saying that some people are built to lead and stand up for others; it is their responsibility to do that for others that don't have the strength to fight. And so, for me, this started a long series of questioning the establishment, systems, and processing and pondering why can't we do better?

I always excelled at math and science and went to a very blue-collar high school where only approximately 10% of the graduates went on to college. The vast majority of those who attended college became engineers, accountants, and math majors. The Math Department head Mr. Thomas LaPena was a fabulous teacher. We learned calculus through osmosis. He always told us that while he loved math, he was more interested that we would live by his class motto: "Cogitation and Tenacity." I am proud to say that I believe I have lived that motto and that frequently when facing a tough challenge, I can hear Mr. LaPena in my head.

My future husband Carl was attending the University of Buffalo for civil engineering when we started dating. And a career choice was born! My parents were very supportive but really didn't understand what would make me want to pursue civil engineering as a career. When I graduated, my dad asked me "Now that you have proven you can do what your brother couldn't, now what?". My response was "Be the best civil engineer I know how to be." And so it has been.

I joined ASCE as a Younger Member and did local committee work and eventually became a local Section Officer. I was pushed by my first true engineering mentor Harry Quinn to see ASCE not only as a way to give back to the profession but also as a way to grow personally and professionally. Harry taught me to "follow the money" if I wanted to lead projects. He was a tremendous mentor and he did so effortlessly. He became like a second dad, encouraging me all along the way. And I did learn to follow the money.

I rose through the ranks in the local Section of ASCE and became Section President. Then I got involved with the New York State Council related to government affairs. We researched various topics related to the profession and advocated for change. We passed resolutions with our opinions on various topics. We also got some legislation passed. One of the things I worked on at that point was to get legislation passed to establish a dedicated highway fund in New York State. At that time, NY was one of only three states that did not have a dedicated fund. Then State Senator Tony Masiello worked with us and sponsored legislation which eventually passed. Tony left the New York Senate to become the Mayor of Buffalo, and we got to work together when I became Commissioner of Public Works for Erie County in 2000. We have been friends ever since.

That experience made me realize how little the general public and elected officials really knew about how the built environment works, how people interact with it, and how we must be stewards of our environment. So in 1993 I ran to serve on the Board of Directors for ASCE representing then District 1: New York, New Jersey, Eastern Canada, and Puerto Rico.

One of the National Committees I was assigned was Public Communications. I remember thinking this is a profession unto itself and asked the Director at the time, Jim Quigley, to teach me Public Relations 101. For 8 years, I served on Public Communications and learned all I could about the subject. In 1996, we moved ASCE headquarters from New York City to Reston, VA, to be closer to the Capitol and to be able to better influence legislation related to the built environment. Jim stayed in New York, and ASCE hired Jane Howell as the Public Communications Director.

2.3.1   Public Relations 101

According to the Public Relations Society of America, the modern definition of public relations is: "Public relations is a strategic communication process that builds mutually beneficial relationships between organizations and their publics." It goes on to say that "At its core, public relations is about influencing, engaging and building a relationship with key stakeholders across numerous platforms in order to shape and frame the public perception of an organization."

The PR process circle starts at a baseline. You need to build awareness in a broad group of people. Then you need people to understand that they need to require action by others and ultimately advocate for your cause. The circle then links up when conditions change and you need to start the process all over again.

With that definition and with what I learned from Jane, I changed my focus. Jane taught me that there is a circle of communications in public relations and that trying to change the opinions of public servants requires having their constituency be educated consumers. The average American did not understand what civil engineering was, why they should even care about it, and even worse why would anyone want to be a "nerd"?

The profession was largely pale and male and something had to change. We thought that first we needed to establish programs for community outreach to help educate people about how critical the profession is to their daily lives and standard of living and then to consider this a great profession for all regardless of race or sex.

In 1997, I became the chair of the National Committee on Public Communications, and we used focus groups across the country to test messages and pathways. Once we were comfortable that we had "the secret sauce," we presented the plan to the ASCE Board, and it was approved. That plan helped hone the message for pre-college outreach in the 1990s, as well as producing age-appropriate hands-on activities that excite kids from kindergarten through high school about the possibility of a career in civil engineering. A formal ASCE pre-college outreach program was born.

2.3.2 ASCE National Involvement and Legislative Engagement

So it is 1993 and I am a 33-year-old woman director on a National Board. I couldn't believe that I was there. More importantly, I was convinced that we needed to make a difference in the profession. I was the third woman to serve as a director on the Board. Barbara Fox was the first and Pat Galloway was the second. Pat would eventually become ASCE's first woman President.

As a director, I was asked to serve on two other national committees: Publications and Environmental Policy. At that time, the public policy committees were very internally focused. The individual policy areas had committees that, working in concert with ASCE technical and geographic units, wrote policies that would identify the position ASCE had on many issues ranging from using specific processes and materials to funding projects to pension portability. The positions need to be refreshed on a 3-year cycle. But only senior staff and presidential officers did "Hill visits" to meet with members of Congress and advocate for the policy positions.

So in 1996, when I was chair of the Environmental Policy Committee, I asked for a 2-hour window on the agenda for a field trip activity. I told my staff contact, Martin Hight, that I would need a couple cabs and I was exercising my chair's privilege. What Martin didn't know is that my congressman, Jack Quinn, who was a friend from his time teaching in the Orchard Park Middle School, was on the House Transportation and Infrastructure Committee and was the chair of the Rail Subcommittee. I had arranged a meeting with Congressman Quinn to have him talk about the importance of technical constituents informing members of Congress on issues. So the committee took the elevator down to the ground floor of the Washington offices and took a couple of cabs to Capitol Hill. It was a great experience and we had fun. And, even better, it was the first unofficial ASCE member fly-in.

2.4   ASCE Report Card for America's Infrastructure

During that same time, there was talk of ASCE leading the charge on a Report Card for America's Infrastructure.

From ASCE's website, the history is documented [8]:

> The concept of a report card to grade the nation's infrastructure originated in 1988 with the congressionally chartered National Council on Public Works Improvement report, *Fragile Foundations: A Report on America's Public Works*. A decade later, when the federal government indicated they would not be updating the report, ASCE used the approach and methodology to publish its first Report Card on America's Infrastructure in 1998. With each new report – in 2001, 2005, 2009, 2013, 2017, and now 2021 – the methodology of the Report Card has been rigorously assessed so as to take into consideration all of the changing elements that affect America's infrastructure.
>
> In 1988, when *Fragile Foundations* was released, the nation's infrastructure earned a "C," representing an average grade based on the performance and capacity of existing public works. Among the problems identified within *Fragile Foundations* were increasing congestion and deferred maintenance and age of the system; the authors of the report worried that fiscal investment was inadequate to meet the current operations costs and future demands on the system. In each of ASCE's seven Report Cards, the Society found that these same problems persist. Our nation's infrastructure is aging, underperforming, and in need of sustained care and action.
>
> Elected officials from both sides of the political aisle and at all levels of government regularly cite the Report Card, beginning with the very first release in 1998, when President Bill Clinton referenced the Report Card's grade for Schools. News reports reference the Report Card on a daily basis, with mentions in *The Wall Street Journal*, *The New York Times*, *USA Today*, *The Washington Post*, and the *Los Angeles Times*, as well as on National Public Radio, NBC's *Today Show*, *60 Minutes*, *CBS Evening News*, and HBO's *Last Week Tonight* with John Oliver, among many others.

Table 2.1 shows the grades by category for each issuance of the ASCE Infrastructure Report Card.[1]

In the beginning, ASCE went it alone. Some of our sister societies were concerned that the Report Card would be viewed as an indictment on the failure of our profession to make positive change. ASCE viewed the grades as a way to shed light on the growing challenges in public infrastructure and the need for elected offices to take notice and supplement funding for public infrastructure.

Through its various iterations, ASCE has improved its public relations strategy surrounding the grades and improved the information contained in the Report Card to get the Society's message across.

The 2001 Report Card was a loose leaf packet with an online pdf and had 12 infrastructure categories. It relied on the original report – the federally issued *Fragile Foundations* as the baseline for the grades. The 2001 report was very detailed and text heavy on national conditions and had recommendations for public policy actions. It also included a cost of implementing the needed improvements as well as the estimated deficit in funding available.

---

[1] Each separate report card is archived on ASCE's infrastructure report card site: https://infrastructurereportcard.org/

Table 2.1 *Fragile Foundations* and ASCE Infrastructure Report Card grades

Category	1988[a]	1998	2001	2005	2009	2013	2017	2021
Aviation	B−	C−	D	D+	D	D	D	D+
Bridges	−	C−	C	C	C	C+	C+	C
Dams	−	D	D	D+	D	D	D	D
Drinking water	B−	D	D	D−	D−	D	D	C−
Energy	−	−	D+	D	D+	D+	D+	C−
Hazardous waste	D	D−	D+	D	D	D	D+	D+
Inland waterways	B−	−	D+	D-	D−	D−	D	D+
Levees	−	−	−	−	D−	D−	D	D
Ports	−	−	−	−	−	C	C+	B−
Public parks and recreation	−	−	−	C−	C−	C−	D+	D+
Rail	−	−	−	C−	C−	C+	B	B
Roads	C+	D−	D+	D	D−	D	D	D
Schools	D	F	D−	D	D	D	D+	D+
Solid waste	C−	C−	C+	C+	C+	B−	C+	C+
Stormwater	−	−	−	−	−	−	−	D
Transit	C−	C−	C−	D+	D	D	D−	D−
Wastewater	C	D+	D	D−	D−	D	D+	D+
GPA	**C**	**D**	**D+**	**D**	**D**	**D+**	**D+**	**C−**
Cost to improve[b]	−	−	**$1.3T**	**$1.6T**	**$2.2T**	**$3.6T**	**$4.59T**	**$5.94T**

[a]The first infrastructure grades were given by the National Council on Public Works Improvements in its report *Fragile Foundations: A Report on America's Public Works*, released in February 1988. ASCE's first Report Card for America's Infrastructure was issued a decade later
[b]The 2017 Report Card's investment needs are over 10 years. The 2013 Report is over 8 years. In the 2001, 2005, and 2009 Report Cards, the time period was 5 years

In 2005, the Report Card used focus groups and ad hoc conversations with our stakeholders to inform the process and outputs. The pivot for that iteration was to report on items that the market was interested in and where those needs existed. Members of Congress consistently asked for state-specific data, as they needed to tie back the large national challenge to specific challenges in their districts. And the lesson from the media is that they were more interested in local stories and anecdotes than in policy-focused news. All politics is local.

So after collecting market feedback on the 2005 Report Card, small investments were made by ASCE for the next cycle in 2009. We made small, low-cost changes to the Report Card to test in the market. Limited data provided included data to address state infrastructure needs, and engineering success stories were added for select states. The minor changes were successful. Thus, it was decided that we needed to go back to focus groups and look at reach metrics, as well as the length of the news cycle. It was decided that we needed web analytics as well as social media metrics.

In 2011, I was fortunate to be asked to join the Committee on America's Infrastructure for the 2013 Report Card. Since we had four Report Cards under our belt, it was time to assess our market presence and grow our audience to effectively push elected officials to support infrastructure investment and sound policy. ASCE focused on innovation, as technology began to take over many aspects of our lives. It was time to refresh the Report Card to meet the current expectations of the public.

The ASCE created a website and a mobile app to replace the hardcopy report format for the 2013 Report Card for America's Infrastructure. A new branded logo was created and four new categories were added. The grading methodology was formalized into a much more rigorous process. Most importantly, new sections were added that had the estimated cost and funding deficit in every category graded. There was information on state infrastructure in all 50 states, and the narrative included local success stories. There were sections which were labeled as "solutions that work now." Social media was used to post articles on Facebook and tweets by and about ASCE that were updated in real time. Much of the heavy text was replaced with infographics, interactive charts, pictures, and YouTube videos from ASCE and our partner organizations.

The results were that we had more interest from the general public, the press, and local ASCE Sections. We observed a longer news cycle for local content, while getting more interest among local audiences, and therefore we increased our partnership with local ASCE groups.

As ASCE evaluated the metrics, the small investment was very successful; so ASCE allocated more resources to build out the Report Card components as permanent byproducts of the national report card. Facts on infrastructure systems and success stories were supplied for all 50 states. Supplemental videos were added with heavier local focus.

ASCE used its grassroots network effectively but still needed a broader reach so several cooperative organizations with "multiplier" qualities were added to access larger audiences, provide greater media reach, and acquire even more local examples. The Report Card went from a short paper technical product to a full media outreach program including spokesperson training, press packets, and social media calendars. Relationship models were developed for both historical and new partners that focused on group characteristics and tools to support partner engagement. All these elements were added to the 2017 Report Card.

2.4.1   The 2021 Report Card

The 2021 Report Card committee was populated in 2019, and I was fortunate enough to be asked to participate again as 1 of the 32 experts. We started our work in the fall with no clue as to what the future would hold. Committees were formed for each of the Report Card categories to start the research. Just as we were really getting into the data and our analysis, the COVID-19 pandemic hit. Our meetings went virtual, and we now saw how much more important our roles were. Everyone around the country was trying to keep informed about health and safety protocols to keep critical infrastructure working to ensure essential employees could get to and from work. State and local governments were extremely hard hit by supplemental costs in keeping people safe as well as by draconian losses in revenue. We were really challenged on how to present data on needs and status when we were in the middle of a once-in-a-century pandemic.

On one of our committee TEAMS calls, we were discussing the need to get our message to Congress on the dire impacts COVID-19 was having on our physical infrastructure. An ASCE position paper was being proposed when a light bulb went off in my head. Everyone was sending Congress white papers related to the pandemic. How could our thoughts rise to the top of the pile? ASCE had great brand recognition with the Report Card with both the media and elected officials. ASCE could use the brand and publish an interim report without grades, just speaking to the impacts COVID was having. The team loved the idea, and in June of 2020, ASCE published the "STATUS REPORT: COVID-19's Impacts on America's Infrastructure."[2]

The message was successfully delivered, and the interim report was part of the pitch which yielded much needed support to the owners and operators of our infrastructure. Ultimately, Congress provided emergency funding to state departments of transportations, airports, and water systems through the various COVID-19 relief packages.

Turning back to the development of the 2021 Report Card for America's Infrastructure, we again used the standard methodology that had been used to develop the 2017 Report Card. The methodology we used looked at the following items for researching the data, developing the narrative, and ultimately assigning the grade for all 17 categories:

- Capacity
- Condition
- Funding
- Future need
- Operations and maintenance
- Public safety
- Resilience
- Innovation

Once subcommittees had pulled the data together, each subcommittee graded the category as follows:

A: Exceptional, fit for the future, in excellent condition.
B: Good, adequate for now, but some elements show signs of general deterioration that require attention.
C: Mediocre, requires attention; some elements exhibit significant deficiencies, with increasing vulnerability.
D: Poor, at risk, with many elements approaching the end of their service life. Condition and capacity are of serious concern, with strong risk of failure.
F: Failing/critical, unfit for purpose, with signs of imminent failure.

---

[2] https://www.infrastructurereportcard.org/wp-content/uploads/2020/06/COVID-19-Infrastructure-Status-Report.pdf

Then each individual subcommittee presented to the entire committee over many virtual meetings and had to defend the grades much like you would defend a dissertation. In several cases, more data were needed for the overall committee to accept the grades, and also in several cases, the subcommittees were required to come back and redefend or brought supplemental information to change the overall committee's previous decision. It was an incredibly rigorous process.

The trends that were identified included as follows: maintenance backlogs continue to be an issue, but asset management helps prioritize limited funding, federal investments have moved the needle, and many state and local governments continue to prioritize infrastructure investments to help us keep pace with our growing needs, but there are still infrastructure sectors where data are scarce or unreliable. The resulting grades are shown in Fig. 2.1 [9].

All of the data used for grading each category of infrastructure are based on publicly sourced data. That has been challenging with several categories as there are limited public databases available. Schools are one example. School facilities represent the second largest sector of public infrastructure spending after highways, and yet there is no comprehensive national data source on K-12 public school infrastructure. The limited data that are available indicate that 54% of public school districts report the need to update or replace multiple building systems including heating, ventilation, and air conditioning systems. More than one-third of public schools have portable buildings due to capacity constraints, and 45% of these buildings are in poor or fair condition. Meanwhile, as a share of the economy, state capital funding for schools was down 31% in fiscal year 2017 as compared to 2008. That is the equivalent of a $20 billion cut.

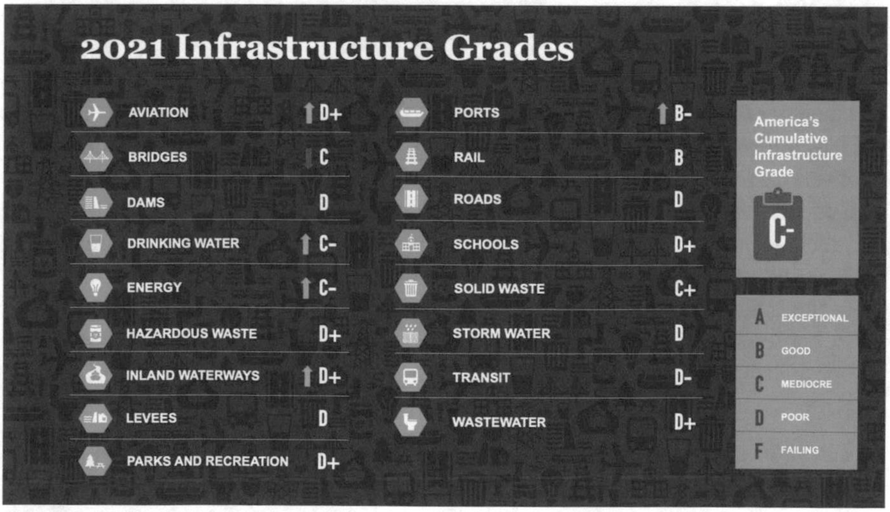

2021 Infrastructure Grades

✈ AVIATION	↑ D+	🚢 PORTS	↑ B-		America's Cumulative Infrastructure Grade	
🌉 BRIDGES	C	🚂 RAIL	B			
🏗 DAMS	D	🚧 ROADS	D		**C-**	
💧 DRINKING WATER	↑ C-	🏫 SCHOOLS	D+			
💡 ENERGY	↑ C-	🗑 SOLID WASTE	C+			
☢ HAZARDOUS WASTE	D+	💦 STORM WATER	D		A EXCEPTIONAL	
🚮 INLAND WATERWAYS	↑ D+	🚌 TRANSIT	D-		B GOOD	
🌊 LEVEES	D	💧 WASTEWATER	D+		C MEDIOCRE	
🌲 PARKS AND RECREATION	D+				D POOR	
					F FAILING	

Fig. 2.1   2021 ASCE Infrastructure Report Card

Broadband is a category that has touched us during the pandemic. ASCE added it as a spotlight area but not a true category as much of broadband is private and as such there are very limited data in the public realm. The importance of broadband infrastructure has grown exponentially as we increasingly rely on it to support our connected lives. Meanwhile, civil engineers play a growing role in broadband installation, and high-speed internet is increasingly critical to the operation and modernization of our legacy infrastructure systems. Because of this, the ASCE Committee on America's Infrastructure felt it important to make recommendations on how to improve broadband infrastructure. However, the committee determined there was insufficient information on broadband infrastructure to justify a category grade.

Overall, the cumulative infrastructure grade increased from a D+ to a C−.

Each individual category has a bibliography of sources at the end of their chapter.

The grades are only one piece of the puzzle. The executive summary and full report include key findings, discussion on how investment pays and what the cumulative investments need to be, as well as recommendations on how to raise the grades.

As for recommendations to raise the grades, these fell into three categories: Leadership and Action, Investment, and Resilience. Specifically under Leadership and Action, ASCE recommends the following:

Smart investment will only be possible **with strong leadership, decisive action, and a clear vision for our nation's infrastructure**. Leaders from all levels of government, business, labor, and nonprofit organizations must come together to:

- Incentivize asset management and encourage the creation and utilization of infrastructure data sets across classes.
- Streamline the project permitting process across infrastructure sectors while ensuring appropriate safeguards and protections are in place.
- Ensure all investments are spent wisely, prioritizing projects with critical benefits to the economy, public safety, environment, and quality of life (e.g., sustainability).
- Leverage proven and emerging tech to make use of limited available resources.
- Consider life cycle costs when making project decisions. Life cycle cost analysis determines the cost of building, operating, and maintaining the infrastructure for its entire life span.
- Support research and development of innovative materials, technologies, and processes to modernize and extend the life of infrastructure, expedite repairs or replacements, and promote cost savings. Innovation should include a component of integration and utilization of big data, as well as the "internet of things."
- Promote sustainability, or the "triple bottom line" in infrastructure decisions, by considering the long-term economic, social, and environmental benefits of a project.

Under Investment, if the United States is serious about achieving an infrastructure system fit for the future, some specific steps must be taken, beginning with **increased, long-term, consistent investment**. To close the $2.59 trillion 10-year investment gap, meet future needs, and restore our global competitive advantage, we must **increase investment from all levels of government and the private**

sector from 2.5% to 3.5% of US gross domestic product (GDP) by 2025. This investment must be consistently and wisely allocated and must begin with the following steps:

- Congress should fully fund authorized programs.
- Infrastructure owners and operators must charge, and Americans must be willing to pay, rates reflecting the true cost of using, maintaining, and improving infrastructure.
- The surface transportation investment gap is the largest deficit among the categories of infrastructure ASCE examines. Continuing to defer maintenance and modernization is impacting our ability to compete in a global marketplace and maintain a high quality of living domestically. Congress must fix the Highway Trust Fund.
- All parties should strive to close the rural/urban and underserved community resource divide by ensuring adequate investment in these areas through programmatic set asides.
- All parties should make use of public-private partnerships, where appropriate.

Resilience is paramount as future investments are made. We must **utilize new approaches, materials, and technologies to ensure our infrastructure** can withstand or quickly recover from natural or man-made hazards. Advancements in resilience across all infrastructure sectors can be made by:

- Enabling communities, regardless of size, to develop and institute their own resilience pathway for all their infrastructure portfolios by streamlining asset management, implementing life cycle cost analysis into routine planning processes, and integrating climate change projections into long-term goal-setting and capital improvement plans
- Incentivizing and enforcing the use of codes and standards, which can mitigate risks of major climate or man-made events such as hurricanes, fires, sea level rise, and more
- Understanding that our infrastructure is a system of systems and encouraging a dynamic, "big picture" perspective that weighs tradeoffs across infrastructure sectors while keeping resilience as the chief goal
- Prioritizing projects that improve the safety and security and systems and communities, to ensure continued reliability and enhanced resilience
- Improving land use planning across all levels of decision-making to strike a balance between the built and natural environments while meeting community needs, now and into the future
- Enhancing the resilience of various infrastructure sectors by including or enhancing natural or "green" infrastructure

The Report Card includes examples of game changers in all the categories. These are project innovations that deliver innovation to the solution. Examples of projects moving the needle are listed in the report and are shown prominently on the website and app, so that people see examples in each state.

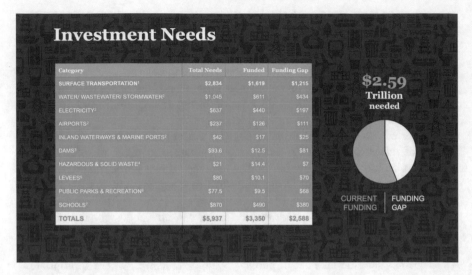

Fig. 2.2 2021 ASCE Infrastructure Report Card investments required. Much of these data were taken from the ASCE Failure to Act study completed in 2021

The Report Card also included a table of cumulative investment needed in the major categories as shown in Fig. 2.2 [10].

2.5   ASCE's Failure to Act Series

The evergreen question about the Report Card is "So what does this mean to me?"

Concurrent with the Report Card, ASCE began a series of independent economic reports that take a deep dive into what the status quo means to the average American. It seeks to quantify the fact that a healthy economy can't be supported by a crumbling infrastructure. And what are the future implications?

As described on the Infrastructure Report Card website [11]:

> **The Failure to Act report series answers this key question—how does the nation's failure to act to improve the condition of U.S. infrastructure systems affect the nation's economic performance?** In 2020 and 2021, ASCE released five Failure to Act reports in a series covering 11 infrastructure sectors that are critical to the economic prosperity of the U.S.
>
> These reports were followed by a fifth, comprehensive final report, *Failure to Act: Economic Impacts of Status Quo Investment Across Infrastructure Systems*, which addressed the aggregate economic impact of failing to act in more than one sector. This report addresses the current infrastructure gaps between today's needs and investment and how they will affect the future productivity of industries, national competitiveness, and the future costs to households.
>
> Recent Failure to Act reports include:

FAILURE TO ACT

Economic Impacts of Status Quo
Investment Across Infrastructure Systems

Our report finds that the over the next 20 years, **the average American household will spend $3,300 a year** due to infrastructure deficiencies. This lost disposable income comes from the disruptions of inadequate infrastructure **like sitting in traffic, hitting a pothole, power outages, and water main breaks.**

Fig. 2.3 Failure to Act summary

Ports and Inland Waterways (2021) — Underinvestment in ports and inland waterways stand to increase waterborne shipping costs from 8% to 22%, on average, by 2039. Manufacturing, agriculture, and production and extraction are most impacted.

Airports (2021) — A recent uptick in airport infrastructure investment is paying dividends for the residents and businesses in the U.S. However, as spending returns to previous averages, the economy will suffer. Specifically, airport congestion will cost U.S. industries and households approximately $28 billion in 2029 and $41 billion in 2039.

Electricity (2020) — An additional investment of $16.9 billion per year between now and 2039 in our electricity infrastructure can protect 540,000 jobs and $5,800 per household in personal income.

Water and Wastewater (2020) — By investing in our water infrastructure to make it more reliable, we can prevent $250 billion in increased costs to businesses by 2039.

Surface Transportation (2021) — These findings show that if industry costs are passed on to customers, costs per household could be as high as $12,500 over 20 years, or $625 dollars per year. Losses to households and industries will amount to $677 billion over the 2020–2029 period and $1.3 trillion during the 2030–2039 decade.

The comprehensive report can be summarized as shown in Fig. 2.3.

2.6   Impact of the 2021 Report Card

United for Infrastructure, a coalition for which ASCE is a Steering Committee member, convened an Infrastructure Candidate Forum in Las Vegas on February 16, 2020. The Forum elevated infrastructure as a top issue of the public agenda. Joe Biden, who was then a Presidential candidate, addressed an issue that 90 percent of swing state voters saw as a top priority for elected officials. The host committee consisted of the International Union of Operating Engineers, Transportation Trades Department, AFL-CIO, North America's Building Trades Unions, Transport Workers Union of America, American Public Transportation Association, ASCE,

Value of Water Coalition, American Council of Engineering Companies, American Road and Transportation Builders Association, Association of Equipment Manufacturers, Airports Council International-North America, and Build Together.

I had the honor and privilege to be there for the candidate discussions and Q&A on their views regarding infrastructure. It was especially poignant to see that the items that President Biden and Secretary Pete Buttigieg spoke about almost a year before they became President and Transportation Secretary were part of the President's plan for the future of infrastructure for a prosperous US future. Much of the language in the ASCE Report Card, the Solutions Summit held immediately following the reveal of the Report Card, and the Failure to Act reports have been not only cited, but used as part of the development of the resulting infrastructure plans and bills of the Biden administration. While ASCE has seen success with the Report Card, it has never hit a grand slam home run like it did in 2021.

As of the end of June 2021, 4 months after the release, the media hits have been higher than for any previous Report Card. There have been 2091 print and online placements that have yielded 1.6 billion print and online impressions. There have been more than 400 original articles written and 1520 broadcast clips with a viewership of more than 65 million. 57 million total stations were reached, including 33 radio interviews with 5 million listener impressions heard on more than 2225 radio stations. There have been 85 press release pickups with a potential audience of 99 million. And ASCE members have really engaged in spreading the news as there have been 468 engagement actions by members.

So by any measure, the Report Card has been a resounding success in getting the message out that our work as civil engineers is critical to our quality of life and we need our political leadership to get it done.

2.7 Personal Thoughts

As I look back to my professional and ASCE life, I see that the activity around public awareness and active advocacy has really been a uniting thread throughout my career. It has been a very exciting journey, and I am very thankful for the thousands of people who have been engaged throughout the journey. It proves that if you have the facts and the passion, amazing results can be achieved.

References

1. Ambrose, Susan A., Kristin L. Dunkle, Barbara A. Lazarus, Indira Nair and Deborah A. Harkus, Editors. *Journeys of Women in Science and Engineering: No Universal Constants.* Temple University Press: Philadelphia, Pennsylvania. 1997.
2. "John Smeaton", Wikipedia, https://en.wikipedia.org/wiki/John_Smeaton, accessed September 26, 2021.

3. "Institution of Civil Engineers", Wikipedia, https://en.wikipedia.org/wiki/Institution_of_Civil_Engineers, accessed September 26, 2021.
4. Institution of Civil Engineers, "Our history," https://www.ice.org.uk/about-ice/our-history, accessed September 26, 2021.
5. "Civil Engineering Defined," http://www.smweng.com/smw-civil-engineering-defined, SMW Engineering Group, Inc., accessed September 26, 2021.
6. Kass-Simon, G. and Patricia Farnes, Editors. *Women of Science: Righting the Record*. Indiana University Press: Bloomington, Indiana. 1990.
7. "History," About ASCE, American Society of Civil Engineers, https://www.asce.org/about-asce, accessed September 26, 2021.
8. "Report Card History," 2021, Report Card for America's Infrastructure, https://infrastructurereportcard.org/making-the-grade/report-card-history/, accessed September 26, 2021.
9. "America's Infrastructure Scores a C-", 2021 Report Card for America's Infrastructure, https://infrastructurereportcard.org/, accessed September 26, 2021.
10. "Investment Gap 2020-2029," 2021 Report Card for America's Infrastructure, https://infrastructurereportcard.org/resources/investment-gap-2020-2029, accessed September 26, 2021.
11. "Failure to Act Economic Reports," 2021 Report Card for America's Infrastructure, https://infrastructurereportcard.org/resources/failure-to-act-economic-reports/, accessed September 26, 2021.

Maria Lehman, P.E, F.ASCE, ENV SP, is GHD's Infrastructure Market Leader for the United States since May of 2020. Maria was the former Vice President for Critical Infrastructure for Parsons, COO, and Acting Executive Director of the New York State Thruway Authority and Commissioner of Public Works for Erie County, NY. She has 40 years of diverse, increasingly responsible, multi-disciplinary technical and leadership experience, both in the private and public sectors and in traditional and alternative delivery. She received her BS in Civil Engineering at the State University of New York at Buffalo, Magna Cum Laude, and is a licensed Professional Engineer in several states. She is currently the National President Elect of the American Society of Civil Engineers (ASCE).

Maria has won numerous national, statewide, and local awards including the ASCE President's Medal, UB's School of Engineering Alumna of the Year, and the New York State Society of Professional Engineers Engineering Manager of the Year.

Chapter 3
Infrastructure Pioneers

Jill S. Tietjen

Abstract Women have contributed to the development of infrastructure in many ways over the years. Elsie Eaves made many contributions in the field of information gathering and publishing about construction and infrastructure. Olive Ann Beech, Anne Morrow Lindbergh, Mabel MacFerran Rockwell, and Elsie Gregory MacGill were aviation pioneers. Bridges pioneer Emily Warren Roebling ensured the completion of the Brooklyn Bridge. Computer pioneers Ada Byron Lovelace, Admiral Grace Murray Hopper, and Anita Borg laid the foundation for the internet and broadband. Ruth Patrick and Ellen Henrietta Swallow Richards worked to ensure clean water for drinking. Energy pioneers include Edith Clarke, Maria Telkes, Ivy Parker, and Ada Pressman. Joan Berkowitz has been a hazardous waste pioneer. Hydraulic engineer Margaret Petersen led the way for women in inland waterways. Public parks pioneers include Mary Colter, Julia Morgan, Marjory Stoneman Douglas, Margaret "Mardy" Murie, and Gale Norton. Railroads benefit from the efforts of Mary Engle Pennington and Olive Dennis. Women road pioneers include Bertha Benz, Alice Huyler Ramsey, Marilyn Jorgenson Reece, and Janet Bonnema. Helen Schultz worked to ensure transit options. These pioneers worked to improve our quality of life and standard of living. We celebrate and honor them.

Keywords Women in infrastructure · ASCE Infrastructure Report Card

3.1 Introduction

Women have contributed to the development of infrastructure in many ways over the years – from proving the value of the automobile to developing the first water quality tables to demonstrating the safety of drinking water. Women pioneered in developing refrigerated railroad cars for the transport of perishables and made railroad cars more comfortable and inviting. Women built airplanes and ensured that

J. S. Tietjen (✉)
Technically Speaking Inc, Greenwood Village, CO, USA

© The Author(s), under exclusive license to Springer Nature
Switzerland AG 2022
P. Layne, J. S. Tietjen (eds.), *Women in Infrastructure*, Women in Engineering
and Science, https://doi.org/10.1007/978-3-030-92821-6_3

bridges got built. They designed the algorithms used to evaluate long-distance electric transmission lines and pioneered in the usage of and development of solar energy. They advocated for the establishment of national parks and designed interstate interchanges. Within this chapter, the featured infrastructure pioneers are presented chronologically by infrastructure category per the American Society of Civil Engineers' (ASCE) Infrastructure Report Card within which they fall. They each helped make our standard of living and quality of life better.

3.2 General

The ASCE Infrastructure Report Card does not have a general category, but Elsie Eaves, who is written about in this section of the chapter, made many contributions in the field of information gathering about and publishing about construction and infrastructure. She was truly a pioneering civil engineer.

3.2.1 Elsie Eaves (1898–1983)

After graduating from the University of Colorado at Boulder in 1920 with a BS in civil engineering (with honors), Elsie Eaves (Fig. 3.1) had a series of jobs in Colorado before heading to New York. There, she began her employment with McGraw-Hill, the publishing company. Colonel Willard T. Chevalier hired Eaves (after an editor of an undisclosed organization told her "a woman's place, if not in the home, is in the department store") and created her job as assistant on market surveys for *Engineering News-Record* in 1926. She became Director of Market Surveys for *Engineering News-Record* and *Construction Methods and Equipment* shortly thereafter. In 1932, Eaves moved to the position of Manager of Business News Department, where she directed the activities of 100 staffers throughout the United States and Canada.

Her career in the publishing field was a series of "firsts." In 1929, Eaves originated and compiled the first national inventory of municipal and industrial sewage disposal facilities – an analysis that she recompiled at regular intervals. A few years later, she compiled statistics on needed construction, which aided the passage of the Federal Loan-Grant legislation used to revitalize the construction industry during the 1931–1935 depression. In 1945, she organized and directed the *Engineering News-Record*'s measurement of Post War Planning by the Construction Industry that was used by the Committee for Economic Development and the ASCE as the official progress report of the industry. This index was unprecedented in the field of engineering analysis. Under Eaves' direction, the "Post War Planning" statistics were converted into a continuous inventory of planned construction. This has become the *Engineering News-Record*'s "Backlog of Proposed Construction," an index to more than $100 billion of construction activity. Another of her unique

Fig. 3.1 Elsie Eaves.
(Courtesy of the Society of
Women Engineers
Photograph Collection,
Walter P. Reuther Library,
Wayne State University)

"firsts" was defining the limits and editing the pilot issues of the *Construction Daily*,
a nationwide service.

Eaves' additional firsts and awards include:

– First woman to be licensed as a professional engineer in New York State
– First woman member of the ASCE (as a corporate member in 1927)
– First woman to be a life member of the ASCE (1962, at which time there were 54
  women among 48,000 members)
– First woman elected to honorary membership of the ASCE (1979); first woman
  to be elected Associate Member, Fellow of ASCE
– First and, for a long time, the only, woman member of the American Association
  of Cost Engineers (1957) as well as the first civil engineer
– First woman to receive the Honorary Life Membership Award from the American
  Association of Cost Engineers (1973)
– First woman to receive the International Executive Service Corporation "Service
  to the Country" award
– First woman to receive the American Association of Cost Engineer's Award of
  Merit (1967) [1–4]

3.3 Aviation

The women pioneers in aviation include Olive Ann Beech, who helped establish and then ran an airplane manufacturing company; Anne Morrow Lindbergh, who with her husband Charles Lindbergh laid the groundwork for commercial airline routes; Mabel MacFerran Rockwell, who helped improve manufacturing processes for aircraft; and Elsie Gregory MacGill – the first woman to design, build, and test an airplane.

3.3.1 Olive Ann Beech (1903–1993)

At age 7, Olive Ann Mellor demonstrated such significant financial acumen that she had her own bank account. At age 11, she was handed the responsibility of keeping the family checkbook. She completed secretarial and business college and began working at an electrical contracting firm when she was 18. In 1924, she joined the Travel Air Company in Wichita, Kansas, as a secretary and the only woman out of the company's 12 employees. In addition to being the only female employee, she was also the only staff member without a pilot's license and endured tremendous teasing because of this status. To combat the teasing, she asked for and got a copy of airplane drawings with parts labeled. She and Walter H. Beech, the founder of Travel Air, married in 1930, and Olive Ann Beech relocated to New York City where Walter Beech was serving as president of the Curtiss-Wright Corporation, which had merged with Travel Air.

In 1932, the Beeches returned to Wichita and established Beech Aircraft Company. She served as secretary-treasurer of their new company. In 1940, Walter Beech became ill, and Olive Ann stepped into the leadership role of the company. He was hospitalized for nearly a year, during which she arranged for the needed loans to retool the company for military production of aircraft in support of World War II. Walter was able to return to the company only to die in 1950 of a heart attack. At that point, Olive Ann was elected president and chairman, at age 47, the first woman to head a major aircraft company.

Beech then guided the Beech Aircraft Company through the growth necessary for it to regain its leadership in the commercial aircraft industry and stayed in the leadership role for more than 30 years. When Beech Aircraft Company merged with the Raytheon Company in 1980, she was elected to the board of directors of Raytheon. She retired in 1982 from the position of chairman of Beech Aircraft to become the company's first Chairman Emeritus. She has been deemed "The First Lady of Aviation." Beech's many honors include being the first person inducted into the Kansas Business Hall of Fame and induction into the National Aviation Hall of Fame and the American National Business Hall of Fame [5, 6].

3.3.2 Anne Morrow Lindbergh (1906–2001)

Writer and aviator Anne Morrow Lindbergh (Fig. 3.2), wife of America's hero Charles Lindbergh, was her husband's co-pilot, navigator, and radio operator during the early years of their marriage while the pair flew all over the world to chart potential air routes for commercial airlines. Their work across the North American continent and the Caribbean laid the groundwork for Pan American's air mail service. Her first book, *North to the Orient*, resulted from the flights they took on uncharted routes from Canada and Alaska to China and Japan in 1931. Another of her books, *Listen! The Wind*, resulted from their chronicling of North and South American potential air routes. She received the Hubbard Gold Medal from the National Geographic Society in 1934 for her contributions to 40,000 miles of exploratory flying with her husband over five continents. She was the first woman licensed glider pilot in the United States.

Lindbergh exhibited writing skills from a very young age. When she graduated from Smith College, she won two prizes for her writing – the Mary Augusta Jordan Prize for the most original literary piece and the Elizabeth Montagu Prize for the best essay on women of the eighteenth century. In addition to the two books noted above, Lindbergh would write 11 other books including the 1955 *Gift from the Sea*, which spent many weeks on the bestseller list. Among her many honors, she has been inducted into the National Aviation Hall of Fame and the National Women's Hall of Fame [7–9].

Fig. 3.2 Anne Morrow Lindbergh with her husband. (Courtesy of the Library of Congress)

Col and Mrs Chas A Lindberg
snaped on their arrival in

3.3.3   Mabel MacFerran Rockwell (1925–1981)

Credited as possibly the first female aeronautical engineer in the United States, Mabel MacFerran Rockwell (Fig. 3.3) received her BS from MIT in 1925 in science, teaching, and mathematics and a BS from Stanford University in electrical engineering. Before World War II, she served as a technical assistant with the Southern California Edison Company, where she was a pioneer in the application of symmetrical components to transmission relay problems in power systems. Through this work, she made it easier to diagnose system malfunctions and to enhance the reliability of multiple-circuit lines. Rockwell then worked for the Metropolitan Water District in Southern California where she was a member of the team that designed the Colorado River Aqueduct's power system and the only woman to participate in the creation of the electrical installations at the Hoover Dam.

Later, Rockwell joined Lockheed Aircraft Corporation and worked to improve the manufacturing operations of aircraft. Her many innovations included refining the process of spot welding and developing techniques for maintaining cleaner working surfaces so that the welds completely fused. After the war, Rockwell went to work for Westinghouse where she designed the electrical control system for the Polaris missile launcher. At Conair, she developed the launching and ground controls for the Atlas guided missile systems. In 1958, President Eisenhower named her Woman Engineer of the Year. Also in 1958, she received the Society of Women Engineers' (SWE) Achievement Award "in recognition of her significant contributions to the field of electrical control systems" [10, 11].

Fig. 3.3 Mabel MacFerran
Rockwell. (Courtesy of the
Society of Women
Engineers Photograph
Collection, Walter
P. Reuther Library, Wayne
State University)

3.3.4 Elsie Gregory MacGill

Elsie Gregory MacGill (Fig. 3.4) overcame polio, with which she was stricken while in graduate school, to become the first woman to design, build, and test an airplane as well as the first female chief aeronautical engineer of a firm in North America. When she graduated with her master's degree in aeronautical engineering degree, after taking her examinations from her hospital bed – unable to walk due to the polio – she became the first woman to complete a master's degree at the University of Michigan. She would continue to trailblaze throughout her life: she was the first female Fellow of the Canadian Aeronautics and Space Institute and the first female member of the Association of Consulting Engineers of Canada.

Elsie MacGill grew up in Vancouver, British Columbia, Canada, and was interested in radios from a young age. She pursued that interest in her college education at the University of Toronto, where she was the first woman to obtain a degree in electrical engineering from that university. Her first job was with the Austin Aircraft Company in Pontiac, Michigan, and she pursued her master's degree at the University of Michigan in aeronautical engineering concurrently. During her polio convalescence, she was home in Vancouver writing articles on the topics of planes and flying. After she was able to walk with the aid of crutches, MacGill went to the Massachusetts Institute of Technology (MIT) and spent 2 years of post-graduate work there studying air currents.

When MacGill left MIT, she went to work for the Fairchild Aircraft Company performing stress analysis. Subsequently, she became Chief Aeronautical Engineer

Fig. 3.4 Elsie Gregory MacGill. (Courtesy of the Society of Women Engineers Photograph Collection, Walter P. Reuther Library, Wayne State University)

with the Canadian Car and Foundry Company, Ltd., in Fort William, Ontario. Her responsibilities there included design, building, and testing the "Maple Leaf Trainer" for the Mexican Air Force and then tooling up to make Hurricane Fighter Planes that had interchangeable parts with those manufactured in the United Kingdom. She not only converted a factory that had previously made box cars into an aircraft manufacturing facility, but she also oversaw the training of unskilled labor to manufacture aircraft. No wonder this effort made her famous and led to a comic book about her that dubbed her the "Queen of the Hurricanes."

After World War II, MacGill started a consulting business. She became the first woman to serve as a Technical Advisor to the International Civil Aviation Organization. In 1974, she became the chair of the United Nations' Stress Analysis Committee, the first woman to chair a UN committee. An advocate for equal pay for equal work, she was appointed to the Royal Commission on the Status of Women. MacGill received SWE's Achievement Award in 1953 "in recognition of her meritorious contributions to aeronautical engineering." She said, "I have received many engineering awards, but I hope I will also be remembered as an advocate for the rights of women and children" [1, 10, 12].

3.4 Bridges

Our bridges pioneer is Emily Warren Roebling – without whom the Brooklyn Bridge as we know it would not have been completed.

3.4.1 Emily Warren Roebling (1844–1903)

Emily Warren Roebling (Fig. 3.5), generally considered the first US female field civil engineer and construction manager, is remembered for her significant accomplishments in the construction of the Brooklyn Bridge. The inscription on the East Tower of the bridge reads:

THE BUILDERS OF THE BRIDGE
DEDICATED TO THE MEMORY OF
EMILY WARREN ROEBLING
1843–1903
WHOSE FAITH AND COURAGE HELPED HER STRICKEN HUSBAND
COL. WASHINGTON A. ROEBLING, C.E.
1837–1926
COMPLETE THE CONSTRUCTION OF THIS BRIDGE
FROM THE PLANS OF HIS FATHER
JOHN A. ROEBLING, C.E.
1805–1869
WHO GAVE HIS LIFE TO THE BRIDGE

BACK OF EVERY GREAT WORK WE CAN FIND
"THE SELF-SACRIFICING DEVOTION OF A WOMAN."

Fig. 3.5 Emily Warren
Roebling. (Courtesy of the
Brooklyn Museum)

THIS TABLET ERECTED 1931 BY
THE BROOKLYN ENGINEERS CLUB
WITH FUNDS RAISED BY POPULAR SUBSCRIPTION

Without Emily Warren Roebling, the Brooklyn Bridge (Fig. 3.6) – one of the greatest engineering projects of the nineteenth century – might not have been completed on May 24, 1883. At a time when most women did not pursue higher education and their proper place as defined by society was wife and mother, Roebling learned engineering through the study of higher mathematics, strength of materials, stress analysis, the calculation of catenary curves, bridge specifications, and the intricacies of cable construction. She was assisted in these studies by her husband and brother. Her engineering skills allowed her to become the principal assistant and inspector of the bridge as her husband, Washington Roebling, could no longer visit the site because he had "Bends" disease, what today we know as caissons disease that results from improper decompression when divers ascend from depths. She was able to discuss structural steel requirements with representatives from steel mills and assisted them with designs and shapes never before fabricated.

She said, "… I have more brains, common sense, and know-how generally than any two engineers civil or uncivil that I have ever met …" The bridge, with a span of 1595 feet, was the largest suspension bridge in the world when it was completed

Fig. 3.6 Brooklyn Bridge circa 1915. (Courtesy of the Library of Congress)

and remains functional today. She took the first ride over the bridge carrying a live rooster as a symbol of victory [2, 13–15].

3.5   Broadband

Broadband – and the internet – requires the use of computers. Computers are no longer the province solely of individuals with graduate degrees in mathematics and physics because of the work of pioneering women including Ada Byron Lovelace, who foresaw that computers would be programmed with computer software; Admiral Grace Murray Hopper, who developed the computer compiler that translates human languages into the zeros and ones that computers understand; and Anita Borg – who not only made technical contributions to computer hardware but also encouraged women in computer science through the Systers email list and the annual Grace Hopper Conference.

3.5.1   Ada Byron Lovelace (1815–1852)

The daughter of the English poet Lord George Byron, Ada Lovelace now has a computer language named (Ada) after her. A somewhat sickly child, Lovelace was tutored at home and was competent in mathematics, astronomy, Latin, and music by the age of 14. Totally enthralled by Charles Babbage's Difference Engine (an early computer concept), at 17 years old, Lovelace began studying differential equations. As proposed, Babbage's second machine, the analytical engine, could add, subtract, multiply, and divide directly, and it would be programmed using punched cards, the same logical structure used by the first large-scale electronic digital computers in the twentieth century.

In 1842, the Italian engineer L.F. Menabrea published a theoretical and practical description of Babbage's analytical engine. Lovelace translated this document adding "notes" in the translation. Her notes constitute about three times the length of the original document, and, as explained by Babbage, the two documents together show "That the whole of the development and operations of analysis are now capable of being executed by machinery." These notes include a recognition that the engine could be told what analysis to perform and how to perform it – the basis of computer software. Her notes (Fig. 3.7) were published in 1843 in *Taylor's Scientific Memoirs* under her initials, because although she wanted credit for her work, it was considered undignified for aristocratic women to publish under their own names.

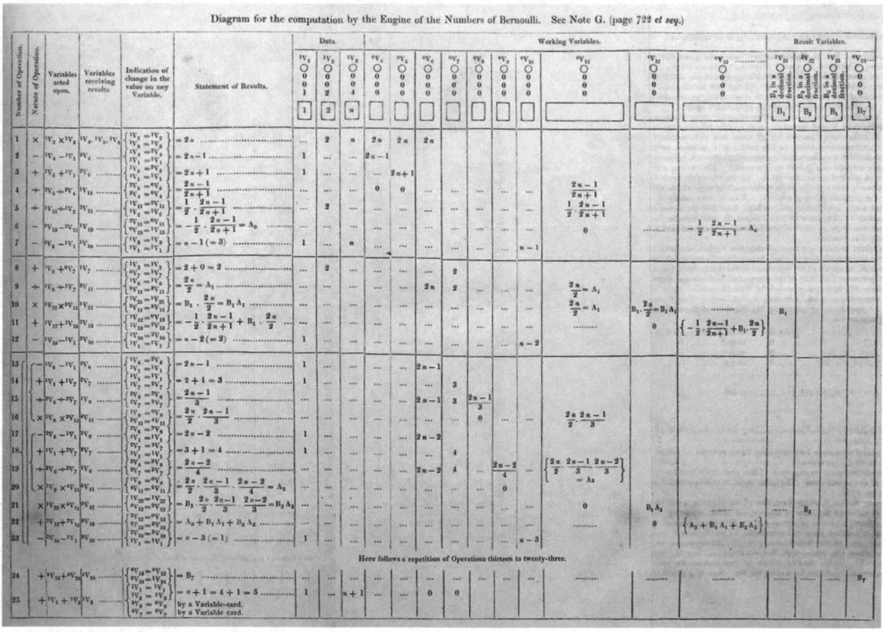

Fig. 3.7   Ada Byron Lovelace – Note G – Wikipedia

Ada Lovelace is considered to be the first person to describe computer programming [16, 17].

3.5.2 *Grace Murray Hopper (1906–1992)*

Admiral Grace Murray Hopper (Fig. 3.8) was famous for carrying "nanoseconds" around with her. These lengths of wire – just less than one foot – represented the distance light traveled in a nanosecond, one billionth of a second. She was renowned for trying to convey scientific and engineering terms clearly and coherently to non-technical people.

Hopper, also known as "Amazing Grace" and "The Grandmother of the Computer Age," helped develop languages for computers and developed the first computer compiler – software that translates English (or any other language) into the zeroes and ones that computers understand (machine language). Actually, her first compiler translated English, French, and German into machine language, but the Navy told her to stick with English because computers didn't understand French and German! Computers truly only understand numbers, but humans can translate those numbers now into English, French, German, and even Chinese and Japanese. She was also part of the group that found the first computer "bug" – a moth that had gotten trapped in a relay in the central processor. When the boss asked why they weren't making any numbers, they responded that they were "debugging" the computer.

Fig. 3.8 Admiral Grace Murray Hopper. (Courtesy of the Library of Congress)

Although Admiral Hopper loved to lay claim to the discovery of this first computer "bug" – and it is in the Smithsonian's National Museum of American History – the term bug had been in use for many years by then.[1]

Hopper received SWE's Achievement Award in 1964 "in recognition of her significant contributions to the burgeoning computer industry as an engineering manager and originator of automatic programming systems." She was the first woman to attain the rank of Rear Admiral in the US Navy. The Arleigh Burke-class guided missile destroyer USS Hopper (DDG-70) was commissioned by the US Navy in 1997. Hopper received the National Medal of Technology from President Bush in 1991, the first individual woman to receive the medal: "For her pioneering accomplishments in the development of computer programming languages that simplified computer technology and opened the door to a significantly larger universe of users." She was inducted into the National Women's Hall of Fame in 1994.

Hopper said she believed it was always easier to ask for forgiveness than permission. "If you ask me what accomplishment I'm most proud of, the answer would be all of the young people I've trained over the years; that's more important than writing the first compiler" [2, 18–23].

3.5.3   Anita Borg (1949–2003)

Anita Borg earned her BS, MS, and PhD (1981) degrees in computer science from Courant Institute of Mathematical Sciences, New York University. Early in her career, Borg was the lead designer and co-implementer of a fault-tolerant microprocessor, message-based UNIX system. The system provided users the ability to run programs that would automatically recover from hardware failures. She designed and built the first software system for generating and analyzing extremely long address traces (Fig. 3.9 shows the first page of her patent related to this system). The knowledge gained from this effort was used in the development of Digital Equipment Corporation's Alpha technology. Later, she designed and managed the implementation of Mecca, a web-based email system used by thousands of people.

Borg was known for much more in the computer industry than her significant technical accomplishments. In 1987, she founded the "Systers" email list linking technical women in computing when email was in its infancy. Borg founded the Grace Hopper Celebration of Women in Computing in 1994. After joining Xerox in 1997, she created a center to find ways to apply information technology to assure a positive future for the world's women. The Institute for Women and Technology (renamed the Anita Borg Institute after her death) researches, develops, and deploys useful, usable technology in support of women's communities.

---

[1] Zuckerman reports that Thomas Edison referred to a "bug" in his phonograph as early as 1889. Edison is reported to have defined a bug as "an expression for solving a difficulty, and implying that some imaginary insect has secreted itself inside and is causing all the trouble."

US005274811A

United States Patent [19]

Borg et al.

[11] **Patent Number:** **5,274,811**

[45] **Date of Patent:** **Dec. 28, 1993**

[54] **METHOD FOR QUICKLY ACQUIRING AND USING VERY LONG TRACES OF MIXED SYSTEM AND USER MEMORY REFERENCES**

[75] Inventors: **Anita Borg; David W. Wall**, both of San Mateo County, Calif.

[73] Assignee: **Digital Equipment Corporation**, Maynard, Mass.

[21] Appl. No.: **368,273**

[22] Filed: **Jun. 19, 1989**

[51] Int. Cl.⁵ G06F 9/44

[52] U.S. Cl. 395/700; 364/DIG. 1; 364/280; 364/280.4; 364/261.6; 364/262.4; 364/262.5; 364/261.3

[58] Field of Search 364/200, 900; 395/575, 395/375, 250, 700

[56]   **References Cited**

U.S. PATENT DOCUMENTS

3,673,573	6/1972	Smith	364/200
4,205,370	5/1980	Hirtle	364/200
4,462,077	7/1984	York	364/200
4,574,351	3/1986	Dang et al.	364/200
4,598,364	7/1986	Gum et al.	364/200
4,812,964	3/1989	Kiya et al.	364/200
4,819,233	4/1989	Delucia et al.	364/200

Primary Examiner—Kevin A. Kriess
Attorney, Agent, or Firm—Flehr, Hohbach, Test, Albritton & Herbert

[57]   **ABSTRACT**

The present invention utilizes link time code modification to instrument the code which is to be executed, typically comprising plurality of kernel operations and user programs. When the code is instrumented, wherever a data memory reference appears, the linker inserts a very short stylized subroutine call to a routine that logs the reference in a large, trace buffer. The same call is inserted at the beginning of each basic block to record instruction references. When the trace buffer fills up with recorded memory references, the contents of the buffer are processed, either by dumping the contents to an output device emptying the trace buffer, or a cache simulation routine is run to analyze the data. The results of the analysis are stored rather than storing the entire results of the tracing program.

16 Claims, 9 Drawing Sheets

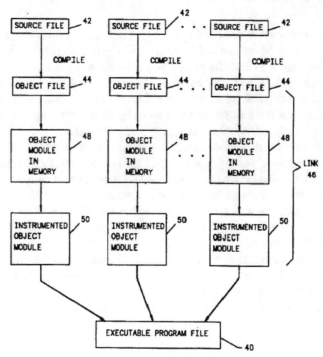

Fig. 3.9 Anita Borg's Patent No. 5,274,811. Method for Quickly Acquiring and Using Very Long Traces of Mixed System and User Memory References

Borg received many honors and recognitions including induction into the Women in Technology International Hall of Fame, the Melitta Bentz Women of Innovation and Invention Award, the Pioneer Award from the Electronic Frontier Foundation, the Augusta Ada Lovelace Award from the Association of Women in Computing, and the Heinz Award for Technology, the Economy and Environment. She was a fellow of ACM and holds two patents. In 1999, she was the presidential appointee (by President Clinton) to the Commission on the Advancement of Women and Minorities in Science, Engineering, and Technology [24–33].

3.6   Drinking Water

The average life expectancy of the population in the United States increased from 45 years of age in 1900 to 77 years of age in 2000. Half of the reason was the availability of clean water. The drinking water pioneers include Ellen Henrietta Swallow Richards who developed the first water quality tables and Ruth Patrick who established a means of identifying levels of pollution in freshwater bodies.

3.6.1   Ellen Henrietta Swallow Richards (1842–1911)

Engineering News-Record called Ellen Henrietta Swallow Richards (Fig. 3.10) "the first female environmental engineer." Her immense legacy includes seminal contributions in the areas of environmental and sanitary engineering, groundbreaking contributions toward water and air purification, leading edge analysis of food and human diet, and the design of healthier and safe buildings. Richards produced the first water purity tables in 1887, helped establish the first systematic course in sanitary engineering at MIT, founded the science of home economics, is called the "mother of ecology," and was one of the founders of what today is called the American Association of University Women.

Richards grew up in Massachusetts and graduated from the Westford Academy in 1863. She attended Vassar College as a special student, joining the senior class in her second year, and earned a bachelor's degree in chemistry in 1870. In the fall of 1870, Richards was accepted to MIT as a "special student" labeled the "Swallow Experiment." In 1872, she began her work on water testing through her association with Professor William Nichols, whose consulting work with the Massachusetts Board of Health included testing public water supplies. Nichols, who had objected to admitting women to MIT, said that the studies Richards conducted "made her a preeminent international water scientist even before her graduation" (and she was his best student!). In 1873, she was awarded a BS degree from MIT as well as an MA degree from Vassar College.

In 1876, the Women's Education Association of Boston provided the funding she had requested to open the Woman's Laboratory at MIT. Her new Laboratory, in

Fig. 3.10 Ellen Henrietta
Swallow Richards –
Library of Congress

space provided by MIT and with apparatus, books, and scholarships provided by the Women's Education Association, provided women with training in chemical analysis, industrial chemistry, biology, and mineralogy. The Laboratory was ultimately so successful that it was closed in 1883 as four women had graduated and the rest were admitted to the regular courses at MIT.

In 1882, she was instrumental in organizing the Association of Collegiate Alumnae (today known as the American Association of University Women – AAUW). The organization was founded to fight the prevailing opinion that too much education was dangerous for women as it was deemed to be dangerous to their health. The organization provided fellowships to women in order to ensure that they had the means to fund their education and worked to raise educational standards at the college level for women. Richards was a leader in efforts to improve physical education in colleges and to widen educational opportunities for women, particularly for graduate education.

Richards was appointed as an instructor of sanitary chemistry after the Woman's Laboratory closed – a position she would hold for the rest of her life. This position was located in MIT's newly established laboratory for the study of sanitation chemistry and engineering with Professor Williams Nichols as the director and Ellen Swallow Richards as his assistant. She taught air, water, and sewage analysis and treatment and introduced biology to MIT's curriculum after MIT established the first program in sanitary engineering anywhere in the United States. Her textbook *Air, Water and Food for Colleges* was published in 1900 with A.G. Woodman.

She supervised a study in 1887–1889 of the quality of Massachusetts' inland waters for the Massachusetts State Board of Health. Over 40,000 samples of water were analyzed which represented the water supply consumed by 83 percent of the state's population. In total, 100,000 analyses were conducted – a study that was unprecedented in its scale. This effort and her involvement with environmental chemistry were significant contributions to the new science of ecology, for which she is often referred to as the "mother." She helped invent new laboratory techniques and apparatuses which were needed to conduct these new analyses. Richards plotted the samples on a map of the state of Massachusetts. Through these efforts, she was able to detect geographic patterns in the chlorine data. Thus, she developed the "Normal Chlorine Map" (Fig. 3.11) showing levels of similar chlorine throughout the state via isochlors (imaginary lines linking the places that have the same levels of chlorine) that served as an early warning system for inland water pollution.[2] She also developed "water purity tables" that were the first water quality standards developed in the United States. She was the official water analyst for the Massachusetts State Board of Health for the next 10 years.

She organized the first of a series of conferences in 1899 at Lake Placid, New York, "for the betterment of the home." These conferences are credited with formally establishing the profession of home economics. Richards was instrumental in founding the American Home Economics Association established for the "improvement of living conditions in the home, the institutional household and the

Fig. 3.11 Ellen Swallow Richards' Normal Chlorine Map [16]

---

[2] Significant deviations from normal levels of chlorine are an indication of sewage contamination.

community." She believed that "One of the most serious problems of civilization is maintaining clean water and clean air, not only for ourselves but for the Planet."

Richards was a prolific author and speaker. She authored more than 30 books and pamphlets and published numerous papers in addition to the nutrition bulletins she produced for the US Department of Agriculture. Richards believed that her ideas could be called euthenics – the science of controlled environment for right living; thus, many call her the developer of sanitary engineering. A generous woman with many interests, her sister-in-law referred to her as "Ellencyclopedia."

Ellen Swallow Richards was listed in the first edition of *American Men and Women of Science* and was elevated to the rank of Fellow of the American Association for the Advancement of Science (AAAS) in 1878. She has been inducted into the National Women's Hall of Fame. She was driven to serve society and once rued that there were only 24 hours in a day: "I wish I were triplets." The Ellen Swallow Richards Professorship Fund, established in 1973 at MIT on the 100-year anniversary of her graduation, honors her achievements and is intended to strengthen the role of women on the faculty at MIT [2, 10, 34–40].

3.6.2   Ruth Patrick (1907–2013)

Dr. Ruth Patrick (Fig. 3.12) is credited with laying the groundwork for modern water pollution control efforts. Over her 60-year career, Ruth Patrick advanced the field of limnology, which is the study of freshwater biology. Patrick is recognized, along with Rachel Carson, as having ushered in the current concern for the environment and ecology.

Patrick studied botany, receiving her undergraduate degree from Coker College and both her MS and PhD degrees at the University of Virginia. She was originally hired as a "volunteer" (without pay – as women scientists at the time were not paid) in 1933 at the Academy of Natural Sciences in Philadelphia, Pennsylvania. Patrick's initial efforts were in microscopy to work with their collection of diatoms, considered to be one of the best collections in the world. Diatoms are microscopic, symmetric single-celled algae with silica cell walls. They are an important part of the food chain of freshwater ecosystems and indicators of water quality. She continued without pay until 1945 while supporting herself through part-time teaching at the Pennsylvania School of Horticulture and making chick embryo slides for Temple University.

She progressed through several positions at the Academy of Natural Sciences. In 1947, she became the curator and chairwoman of the Academy's limnology department, which she founded, today called the Patrick Center for Environmental Research.[3] In 1973, Patrick was named the Francis Boyer Research Chair of Limnology at the Academy. From 1973 to 1976, she served as chairwoman of the

---

[3] In 2011, the Academy of Natural Sciences became affiliated with Drexel University.

Fig. 3.12 Ruth Patrick.
(Courtesy of the Academy
of Natural Sciences,
Philadelphia)

Academy's board, the first woman to hold that position. Concurrently, she taught at the University of Pennsylvania. Her courses included limnology, pollution biology, and phycology. Her research included taxonomy, ecology, the physiology of diatoms, the biodynamic cycle of rivers, and the diversity of aquatic ecosystems.

Patrick gave a paper in the late 1940s at a scientific conference on her diatom research. An oil company executive in the audience was so impressed with the ability of diatoms to predict the health of a body of water that he provided the funds to support her research. With these funds, in 1948, Patrick undertook a survey of the then severely polluted Conestoga Creek in Pennsylvania: the first study of its kind. The Creek contained many types of pollution including fertilizer runoff, sewage, and waste products (some of them toxic) from industries in the area. She matched the types and numbers of diatoms in the water to the type and extent of pollution. This procedure is today used universally but was groundbreaking at the time. To aid in the effort, she invented the diatometer, a clear acrylic device that holds glass microscope slides. The diatometer collects the diatoms from bodies of water: they attach to the slides and grow there. Her research showed that healthier bodies of water contain many species of organisms. The belief that biodiversity (the number and kinds of species) is the key indicator of water health is today known as the Patrick Principle in her honor. The Patrick Principle is the foundation of all current environmental assessments.

> My great aim has been to be able to diagnose the presence of pollution and develop means of cleaning things up.

Patrick was actively involved in the drafting of the federal Clean Water Act, passed in 1972. She was called the foremost authority on America's river systems. Patrick estimated at one point that she had waded into 850 different rivers around the globe including the Amazon River. Dr. Patrick was the first woman to serve on the board of directors of the DuPont Corporation and was its first environmental activist. She also served on the board of directors of Pennsylvania Power and Light and advised Presidents Lyndon B. Johnson (on water pollution) and Ronald Reagan (on acid rain) as well as several Pennsylvania governors on water quality issues. She served on water pollution and water quality panels for the National Academy of Sciences and the US Department of Interior as well as other federal advisory groups.

Patrick was elected to the National Academy of Sciences in 1970, as the 12th woman to receive this form of recognition. Patrick received the National Medal of Science in 1996 from President Bill Clinton "for her algal research, particularly the ecology and paleoecology of diatoms, and for elucidating the importance of biodiversity of aquatic life in ascertaining the natural condition of rivers and the effects of pollution." She was elected as a Fellow of the American Academy of Arts and Science in 1976 and was the recipient of over 25 honorary degrees. Dr. Patrick has been inducted into the National Women's Hall of Fame [10, 36, 38, 41–47].

3.7 Energy

The energy pioneers include Edith Clarke, who literally wrote the book on how to evaluate transmission lines for electric grids; Maria Telkes, who was an advocate for solar energy; Ivy Parker, who investigated corrosion in pipelines; and Ada Pressman, who helped develop control systems for electric utility power plants.

3.7.1 Edith Clarke (1883–1959)

A woman engineer with many firsts to her name, Edith Clarke (Fig. 3.13) grew up in Maryland without any intentions of even going to college. After graduating from Vassar with an AB in mathematics and astronomy in 1908 (Phi Beta Kappa), Clarke taught math and science for 3 years in San Francisco and West Virginia. But, teaching was not holding her interest, and she decided to pursue becoming an engineer instead. She enrolled as a civil engineering undergraduate student at the University of Wisconsin and remained there for a year. Then, she went to work for American Telephone and Telegraph Company (AT&T) as a computing assistant. She intended to return to the University of Wisconsin to complete her engineering studies but found the work so interesting at AT&T that she stayed for 6 years.

During World War I, she supervised the women at AT&T who did computations for research engineers in the Transmission Department. She simultaneously studied radio at Hunter College and electrical engineering at Columbia University at night.

Fig. 3.13 Edith Clarke.
(Courtesy of the Society of
Women Engineers
Photograph Collection,
Walter P. Reuther Library,
Wayne State University)

Eventually, she enrolled at MIT and received her master's degree in electrical engineering in 1919: the first woman awarded that degree from MIT. Upon graduation, she wanted to work for either General Electric (GE) or Westinghouse. But even with her stellar credentials, no one would hire her as an engineer because of her gender – they had no openings for a woman engineer! In 1920, after a long job search, GE offered Clarke a computing job, directing women computers who were calculating the mechanical stresses in turbines for the turbine engineering department at GE.

But, Clarke wanted to be an electrical engineer! Since that was not the job she was offered and since she wanted to travel the world, she left GE in 1921 to teach physics at the Constantinople Women's College (now Istanbul American College) in Turkey. A year later GE did offer her a job as an electrical engineer in the central station engineering department. When she accepted this job, she became the first professionally employed female electrical engineer in the United States.

Clarke's area of specialty was electric power systems and problems related to its operation. She made innovations in long-distance power transmission and the development of the theory of symmetrical components and circuit analysis. Symmetrical components are a mathematical means by which engineers can study and solve problems of power system losses and performance of electrical equipment. Clarke literally wrote the textbook *Circuit Analysis of AC Power Systems; Symmetrical and Related Components* (1943) and a second volume in 1950. This textbook, in its two volumes, was used to educate all power system engineers for many years.

She published 18 technical papers during her employment at GE reflecting her status as an authority on the topics of hyperbolic functions, equivalent circuits, and graphical analysis within electric power systems. "Simplified Transmission Line Calculations," which appeared in the *General Electric Review* in May 1926, provided charts for transmission line calculations. She was also involved in the design of hydroelectric dams in the Western United States.

Clarke received a patent in 1925 (1,552,113) for her "graphical calculator" – a method of considering the impacts of capacity and inductance on long electrical transmission lines. It greatly simplified the calculations that needed to be done. In 1926, she was the first woman to address what is today the Institute of Electrical and Electronics Engineers (IEEE) – at the time, it was the American Institute of Electrical Engineers (AIEE). Her topic was "Steady-State Stability in Transmission Systems." In 1932, Clarke became the first woman to present a paper before the AIEE; her paper, "Three-Phase Multiple-Conductor Circuits," was named the best paper of the year in the northeastern district. This paper examined the use of multiple conductor transmission lines with the aim of increasing the capacity of the power lines. In 1948, Clarke was named one of the first three women fellows of IEEE. She had previously become the first female full voting member of IEEE. Clarke was one of the few women who were licensed professional engineers in New York State.

A year after her retirement from GE in 1945, Clarke became an associate professor of electrical engineering at the University of Texas. In 1947, she rose to full professorship becoming the first woman professor of electrical engineering in the United States. She served on numerous committees and provided special assistance to graduate students through her position as graduate student advisor.

In 1954, Clarke received the SWE Achievement Award "in recognition of her many original contributions to stability theory and circuit analysis." In 2015, she was posthumously inducted into the National Inventors Hall of Fame for her invention of the graphical calculator [1, 10, 15, 36, 48, 49].

3.7.2 Maria Telkes (1900–1995)

A celebrated innovator in the field of solar energy, one of the first people to research practical ways for humans to use solar energy, and the so-called Sun Queen, Maria Telkes (Fig. 3.14) was born in Budapest, Hungary. She built her first chemistry laboratory when she was 10 years old. Educated at Budapest University as a physical chemist (BA in 1920 and PhD in 1924), she became interested in solar energy as early as her freshman year in college when she read a book titled *Energy Sources of the Future* by Kornel Zelowitch, which described experiments with solar energy that were taking place, primarily in the United States.

Telkes served as an instructor at Budapest University after receiving her PhD. Her life changed significantly, however, when she traveled to Cleveland, Ohio, to visit her uncle who was the Hungarian consul. During her lengthy visit, she was offered a position as a biophysicist at the Cleveland Clinic Foundation working with American surgeon George Washington Crile. She accepted in 1925. Telkes would spend her entire professional career in the United States.

In 1937, the same year she became a naturalized citizen, Telkes began her employment with Westinghouse Electric where for 2 years she developed and patented instruments for converting heat energy into electrical energy, so-called thermoelectric devices. In 1939, she began her work with solar energy as part of the

Fig. 3.14 Maria Telkes receiving the Society of Women Engineers' Achievement Award. (Courtesy of the Society of Women Engineers Photograph Collection, Walter P. Reuther Library, Wayne State University)

Solar Energy Conversion Project at MIT. Initially, her role was working on thermo-electric devices that were powered by sunlight. During World War II, Telkes served as a civilian advisor to the US Office of Scientific Research and Development (OSRD) where she was asked to figure out how to develop a device to convert salt-water into drinking water.

This assignment resulted in one of her most important inventions, a solar distiller that vaporized seawater and then recondensed it into drinkable water. Its significant advancement used solar energy (sunlight) to heat the seawater so that the salt was separated from the water. This distillation device (also referred to as a solar still) was included in the military's emergency medical kits on life rafts and saved the lives of both downed airmen and torpedoed sailors. It could provide one quart of freshwater daily through the use of a clear plastic film and the heat of the sun and was very effective in warm, humid, tropical environments. Later, the distillation device was scaled up and used to supplement the water demands of the Virgin Islands. For her work, Telkes received the OSRD Certificate of Merit in 1945.

Telkes was named an associate research professor in metallurgy at MIT in 1945. During her years at MIT, she created a new type of solar heating system – one that converted the solar energy to chemical energy through the crystallization of a sodium sulfate solution (Glauber's salt). In 1948, Telkes and architect Eleanor Raymond developed a prototype five-room home built in Dover, Massachusetts.

Called the Dover Sun House, this was the world's first modern residence heated with solar energy, and it used Telkes's solar heating system. The system was both efficient and cost-effective.

She next spent 5 years at New York University (NYU) (1953–1958) as a solar energy researcher. At NYU, Telkes established a laboratory dedicated to solar energy research and continued working on solar stills, heating systems, and solar ovens. Her solar ovens proved to be cheap to make and simple and easy to build and could be used by villagers worldwide. Her work also led her to the discovery of a faster way to dry crops. In 1954, she received a $45,000 grant from the Ford Foundation to further develop her solar ovens.

After NYU, she worked for Curtiss-Wright Company as director of research for their solar energy laboratory (1958–1961). Here, she worked on solar dryers as well as the possible use of solar thermoelectric systems in outer space. She also designed the heating and energy storage systems for a laboratory building constructed by her employer in Princeton, New Jersey. This building included solar-heated rooms, a swimming pool, laboratories, solar water heaters, dryers for fruits and vegetables, and solar cooking stoves.

In 1961, she moved to Cryo-Therm where she spent 2 years as a researcher working on space-proof and sea-proof materials for use in protecting sensitive equipment from the temperature extremes that would be experienced in those environments. Her work at Cryo-Therm was used on both the Apollo and Polaris projects. Subsequently, she served as the director of Melpar, Inc.'s solar energy laboratory looking at obtaining freshwater from seawater (1963–1969) before returning to academia at the University of Delaware. At the University of Delaware, Telkes served as a professor and research director for the Institute of Energy Conversion (1969–1977) and emerita professor from 1978. Here she worked on materials used to store solar energy as well as heat exchangers that could efficiently transfer energy. The experimental solar-heated building constructed at the University of Delaware, known as Solar One, used her methods. In addition, she researched air-conditioning systems that could store coolness during the night to be used during the heat of the following day.

After her retirement, she continued to serve as a consultant on solar energy matters. In 1980, after the 1970s oil crisis and a renewed interest nationwide in solar energy, Telkes was involved with a second experimental solar-heated house, the Carlisle House, which was built in Carlisle, Massachusetts.

In 1952, Telkes was the first recipient of SWE's Achievement Award. The citation reads "In recognition of her meritorious contributions to the utilization of solar energy." In 1977, she received the Charles Greeley Abbot Award from the American Section of the International Solar Energy Society which was in recognition of her being one of the world's foremost pioneers in the field of solar energy. In that same year, she was honored by the National Academy of Sciences Building Research Advisory Board for her work in solar-heated building technology. The holder of more than 20 patents (shown in Table 3.1), in 2012, Telkes was inducted into the National Inventors Hall of Fame. In addition to her patents, Telkes also had many publications on the topics of using the sunlight for heating, thermoelectric/solar

Table 3.1 Maria Telkes patents

Number	Date	Title
2,229,481	January 21, 1941	Thermoelectric couple
2,229,482	January 21, 1941	Thermoelectric couple
2,246,329	June 17, 1941	Heat absorber
2,289,152	July 7, 1942	Method of assembling thermoelectric generators
2,366,881	January 9, 1945	Thermoelectric alloys
2,595,905	May 6, 1952	Radiant energy heat transfer device
2,677,243	May 4, 1954	Method and apparatus for the storage of heat
2,677,367	May 4, 1954	Heat storage unit
2,677,664	May 4, 1954	Composition of matter for the storage of heat
2,808,494	October 1, 1957	Apparatus for storing and releasing heat
2,856,506	October 14, 1958	Method for storing and releasing heat
2,915,397	December 1, 1959	Cooking device and method
2,936,741	Mary 17, 1960	Temperature-stabilized fluid heater and a composition of matter for the storage of heat therefor
2,989,856	June 27, 1961	Temperature-stabilized container and materials therefor
3,206,892	September 21, 1965	Collapsible cold frame
3,248,464	April 26, 1966	Method and apparatus for making large-celled material
3,270,515	September 6, 1966	Dew collecting method and apparatus
3,415,719	December 10, 1968	Collapsible solar still with water vapor-permeable membrane
3,440,130	April 22, 1969	Large-celled material
3,695,903	October 3, 1972	Time/temperature indicators
3,986,969	October 19, 1976	Thixotropic mixture and making of same
4,010,620	March 8, 1977	Cooling system
4,011,190	March 8, 1977	Selective black for absorption of solar energy
4,034,736	July 12, 1977	Solar heating method and apparatus
4,187,189	February 5, 1980	Phase change thermal storage materials with crust forming stabilizers
4,250,866	February 17, 1981	Thermal energy storage to increase furnace efficiency
4,954,278	September 4, 1990	Eutectic composition for coolness storage

generators and distillers, and the electrical conductivity properties of solid electrolytes. She believed so strongly in using solar energy that she said, "Sunlight will be used as a source of energy sooner or later … Why wait?" [10, 50–57].

3.7.3   Ivy Parker (1907–1985)

Dr. Ivy Parker (Fig. 3.15), chemist and research engineer, became a specialist in the causes and prevention of corrosion of pipelines. Parker graduated with her undergraduate degree in chemistry from West Texas State University and earned her master's and PhD degrees in organic chemistry from the University of Texas. She was the first woman to earn a PhD in chemistry from the university. Parker pursued a career initially in academia but then switched to industry.

She worked for Shell Oil in the 1930s and early 1940s and later was employed by the Plantation Pipe Line Company in Atlanta, Georgia. She was elected the first editor of *Corrosion* magazine in 1944, the official publication of the National Association of Corrosion Engineers.

Parker published many papers on the causes and prevention of corrosion in pipelines. The pipelines in the United States expanded after World War II, and Parker was there conducting research on the "quality of product" that flowed in the pipeline. Her corrosion protection work included innovation on both water- and oil-soluble inhibitors. She also researched ways to keep the pipeline clean, filtration,

Fig. 3.15 Ivy Parker at work. (Courtesy of the Society of Women Engineers Photograph Collection, Walter P. Reuther Library, Wayne State University)

tank painting, and tank seals. Parker became a Fellow of the American Institute of Chemists, and an endowed scholarship in her memory was established by SWE [2, 10, 58].

3.7.4   Ada Pressman (1927–2003)

Ada Pressman (Fig. 3.16) was a pioneer in combustion control and burner management for supercritical power plants including the input logic and fuel air mixes associated therewith. She was directly involved in early design efforts toward more automated controls of equipment and systems, the new packaging techniques, and breakthroughs in improved precision and reliability of sensors and controls. As she progressed through the management ranks at Bechtel (earning her MBA during the process), she was recognized as one of the nation's outstanding experts in power plant controls and process instrumentation and worked on fossil-fired and nuclear power plants. Pressman is credited with significantly improving the safety of both coal-fired and nuclear power plants for workers as well as nearby residents.

Planning to become a secretary after she graduated from high school in Ohio, Pressman was encouraged to attend college by her father. She earned her BS in mechanical engineering from The Ohio State University. Pressman characterized her professional experience as including the engineering management of millions of individual hours of power generation plant design and construction and of economic studies and proposals for potential projects. She continually monitored the costs for each project as well as the technical engineering details as the design progressed. A dedicated advocate for women who served as Society President of SWE, Pressman

Fig. 3.16 Ada Pressman at work. (Courtesy of the Society of Women Engineers Photograph Collection, Walter P. Reuther Library, Wayne State University)

received SWE's Achievement Award in 1976 "For her significant contribution in the field of power control systems engineering" [59, 60].

3.8 Hazardous Waste

In the area of hazardous waste, Joan Berkowitz is profiled – who investigated sources and means of handling hazardous wastes, particularly from industrial facilities.

3.8.1 Joan Berkowitz (1931–2020)

The first woman president of the Electrochemical Society, chemist Joan Berkowitz (Fig. 3.17) excelled in science from an early age. After graduating Phi Beta Kappa from Swarthmore College, she wanted to attend Princeton University for graduate school – but women were not accepted at Princeton at the time. Berkowitz attended the University of Illinois at Urbana-Champaign graduating with a PhD in physical chemistry in 1955. After postdoctoral work at Yale University, funded by the

Fig. 3.17 Joan Berkowitz.
(Courtesy of the Society of
Women Engineers
Photograph Collection,
Walter P. Reuther Library,
Wayne State University)

National Science Foundation, she joined the consulting firm of Arthur D. Little in 1957 and was promoted to Vice President of Hazardous Waste Management Consulting in 1980.

Berkowitz characterized the field of hazardous waste management as embryonic when she entered it in 1972. As to how she entered this new field, she says:

> One of my colleagues needed someone to write about hazardous waste generation and disposal in the tanning industry. The work had to be done between Christmas and New Year's Day. I was the only volunteer.... . I was looking for a new field, and hazardous waste had everything.

Her accomplishments in hazardous waste management include:

- Contributed to the First Report to Congress on Hazardous Waste in 1973.
- Developed a system for classifying, sampling, and analyzing industrial hazardous wastes in 1974.
- Participated in the first major field program sponsored by the Environmental Protection Agency in 1975 to evaluate the environmental effects of incineration and pyrolysis of hazardous wastes.
- Served as the principal author of a book on alternatives to landfill and incineration in the management of hazardous wastes.
- Identified hazardous waste streams for which land-farming might be the most economically sound and economic waste management alternative.
- Directed the development of computer simulation to model the transport of chemicals from inactive disposal sites and to assess the environmental benefits and economic costs of alternative remedial actions.
- Conducted one of the first studies that evaluated alternative methods for clean-up of inactive disposal sites. This 1975 study was conducted for the Rocky Mountain Arsenal.
- Directed programs to assess the hazardous waste management facility needs of New England and the State of Maryland.

Berkowitz received the 1983 SWE Achievement Award with a citation that read "For significant contributions in the field of hazardous waste management." She was active in a number of organizations in addition to the Electrochemical Society. Berkowitz served on their boards of directors and edited journals [61–66].

3.9   Inland Waterways

Hydraulic engineering pioneer Margaret Petersen contributed significantly to the category of inland waterways – from the Panama Canal to many projects in the United States for the US Army Corps of Engineers and, later, as a professor and author.

3.9.1   Margaret S. Petersen (1920–2013)

Margaret Petersen (Fig. 3.18) didn't start out to be an engineer. She joined the US Army Corps of Engineers (COE) in Rock Island, Illinois, in June of 1942 and worked as a draftsman for the Rock Island District. Later that year, Petersen was tapped as one of the ten draftsmen selected to go to Panama and work on the Three Locks Project for the Panama Canal. When she returned to the United States, she enrolled at the University of Iowa and earned her BS in civil engineering in 1947 while working part time at a COE sub-office at the University. She was hired by the COE in the Vicksburg, Mississippi office as a hydraulic engineer at the Waterways Experiment Station after her graduation.

There she worked on the design and operation of the Mississippi Basin Model until she returned to the University of Iowa and earned her master's degree in mechanics and hydraulics in 1953. She returned to the COE, working at the Missouri River Division in Omaha, Nebraska. She worked on designs of spillways and other projects on the Missouri River during her time there. Then she went to the Little Rock District where she focused on river engineering for the Arkansas River. In 1961, Petersen became Chief of the Channel Hydraulics Investigation Section. A former student trainee recalls the early days of his career at the Hydraulics Branch of the COE:

> Fifty-one years ago I went to work as a student trainee in the Hydraulics Branch of the Corps of Engineers Little Rock District. More than 1,000 employees of the District were involved in the planning and design work for the 15 locks and dams of the Arkansas River

Fig. 3.18 Margaret S. Petersen, PE. (Courtesy of the US Army Corps of Engineers)

Navigation System to be located in the Little Rock District. Almost all of the student trainees assigned to the Hydraulics Branch spent at least a few weeks working under Margaret Petersen's supervision. There might have been a few of us who had heard that there were female civil engineers; maybe one or two of us had actually seen a female civil engineer; but none of us had ever worked for a female civil engineer (or even dreamed that we would!). We quickly learned, however, that working for Miss Petersen was going to be one of the most valuable educational experiences of our life… . Only Margaret knew that the work we were doing was at the cutting edge of American hydraulic engineering …

In 1964, she returned to Vicksburg as Chief of the Wave Dynamics Section. Later that same year, she moved to the Sacramento District as a project engineer in the planning branch. She became Chief of the Marysville Lake Investigations Section and worked in the Sacramento District until her retirement in 1977.

But her career was not over. Although she had not been interested in teaching and was afraid of public speaking, in 1980, she became a visiting associate professor at the University of Arizona in the Department of Civil Engineering and Engineering Mechanics. During her academic career, Petersen developed four graduate-level courses in hydraulic engineering: Hydropower Engineering, River Engineering, Flow Through Hydraulic Structures, and River Basin and Project Planning. Petersen also authored three books: *Water Resource Planning and Development* (1984), *River Engineering* (1986), and *Water Resources Development in Developing Countries* (co-authored with David Stephenson of South Africa and published in England in 1991). She wrote more than 100 reports and papers over the course of her career. Petersen became an emeritus professor in 1991 but didn't actually fully retire from teaching until 1997.

Petersen was an active mentor and supporter for her students. She endowed two scholarships at the University of Iowa for female engineering undergraduates. Active in ASCE throughout her career, Petersen was a Fellow of ASCE, was only the second female honorary member of the ASCE (after Elsie Eaves), received ASCE's Hunter Rouse Hydraulic Engineering Award, was honored with the Environmental and Water Resources Institute's (EWRI) first Lifetime Achievement Award, and received the University of Iowa's Distinguished Alumni Achievement Award.

In 2010, the EWRI established the Margaret Petersen Outstanding Woman of the Year Award which recognizes women whose commitment to the water resources profession and active engagement in ASCE and EWRI are modeled on Petersen's involvement. ASCE established the Margaret S. Petersen Award in 2013 in honor of her pioneering work in hydraulics and water resources engineering. The Award is for an outstanding woman in environmental and water resources [67–72].

3.10   Public Parks

Public parks across the country have benefited from the efforts of women. Mary Colter designed many of the buildings at the Grand Canyon. Architect Julia Morgan is most known for her work on the Hearst Castle – now a California state park. Marjory Stoneman Douglas dedicated many years of her life to preserving the Everglades. Mardy Murie is considered the grandmother of the modern conservation movement. Gale Norton was the first female Secretary of the Interior.

3.10.1   Mary Colter (1869–1958)

Originally a schoolteacher in St. Paul, Minnesota, Mary Colter became an architect, designer, and decorator for the Fred Harvey Company in 1902. The Fred Harvey Company had operated the gift shops, newsstands, restaurants, and hotels of the Atchison, Topeka, and Santa Fe Railway since 1876. Fred Harvey made an agreement with AT & Santa Fe in which the railway would build and own the station hotels and restaurants, while he would manage them and provide good food and service for reasonable prices. Colter created those Fred Harvey hotels along the railroad and at the Grand Canyon, putting the Southwest on the map.

Colter demonstrated a new style of architecture – one that grew out of the surrounding land. Her buildings pay homage to the early inhabitants of the region as she used Hopi, Navajo, Zuni, and Mexican motifs. Her buildings have the simplicity of the early architecture after which they are patterned. Colter also developed what is today called "National Park Service Rustic." For these buildings, she used local and natural materials. In 1925 at the La Fonda hotel in Santa Fe, New Mexico, her interior design and decorating featured local craftsmen from the pueblos; this created what is today called the Santa Fe Style. At Grand Canyon National Park, her buildings included the Hopi House (1905) (Fig. 3.19), Desert View Watchtower (1932), Hermit's Rest (1914), Lookout Studio (1914), Bright Angel Lodge (1935), and Phantom Ranch (1922). The creator of Mimbreno china and flatware that was used on the trains, Colter also decorated the exteriors of the train stations in St. Louis, Chicago, and Los Angeles [73, 74].

3.10.2   Julia Morgan (1872–1957)

Best known for her work as the architect on the Hearst Castle in San Simeon, California, Julia Morgan was the first woman to be licensed as an architect in California. She graduated from the University of California, Berkeley, with a degree in civil engineering (the only woman in her class) in 1894. Mentored by an architect who was a lecturer in her senior year, she journeyed to Paris with the hopes of

Fig. 3.19 Hopi House. (Courtesy of the Library of Congress)

studying to be an architect. In 1897, the École Nationale supérieure des Beaux-Arts succumbed to pressure from French women artists and decided to admit women. Morgan was admitted a few years later and graduated in 1902.

Morgan's architectural practice in San Francisco, established in 1904, was especially busy after the 1906 San Francisco earthquake. Her many projects for the YWCA included the Asilomar Conference Center in Pacific Grove, California. After accepting more than 450 commissions, she was quite famous and was selected by William Randolph Hearst to serve as the architect for his ranch at San Simeon, California. His simple instructions to her in 1919 were: "Miss Morgan, we are tired of camping out in the open at the ranch in San Simeon and I would like to build a little something." By 1947, Morgan had created an estate of 165 rooms in numerous buildings; there were also 127 acres of gardens, terraces, pools, and walkways. The ornate Roman Pool at the Hearst Castle that she designed is shown in Fig. 3.20.

Today, the project on which she worked for 20 years is called the Hearst Castle and is a California State Park. It is listed as both a National Historic Landmark and a California Historic Landmark. In 2014, Julia Morgan became the first woman awarded the American Institute of Architects' Gold Medal. The award was established in 1907 [75, 76].

3.10.3   Marjory Stoneman Douglas (1890–1998)

In 1947, when Marjory Stoneman Douglas' book *The Everglades: River of Grass* was published, the Everglades were regarded as a breeding ground for disease and were in danger of being paved over. Her book, buttressed by the establishment of the Everglades as a national park, changed the perspective that the Everglades were a swamp that needed to be drained and reclaimed to the understanding of the value and need for its ecosystem.

Fig. 3.20 The Roman Pool at the Hearst Castle designed by Julia Morgan. (Courtesy of the Library of Congress)

Douglas moved to Florida in 1915 to work as a reporter for the *Miami Herald*. Originally a society reporter, she became a crusader for women's rights, conservation, and social justice. Douglas became the public voice for preservation of the Everglades. Her book describing the Everglades opened with the following words:

> There are no other Everglades in the world. They are, they have always been, one of the unique regions of the earth; remote, never wholly known. Nothing anywhere else is like them...

Douglas often found herself opposing the Army Corps of Engineers in her efforts to save the Everglades. In 1970, she established a voting constituency, Friends of the Everglades. A tiny woman, known for her trademark hats and glasses, she commanded attention whenever she spoke. In 1986, the National Parks Conservation Association established the Marjory Stoneman Douglas Award "to honor individuals who often must go to great lengths to advocate and fight for the protection of the National Park System." After her death in 1998, at the age of 108, in a fitting tribute, her ashes were spread in the park she loved so much.

Among her many honors, Douglas received the Presidential Medal of Freedom with the following citation:

> Marjory Stoneman Douglas personifies passionate commitment. Her crusade to preserve and restore the Everglades has enhanced our Nation's respect for our precious environment, reminding all of us of nature's delicate balance. Grateful Americans honor the 'Grandmother

of the Glades' by following her splendid example in safeguarding America's beauty and splendor for generations to come.

Douglas has been inducted into the National Women's Hall of Fame. President Clinton, who had presented her with the Presidential Medal of Freedom, said upon her death "Long before there was an Earth Day, Mrs. Douglas was a passionate steward of our nation's natural resources, and particularly her Florida Everglades" [77–80].

3.10.4 Margaret "Mardy" Murie (1902–2003)

The first woman to graduate from the University of Alaska, Margaret "Mardy" Murie was known as the grandmother of the modern conservation movement. She and her husband Olaus were instrumental in the establishment of the Grand Teton National Park in 1929, and the work of Mardy and her husband Olaus was pivotal to the designation of the Arctic National Wildlife Refuge in 1960. She worked to sow the seeds for the 1964 Wilderness Act and was in the Rose Garden when President Lyndon Johnson signed it into law. She also worked for passage of the Alaska Lands Act, signed by President Carter in 1980.

Mardy was named an honorary ranger by the National Park Service. She founded the Teton Science Schools in Jackson Hole, Wyoming, to teach students of all ages the value of ecology. Her steadfast and inspiring efforts to safeguard America's wilderness for future generations merited her the Presidential Medal of Freedom in 1998. Murie's 77 acre estate in Moose, Wyoming, was bequeathed to the National Park Service. Her legacy will continue through the Murie Center located there [81–83].

3.10.5 Gale Norton (1954–)

Gale Norton (Fig. 3.21) was the first woman to serve as the Secretary of the US Department of the Interior. She received her BA in political science at the University of Denver and her law degree from the University of Denver, College of Law. Prior to her election as Colorado's Attorney General, she worked as Associate Solicitor of the US Department of the Interior where she oversaw endangered species and public lands' legal issues for the National Park Service and the US Fish and Wildlife Service. Her other positions included Assistant to the Deputy Secretary of Agriculture, National Fellow of Stanford University's Hoover Institution, and Senior Attorney for Mountain States Legal Foundation. She returned to private practice after her years as Attorney General.

President George W. Bush nominated Norton to serve as the Secretary of the US Department of the Interior: the first woman to hold that position. As Secretary from

Fig. 3.21 Gale Norton.
(Courtesy of the National
Park Service)

2001 to 2006, Norton headed a federal agency that managed over 500 million acres
(over 20 percent of the land area of the United States) with an appropriated budget
of over $10 billion and 70,000 employees. Among other major responsibilities, she
oversaw 388 national parks, 545 national wildlife refuges, over 2000 dams, 184
Indian schools, and numerous scientific facilities.

After leaving Interior, Norton became General Counsel for Royal Dutch Shell
Unconventional Oil. She later established Norton Regulatory Strategies. She
remains active in the public interest sphere, having chaired the National Park
Foundation and the Migratory Bird Conservation Commission. Norton serves as a
board member of the University of Colorado Renewable and Sustainable Energy
Institute and as a founding member of the Conservation Leadership Council
[84–86].

3.11 Rail

Women railroad pioneers focused on transporting food and passenger comfort.
Mary Engle Pennington ensured that food could be safely transported. Olive Dennis
designed equipment to ensure passenger safety and comfort. Mary Colter, profiled
in the section on Public Parks, created the Mimbreno china and flatware that was
used on the trains of the Atchison, Topeka, and Santa Fe Railway.

3.11.1   Mary Engle Pennington (1872–1952)

Mary Engle Pennington was the first woman member of the American Society of Refrigerating Engineers. Her picture hangs today at its successor organization – the American Society of Heating, Refrigerating and Air-Conditioning Engineers. She later became the president of the American Institute of Refrigeration. In 1947, she was elected a fellow of the American Society of Refrigerating Engineers and a fellow of the American Association for the Advancement of Science.

Pennington completed the coursework for a bachelor's degree in chemistry, biology, and hygiene at the University of Pennsylvania, but at that time (1892), the University did not grant bachelor's degrees to women. Instead, she received a Certificate of Proficiency in biology. She continued her studies and, in 1895, received a PhD in chemistry from the University of Pennsylvania.

Her work in refrigeration led to her appointment as head of the Department of Agriculture's food research laboratory. As she used the name "M.E. Pennington," not everyone was aware that she was a woman. In 1916, when she had been chief of the Food Research Laboratory for a decade, a railroad vice president on whom she called instructed his secretary "to get rid of the woman," because he had "an appointment with Dr. Pennington, the government expert."

Pennington developed standards of milk and dairy inspection that were adopted by health boards throughout the country. Her methods of preventing spoilage of eggs, poultry, and fish were adopted by the food warehousing, packaging, transportation, and distribution industries. She has six patents associated with refrigeration and spoilage prevention methods (Fig. 3.22). The standards she established for refrigeration railroad cars, which were informed by the time she spent riding freight trains, remained in effect for many years and gained her worldwide recognition as a perishable food expert. Pennington received the Garvan Medal from the American Chemical Society in 1940 and was the first woman elected to the American Poultry Historical Society's Hall of Fame (1947). She has been inducted into the National Women's Hall of Fame [1, 10, 40, 87].

3.11.2   Olive Dennis (1885–1957)

Olive Dennis first studied mathematics and science at Goucher College. After several years as a teacher, she completed a degree in civil engineering from Cornell (1920), with a specialization in structural engineering. That fall, she went to work in the bridge department of the Baltimore and Ohio Railroad. However, the president of the B&O railroad had other ideas about how she could contribute to the organization. After 14 months in the bridge department, Dennis was promoted to the position of Engineer of Service for the railroad, riding the rails and figuring out ways to make the railroad more accommodating to its passengers.

Oct. 11, 1932.   M. E. PENNINGTON   1,882,030

CONDITIONING SYSTEM FOR COLD STORAGE ROOMS

Filed Oct. 18, 1929

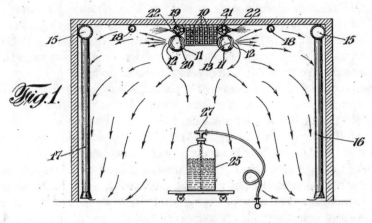

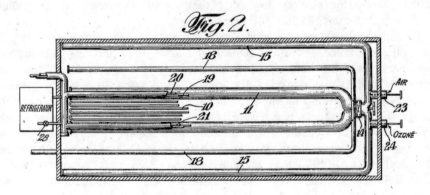

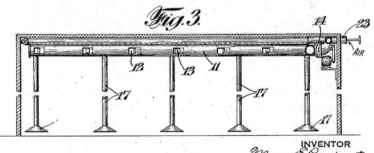

INVENTOR
Mary E. Pennington
BY
Kenyon & Kenyon
ATTORNEYS.

Fig. 3.22 Mary Engle Pennington Patent for Refrigeration

During her years with the railroad, Dennis pushed for better lighting and better seating (cleaner, better fabrics and lower, reclining seats) in the coach cars. She was an advocate for air conditioning in the cars, and she designed and received a patent for an individually operated ventilator (Fig. 3.23). Dennis even designed the blue

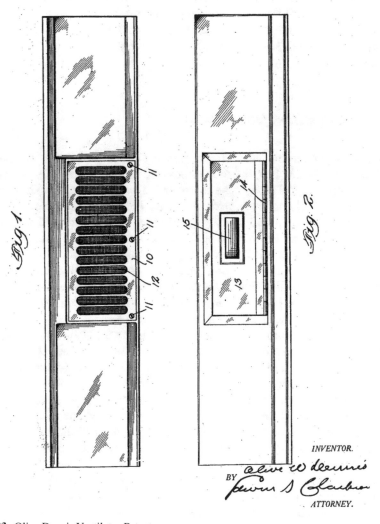

Nov. 27, 1928.   1,693,108

O. W. DENNIS

VENTILATOR

Filed Dec. 13, 1927   2 Sheets-Sheet 1

INVENTOR.

BY

ATTORNEY.

Fig. 3.23 Olive Dennis Ventilator Patent

colonial china provided in the dining car. At the Women's Centennial Congress in New York in 1940, she was named by Carrie Chapman Catt as 1 of the 100 outstanding career women in the United States [1].

3.12   Roads

For more than 100 years now, humans have been able to travel by automobile – and the road system was developed to accommodate those mechanical beasts. But, it took Bertha Benz to demonstrate that we could travel by automobile and that cars had a future. Alice Ramsey demonstrated the feasibility of cross-country travel in the United States at a time before gas stations and interstates. Civil engineer Marilyn Jorgenson Reece is known for her spiral design of the San Diego-Santa Monica freeway interchange (I-10/405). Janet Bonnema fought for the right of women to work on tunnels.

3.12.1   Bertha Benz (1849–1944)

The business partner and wife of automotive pioneer Karl Benz, Bertha Benz (Fig. 3.24) took the first long-distance automobile trip. She proved the feasibility of travel by automobile and brought the company (today Mercedes Benz) its first sales. The 120-mile-long Bertha Benz Memorial Route honors her 1888 trip.

Bertha Benz grew up in a wealthy family in Germany and used part of what would be her dowry when she married to support the iron construction company of Karl Benz. After they married, she no longer controlled her money under German law at the time, and Karl Benz used the money as financial support for his new horseless carriage that he completed in 1885. She was a field tester in the business, invented leather brake pads, and improved the fuel line design. In addition, she understood marketing; Karl was a tinkerer not a marketer.

In 1888, without informing her husband, she and two of their sons left home in the Patent-Motorwagen No. 3. This automobile had one front wheel and two rear wheels and could achieve a speed of 25 miles per hour. This journey took place before there were roads or gas stations or any of the services that we encounter today. Following wagon trails, the intrepid trio progressed. When fuel ran low, they bought ligroin, a petroleum solvent, from a local pharmacy. She used her garter to make an emergency repair to the car's ignition. A fuel line problem was solved with her hairpin. When the wooden brakes began to fail, she had a local cobbler make leather soles – the first brake pads. The 65 miles from their home in Mannheim to her mother's home in Pforzheim took about 12 hours. Although she informed her husband after her arrival of her whereabouts by telegram, the townspeople and press along the way already knew about her exploits. She returned to Mannheim by a different route a few days later, covering 120 miles in total.

Fig. 3.24 Bertha Benz.
(Courtesy of Wikipedia)

The trip had a number of positive outcomes. Karl was convinced to add gearing to the car – as his sons had had to push the car uphill numerous times during the trip. The trip generated a tremendous amount of publicity and orders began to come in. Within a decade, Benz & Cie. was the largest automobile company in the world.

Karl Benz said of Bertha, "Only one person remained with me in the small ship of life when it seemed destined to sink. That was my wife. Bravely and resolutely she set the new sails of hope" [88, 89].

3.12.2   Alice Huyler Ramsey (1886–1983)

Following in the footsteps of Bertha Benz, Alice Ramsey (Fig. 3.25) became the first woman to drive coast to coast in the United States in 1909 when she was 22 years old. She drove 3800 miles from New York to California over the course of 59 days.

The plan for Ramsey to undertake the first female cross-country trip was set in motion after her husband bought her a car with which she entered an endurance race. At this point in our country's history, very few people owned cars, and very few women drove – this is before women had the right to vote. Automaker Maxwell-Briscoe provided the automobile and paid the expenses of the four women who

Fig. 3.25 Alice Ramsey. (Courtesy of Wikipedia)

embarked on this journey. Ramsey was accompanied by three non-drivers, two sisters-in-law and a friend.

At a time when roads hadn't been built for long-distance driving and maps, service stations, and other niceties didn't exist, Ramsey and her crew persevered. Often accompanied by enthusiastic locals who wanted a glimpse of the women, they overcame a host of mechanical problems with the car. When they arrived in California, the San Francisco newspaper headline was "Pretty Women Motorists Arrive After Trip Across the Continent."

Ramsey would make more than 30 cross-country drives during her lifetime. The Automobile Manufacturers' Association named her the "First Lady of Automotive Travel" in 1960 for her 1909 trek. Her account of the trip was published in a 1961 book with the title *Veil, Duster and Tire Iron*. In 2000, she was the first woman inducted into the Automotive Hall of Fame [90, 91].

3.12.3 Marilyn Jorgenson Reece (1926–2004)

Marilyn Jorgenson Reece studied civil engineering at the University of Minnesota because as she later said, "Well, I like mathematics and I didn't want to be a teacher." After her graduation in 1948, she moved to California where she became the first female engineer for California's Division of Highways (now Caltrans). In 1954, she became the first female licensed civil engineer in the state. Later, she would become the state's first woman resident engineer for construction projects.

Reece is known for her spiral design of the San Diego-Santa Monica freeway interchange (I-10/405) which is not only noteworthy for its looks but was specifically designed to accommodate high-speed traffic. Reece spoke of the curvature of the interchange and the effort that was taken to allow drivers to maintain their speed through the curve. She said she put her heart and soul into it and had aesthetics in mind for the design.

"It is very airy. It isn't a cluttered, loopy thing," she said, adding that specifications to keep traffic moving at high speeds necessitated the long, sweeping curves. "That was so you didn't have to slam on the brakes, like you do on some interchanges." Urban critic Reyner Banham, author of *Los Angeles: The Architecture of Four Ecologies*, said the interchange "is a work of art, both as a pattern on the map, as a monument against the sky, and as a kinetic experience as one sweeps through it." Reece received the Governor's Design Excellence Award for the interchange. Today that interchange is named for her.

Reece later worked on construction of the I-605 Freeway, the I-210 extension, and the I-105 Century Freeway. When she worked on the I-210 extension, it was a $40 million project, the highest dollar project that Caltrans had ever awarded. After her retirement, Reece taught engineering classes at Cal State Long Beach. In 1983, the Los Angeles City Council honored her for making significant contributions to the city. ASCE awarded her life membership in 1991 [92–96].

3.12.4 Janet Bonnema (1938–2008)

Janet Bonnema (Fig. 3.26) got a job as an engineering technician with the Colorado Department of Highways for Colorado's Straight Creek Tunnel (now renamed the Eisenhower Tunnel) in 1970 because her name was misspelled on her employment application. CDOH read her name as J-A-M-E-T and, thinking she was male, hired her. When she reported to work at this tunnel project along I-70 under Loveland Pass, her supervisors barred her from working in the tunnel bore because she was female. Superstition held that women in tunnels and mines would bring bad luck. Instead of the job she thought she had been hired for, Bonnema was assigned desk duty. She filed a sexual discrimination law suit in 1972.

The two sides settled out of court in her favor. Despite initially being shunned by fellow workers and labeled a trouble-maker, Bonnema opened up new job opportunities for women in highway construction, mining, and other previously all-male professions. She was allowed to record measurements, collect rock samples, and produce technical drawings.

After the Eisenhower Tunnel's completion in 1973, Bonnema earned her master's degree in civil engineering at the University of Colorado at Denver and traveled the world to work on projects. She has been inducted into the Colorado Women's Hall of Fame [97–99].

Fig. 3.26 Janet Bonnema.
(Courtesy of the Colorado
Women's Hall of Fame)

3.13 Transit

The transit pioneer, Helen Schultz, established a bus transportation company connecting cities.

3.13.1 Helen Schultz (1898–1974)

When bus transportation first emerged in the United States in the 1910s and 1920s, it was primarily small scale and family-owned. Almost all of the early bus pioneers were men. Helen Schultz, however, was not.

In 1922, Schultz, who wanted to be an entrepreneur, established the Red Ball Transportation Company that provided city-to-city transportation by bus. Initially, her bus routes were between Charles City, Iowa, and Waterloo, Iowa. Her primary clientele were women shoppers and traveling salesmen. Dirt roads, such as they were, were often impassable, and the railroads did not want buses as a form of competition. Later, other bus companies became competitors as well. Nevertheless, she overcame many challenges in establishing and running her business.

The Des Moines Register, the local newspaper, nicknamed her the "Iowa Bus Queen." Schultz maintained her business until 1930, when she sold Red Ball and retired from the bus business [95, 100, 101].

These infrastructure pioneers all contributed in significant ways to the quality of life and standard of living that we enjoy around the world today. We celebrate and honor them as we strive to ensure that the country – and the world – has the infrastructure that it needs for the future.

References

1. Goff, Alice C., *Women Can Be Engineers*, Ann Arbor, Michigan: Edwards Brothers, Inc., 1946.
2. G. Kass-Simon and Farnes, Patricia, Editors (1990). *Women of Science: Righting the Record*. Indiana University Press, Bloomington, Indiana.
3. "SWE's first fellow members: their achievements and careers," *U.S. Woman Engineer*, December 1980.
4. "Elsie Eaves Scores Again: Is First Woman Honored with ASCE Life Membership," *McGraw-Hill News-Bulletin*, March 15, 1962.
5. Olive Ann and Walter H. Beech: Partners in Aviation, Special Collections and University Archives – Wichita State University Libraries, http://specialcollections.wichita.edu/exhibits/beech/exhibita.html, accessed April 29, 2020.
6. Olive Ann Beech, https://en.wikipedia.org/wiki/Olive_Ann_Beech, accessed April 29, 2020.
7. Anne Morrow Lindbergh, https://en.wikipedia.org/wiki/Anne_Morrow_Lindbergh, accessed April 29, 2020.
8. Anne Morrow Lindbergh Biography, http://www.charleslindbergh.com/anne/, accessed April 29, 2020.
9. Anne Morrow Lindbergh, https://www.womenofthehall.org/inductee/anne-morrow-lindbergh/, accessed April 29, 2020.
10. Ogilvie, Marilyn and Joy Harvey, Editors. *The Biographical Dictionary of Women in Science: Pioneering Lives from Ancient Times to the Mid-20th Century*. New York, New York: Routledge, 2000.
11. Society of Women Engineers, Historical Record of Policy and Interpretation, in the author's possession, approved November 7, 1990.
12. Elsie MacGill, https://en.wikipedia.org/wiki/Elsie_MacGill, accessed April 29, 2020.
13. Weigold, Marilyn, *Silent Builder: Emily Warren Roebling and the Brooklyn Bridge*, Port Washington, NY: Associated Faculty Press, Inc., 1984.
14. "Landmarks of the World: Brooklyn Bridge," Holiday, June 1959, The Curtis Publishing Company.
15. Layne, Margaret, E., Editor. *Women in Engineering: Pioneers and Trailblazers*. Reston, Virginia: ASCE Press, 2009.
16. Alic, Margaret. *Hypatia's Heritage: A History of Women in Science from Antiquity through the Nineteenth Century*, Beacon Press, Boston: Massachusetts, 1986.
17. Morrow, Charlene and Teri Perl, Editors, *Notable Women in Mathematics: A Biographical Dictionary*, Westport, Connecticut: Greenwood Press, 1998.
18. "Grace Hopper 1906 –1992," https://www.womenofthehall.org/inductee/grace-hopper/, accessed September 1, 1999.
19. www.swe.org/SWE/Awards,achieve3.htm, accessed September 1, 1999.
20. "Grace Hopper," National Medals of Science and Technology Foundation, https://www.nationalmedals.org/laureates/grace-hopper, accessed April 9, 2020.
21. Billings, Charlene W., *Grace Hopper: Navy Admiral and Computer Pioneer*, Hillside, NJ: Enslow Publishers, Inc., 1989.
22. Zuckerman, Laurence, "Think Tank: If There's a Bug in the Etymology, You May Never Get it Out," *The New York Times*, April 22, 2000.
23. Stanley, Autumn, *Mothers and Daughters of Invention: Notes for a Revised History of Technology*, New Brunswick, NJ; Rutgers University Press, 1995.
24. About Anita Borg, https://anitab.org/about-us/about-anita-borg/, accessed April 21, 2020.
25. Mieszkowski, Katharine, "Sisterhood is Digital," *Fast Company*, September 1999.
26. "President Clinton Names Anita Borg to the Commission on the Advancement of Women and Minorities in Science, Engineering, and Technology," White House Press Release, June 29, 1999, https://www.govinfo.gov/content/pkg/WCPD-1999-07-05/pdf/WCPD-1999-07-05.pdf, Accessed August 20, 1999.

27. O'Brien, Tia, "Women on the Verge of a High-Tech Breakthrough," *San Jose Mercury News*, May 9, 1999.
28. Eng, Sherri, "Women's group honors pioneers in technology, *San Jose Mercury News*, June 26, 1998.
29. "Dr. Anita Borg," Women in Technology International Hall of Fame, https://www.witi.com/halloffame/102852/Dr.-Anita-Borg-Member-of-Research-Staff,-Xerox-PARC,-Founding-Director-Institute-for-Women-And-Technology/, Accessed August 17, 1999.
30. "Top 25 Women on the Web – Dr. Anita Borg," http://www.top25.org/ab.shtml, accessed August 17, 1999.
31. Corcoran, Cate T., "Anita Borg wants more scientists to start listening to women," *Red Herring*, March 1999.
32. "Method for quickly acquiring and using very long traces of mixed system and user memory references, Patent 5,274,811 granted 12/28/93. Patent 4,590,554 "Backup fault tolerant computer system, granted May 20, 1986.
33. Hafner, Katie, "Anita Borg, 54, creator of Systers list," *Rocky Mountain News*, April 11, 2003.
34. This Day in Water History: A little bit of water history – one day at a time, https://thisdayin-waterhistory.wordpress.com/tag/ellen-swallow-richards/, accessed May 1, 2020.
35. Weingardt, Richard G., "Engineering Legends: Ellen Henrietta Swallow Richards and Benjamin Wright," *Leadership and Management in Engineering*, ASCE, October 2004.
36. Proffitt, Pamela, Editor, *Notable Women Scientists*, Detroit, Michigan: The Gale Group, 1999.
37. James, Edward, T., Editor. *Notable American Women: A Biographical Dictionary*, Cambridge, Massachusetts: The Belknap Press of Harvard University Press, 1971.
38. Bailey, Martha J., *American Women in Science A Biographical Dictionary*, Denver, Colorado: ABC-CLIO, 1994.
39. Durant, Elizabeth, *MIT Technology Review*, "Ellencyclopedia," https://www.technologyre-view.com/s/408456/ellencyclopedia/, August 15, 2007.
40. Shearer, Benjamin H. Shearer and Barbara S. Shearer, Editors, *Notable Women in the Physical Sciences: A Biographical Dictionary*, Westport, Connecticut: Greenwood Press, 1997.
41. http://www.nytimes.com/2013/09/24/us/ruth-patrick-a-pioneer-in-pollution-control-dies-at-105.html
42. Bott, Thomas L. and Bernard W. Sweeney, A Biographical Memoir, Ruth Patrick 1907–2013, National Academy of Sciences, 2013, http://www.nasonline.org/publications/biographical-memoirs/memoir-pdfs/patrick-ruth.pdf.
43. Shearer, Benjamin H. Shearer and Barbara S. Shearer, Editors, *Notable Women in the Life Sciences: A Biographical Dictionary*, Westport, Connecticut: Greenwood Press, 1996.
44. http://www.ansp.org/about/drexel-affiliation/
45. https://www.nationalmedals.org/laureates/ruth-patrick
46. Zauzmer, Julie, "Ruth Patrick, ecology pioneer, dies at 105," https://www.washingtonpost.com/national/health-science/ruth-patrick-ecology-pioneer-dies-at-105/2013/09/23/2bcde762-245e-11e3-b75d-5b7f66349852_story.html?utm_term=.5d9dd8392428,   September 23, 2013.
47. The Academy of Natural Sciences of Drexel University, A Biography of Ruth Patrick, http://www.ansp.org/research/environmental-research/people/patrick/biography/
48. Ingels, Margaret, "Petticoats and Slide Rules," Western Society of Engineers, September 4, 1952.
49. National Inventors Hall of Fame, *Edith Clarke*, https://www.invent.org/inductees/edith-clarke, accessed November 30, 2018.
50. National Inventors Hall of Fame, *Maria Telkes*, https://www.invent.org/inductees/maria-telkes, accessed November 30, 2018.
51. Rafferty, John P., *Maria Telkes: American Physical Chemist and Biophysicist*, https://www.britannica.com/biography/Maria-Telkes, accessed November 30, 2018.
52. Society of Women Engineers – Philadelphia Section, *Maria Telkes*, http://philadelphia.swe.org/hall-of-fame-m%2D%2D-z.html, accessed November 30, 2018.

53. Maria Telkes: The Telkes Solar Cooker, https://lemelson.mit.edu/resources/maria-telkes, accessed December 1, 2018.
54. Maria Telkes, https://www.encyclopedia.com/history/encyclopedias-almanacs-transcripts-and-maps/telkes-maria, accessed December 1, 2018.
55. Nichols, Burt E., and Steven J. Strong, "The Carlisle House: An All-Solar Electric Residence," DOE/ET/20279-133.
56. Telkes, Maria. Preliminary Inventory of the Maria Telkes Papers 1893-2000 (Bulk 1950s–1980s), http://www.azarchivesonline.org/xtf/view?docId=ead/asu/telkes_acc.xml, accessed December 1, 2018.
57. Boyd, Andrew, Engines of Our Ingenuity: No. 2608, Maria Telkes, https://www.uh.edu/engines/epi2608.htm, accessed December 1, 2018.
58. Ivy Parker, https://en.wikipedia.org/wiki/Ivy_Parker, accessed May 1, 2020.
59. Tietjen, Jill S., "Honoring the Legacy of Ada Pressman, P.E.," *SWE: Magazine of the Society of Women Engineers,* Fall 2008.
60. Oakes, Elizabeth H. "Pressman, Ada Irene." Encyclopedia of World Scientists, Revised Edition. New York, New York: Facts on File. 2007.
61. Society of Women Engineers, Achievement Award: 50+ Years of Achievement & Inspiration. October 2006.
62. Joan Berkowitz, https://en.wikipedia.org/wiki/Joan_Berkowitz, accessed March 9, 2021.
63. "Their Light Lives On," *Swarthmore College Bulletin,* Summer-Fall 2020, Issue 1, Volume CXVIII, https://www.swarthmore.edu/bulletin/archive/their-light-lives-on/230.html, accessed April 14, 2021.
64. Oakes, Elizabeth H. *Encyclopedia of World Scientists.* New York, New York: Facts on File, Infobase Publishing. 2007.
65. Joan Berkowitz, "Don't hide your talents, urges Berkowitz, 1983 recipient of SWE's Achievement Award." *U.S. Woman Engineer.* October 1983.
66. Achievement Award Nomination for Joan Berkowitz. Society of Women Engineers. SWE Archives. Walter P. Reuther Library. Wayne State University. January 14, 1982.
67. U.S. Army Corps of Engineers, *Water resources: Hydraulics and hydrology: Interview with Margaret S. Peterson* (Oral History Collection), https://usace.contentdm.oclc.org/digital/collection/p16021coll4/id/338, accessed April 14, 2021.
68. University of Arizona, College of Engineering – Civil & Architectural Engineering & Mechanics, *In Memoriam: Margaret Petersen, P.E., F.ASCE, Hon.D.WRE,* https://caem.engineering.arizona.edu/news-events/memoriam-margaret-petersen-pe-fasce-hondwre, accessed April 14, 2021.
69. American Society of Civil Engineers, *Margaret Petersen Outstanding Woman of the Year Award C,* https://www.asce.org/templates/membership-communities-committee-detail.aspx?committeeid=000001005036, accessed April 14, 2021.
70. American Society of Civil Engineers, *Margaret S. Petersen Award,* https://www.asce.org/templates/award-detail.aspx?id=1625, accessed April 14, 2021.
71. University of Iowa, Distinguished Alumni Award, *Margaret S. Petersen, 47 BSCE, 53MS,* https://foriowa.org/daa/daa-profile.php?namer=true&profileid=335, accessed April 14, 2021.
72. University of Iowa, College of Engineering, Prof. Margaret S. Petersen, https://www.engineering.uiowa.edu/alumni/awards/honor-wall/prof-margaret-s-petersen, accessed April 14, 2021.
73. Grattan, Virginia L. *Mary Colter: Builder Upon the Red Earth.* Grand Canyon, Arizona: Grand Canyon Natural History Association. 1992.
74. Mary Colter, Grand Canyon National Park, https://www.nps.gov/articles/marycolter.htm, accessed May 1, 2020.
75. Sicherman, Barbara and Carol Hurd Green, Editors. *Notable American Women: The Modern Period.* Cambridge, Massachusetts: The Belknap Press of Harvard University Press. 1980.

76. Hawthorne, Christopher, "2014 AIA Honor Awards, Gold Medal: Julia Morgan," https://www.architectmagazine.com/awards/aia-honor-awards/gold-medal-julia-morgan_o, June 23, 2014.
77. Marjory Stoneman Dougas, https://en.wikipedia.org/wiki/Marjory_Stoneman_Douglas, accessed March 11, 2021.
78. Marjory Stoneman Dougas, https://www.womenofthehall.org/inductee/marjory-stoneman-douglas/, accessed March 11, 2021.
79. Marjory Stoneman Douglas Writer & Conservationist, http://scholar.library.miami.edu/msdouglas/, accessed March 11, 2021.
80. Marjory Stoneman Douglas, https://wilderness.net/learn-about-wilderness/marjory-stoneman-douglas.php, accessed March 11, 2021.
81. Olaus and Mardy Murie, Wilderness Connect, https://wilderness.net/learn-about-wilderness/olaus-mardy-murie.php, accessed May 2, 2020.
82. Happy Birthday to Mardy Murie, "Grandmother of the Conservation Movement," The Wilderness Society, https://www.wilderness.org/articles/blog/happy-birthday-mardy-murie-grandmother-conservation-movement, August 14, 2019.
83. Margaret Murie, https://en.wikipedia.org/wiki/Margaret_Murie, accessed May 2, 2020.
84. Gale Norton, National Ocean Industries Association, https://www.noia.org/gale-norton/, accessed May 2, 2020.
85. Gale A. Norton, Secretary of the Interior, 2001–2006, https://georgewbush-whitehouse.archives.gov/government/norton-bio.html, accessed May 2, 2020.
86. Gale Norton, https://en.wikipedia.org/wiki/Gale_Norton, accessed May 2, 2020.
87. Read, Phyllis J., and Bernard L. Witlieb (1992), *The Book of Women's Firsts,* New York: Random House, 1992.
88. Maranzani, Martha, Bertha Benz Hits the Road, https://www.history.com/news/bertha-benz-hits-the-road, August 5, 2013, updated January 18, 2019.
89. Bertha Benz, https://en.wikipedia.org/wiki/Bertha_Benz, accessed May 2, 2020.
90. Ruben, Maria Koestler, "Alice Ramsey's Historic Cross-Country Drive," *Smithsonian Magazine*, https://www.smithsonianmag.com/history/alice-ramseys-historic-cross-country-drive-29114570/, June 4, 2009.
91. Alice Huyler Ramsey, https://en.wikipedia.org/wiki/Alice_Huyler_Ramsey, accessed May 2, 2020.
92. Briseno, Terri, "Ten Women Who Broke New Ground in Engineering," https://science.howstuffworks.com/engineering/structural/10-women-in-engineering3.htm, accessed May 2, 2020.
93. Marilyn Jorgenson Reece, https://en.wikipedia.org/wiki/Marilyn_Jorgenson_Reece, accessed May 2, 2020.
94. Women in Transportation History, https://transportationhistory.org/2019/03/15/women-in-transportation-history-marilyn-j-reece-civil-engineer/, accessed May 2, 2020.
95. Women in Transportation History, U.S. Department of Transportation, https://www.transportation.gov/womenandgirls/collage, accessed May 2, 2020.
96. Marilyn J. Reece, 77; State's First Licensed Female Civil Engineer, https://www.latimes.com/archives/la-xpm-2004-may-21-me-reece21-story.html, accessed May 2, 2020.
97. Janet Bonnema, https://www.cogreatwomen.org/project/janet-bonnema/, accessed May 2, 2020.
98. Janet Bonnema, https://en.wikipedia.org/wiki/Janet_Bonnema, accessed May 2, 2020.
99. Orabetz, Janet and Jon, Glasgow, Denver woman won right for female engineers to work on Eisenhower Tunnel project, https://www.9news.com/article/news/janet-bonnema/73-2c7fd55c-f025-4d5c-a70c-020025c43207, March 29, 2019.
100. Walsh, Margaret, "Iowa's Bus Queen: Helen M. Schultz and the Red Ball Transportation Company," The Annals of Iowa, State Historical Society of Iowa, Volume 53, Number 4 (Fall 1994), pp. 329–355, https://ir.uiowa.edu/cgi/viewcontent.cgi?article=9845&context=annals-of-iowa
101. Bus Operators, Women in Transportation – Changing America's History, Federal Highway Administration, https://www.fhwa.dot.gov/wit/bus.htm, accessed May 2, 2020.

Jill S. Tietjen, PE, entered the University of Virginia in the fall of 1972 (the third year that women were admitted as undergraduates after a suit was filed in court by women seeking admission) intending to be a mathematics major. But midway through her first semester, she found engineering and made all of the arrangements necessary to transfer. In 1976, she graduated with a BS in Applied Mathematics (minor in Electrical Engineering) (Tau Beta Pi, Virginia Alpha) and went to work in the electric utility industry.

Galvanized by the fact that no one, not even her PhD engineer father, had encouraged her to pursue an engineering education and that only after her graduation did she discover that her degree was not ABET-accredited, she joined the Society of Women Engineers (SWE) and for more than 40 years has worked to encourage young women to pursue science, technology, engineering, and mathematics (STEM) careers. In 1982, she became licensed as a professional engineer in Colorado.

Tietjen started working jigsaw puzzles at age 2 and has always loved to solve problems. She derives tremendous satisfaction seeing the result of her work – the electricity product that is so reliable that most Americans just take its provision for granted. Flying at night and seeing the lights below, she knows that she had a hand in this infrastructure miracle. An expert witness, she works to plan new power plants.

Her efforts to nominate women for awards began in SWE and have progressed to her acknowledgement as one of the top nominators of women in the country. Her nominees have received the National Medal of Technology and the Kate Gleason Medal; they have been inducted into the National Women's Hall of Fame and state Halls including Colorado, Maryland, and Delaware and have received university and professional society recognition. Tietjen believes that it is imperative to nominate women for awards – for the role modeling and knowledge of women's accomplishments that it provides for the youth of our country.

Tietjen received her MBA from the University of North Carolina at Charlotte. She has been the recipient of many awards including the Distinguished Service Award from SWE (of which she has been named a Fellow and is a Society Past President) and the Distinguished Alumna Award from both the University of Virginia and the University of North Carolina at Charlotte. She has been inducted into the Colorado Women's Hall of Fame and the Colorado Authors' Hall of Fame. Tietjen sits on the board of Georgia Transmission Corporation and spent 11 years on the board of Merrick & Company. Her publications include the bestselling and award-winning books *Her Story: A Timeline of the Women Who Changed America* for which she received the Daughters of the American Revolution History Award Medal and *Hollywood: Her Story, An Illustrated History of Women and the Movies* which has received numerous awards.

Part I
Moving People and Things

Chapter 4
Airport Infrastructure

Sandra Scanlon

Abstract Major airports encompass property and utilities large enough to dwarf some small cities. Heating and cooling systems, reliable electrical services, and numerous interconnected technology systems are the unseen lifeblood of an efficient and successful airport operation. Master planning for upgrades and modernization are critical for the stability of the airline industry.

Keywords Airport · Runway · Taxiway · Terminal · Concourse · Gate · Checkpoint · Roadway · Utilities · Parking · Airline

4.1 Introduction

Major airports encompass property and utilities large enough to dwarf some small cities. Heating and cooling systems, reliable electrical services, and numerous interconnected technology systems are the unseen lifeblood of an efficient and successful airport operation. Master planning for upgrades and modernization are critical for the stability of the airline industry. Airports are also essential to America's economic success. They have a footprint in every community in America, supporting $1.4 trillion in annual economic output and 11.5 million jobs each year [1].

In 2021, the American Society of Civil Engineers (ASCE) gave America's aviation system a D+, largely because airports' basic inefficiencies and lack of space lead to problems like delays and overcrowding. The airport grade was worse than those of other, oft-maligned parts of US transportation infrastructure, like bridges, which earned a C, and roads, which were given a D. The Colorado Section of the American Society of Civil Engineers (ASCE) released its 2020 Report Card for

S. Scanlon (✉)
Denver International Airport, Denver, CO, USA
e-mail: sandra@cooljest.com

Colorado's Infrastructure, giving 14 categories of infrastructure in the state an overall grade of a C-. Of the individual infrastructure categories, the association's civil engineers graded Colorado's aviation infrastructure a B-, one of the highest grades of the categories [2].

4.2   National Aviation Infrastructure

US airports need \$115.4 billion in infrastructure investments over the next 5 years to address critical needs, an industry trade group found in a report released in March of 2021. The Airports Council International-North America (ACI-NA) found that these investments are needed to improve the air passenger experience, increase convenience, enhance security, expand competition, and ultimately lower prices for travelers [3]. But the long list of projects is mostly for much needed infrastructure long overdue for repairs, upgrades, or enhancements in addition to projects needed to increase capacity. The COVID-19 pandemic further delayed many projects from 2020 until the economy recovers – exacerbating the problem further.

But even before the pandemic, airports have been unable to fund all the large-scale projects they need to meet passenger needs because Congress has not modernized one of the main funding mechanisms for airports in more than two decades. One of the main sources of airport infrastructure funding is the federally capped passenger facility charge, or the PFC: a modest user fee on tickets. Congress last raised the maximum statutory PFC cap 20 years ago – before 9/11/2001 – from \$3.00 to just \$4.50. In the two decades since then, construction and related costs have risen steadily, meaning that the real value of the PFC – what it's actually able to purchase – has declined by 40 percent [4].

Many airports, including the youngest – Denver International Airport – are near, at, or even exceeding their design capacities, causing congestion at security checkpoints, lower levels of service at ticket counters and concessions, and frustrated passengers overall. Existing airport infrastructure cannot handle the expected growth, despite the delay caused by the pandemic. This is what the industry refers to as capacity crunch. Given the complexities involved in planning, financing, design, getting permit approvals, and constructing new infrastructure within an operating airport, by the time the new infrastructure is ready for use, the passenger traffic could easily be twice what it was when the project was conceived. This predicament of infrastructure shortage has negative consequences for the regional and national economies as well as the key industry stakeholders.

Fig. 4.1 Colorado's real gold mine – DEN. (Courtesy of Denver International Airport)

4.3   Denver International Airport (DEN)

Denver International Airport (DEN) (Fig. 4.1), the only major new US airport since the 1974 completion of Dallas-Fort Worth, was the largest public works project in Colorado history. It is the largest airport in geographical size in the USA and No. 3 in the world. It led the Department of Transportation's measure of domestic origin-and-destination traffic in 2020 for the first time. It is also the top money maker for the state, pumping $33.5 billion into the economy with 259,084 jobs [5].

The airport is truly like a small city within a city. There are a total of almost 30,000 badged employees including City and County of Denver employees, Denver Fire Department employees, Denver Police Department employees, Denver Health Paramedics, airlines and cargo carriers, vendors and contractors, federal officers, and numerous tenants such as restaurants and retail stores.

DEN has approximately 53 square miles (34,000 acres) of land area, about half of which will not be needed for aviation purposes. That means about half of DEN's land has the potential for development. DEN has the largest commercial development opportunity connected to any airport in the USA.

4.4   DEN Statistics

- Approximately 53 square miles (34,000 acres) of land area – the largest airport in geographical size in the USA.
- Runways: Six total active; five 12,000 feet in length (3600 meters) and 150 feet in width (45.72 meters); one 16,000 feet in length (4800 meters) and 200 feet wide (60.96 meters). This is the longest commercial runway in North America and the seventh longest in the world. Future capacity for six more runways.
- Just over six million square feet spread out over a large area with multiple buildings, heated and cooled from a single central utility plant.

- – 8 × 20MMBTU boilers, ultra-low NOx.
- – 8 × 2500 ton VFD chillers, R22 refrigerant phased out already.
- – Subsurface, the main floor of the plant is located about 40 feet below grade.
- – Completely land locked and cooling towers in top.

- Jeppesen Terminal:

  - – 2.6 million gross square footage (not including the 10-story airport office building or hotel and transit center).
  - – The approximately 240,000 square foot tent roof is held up by 34 masts and is large enough to cover more than 4 football fields.
  - – The Great Hall Project within the Jeppesen Terminal is necessary to prepare DEN for the future and to increase capacity; however, enhanced security is the biggest catalyst for the project. Moving the checkpoints up to Level 6 will increase safety and make the screening process more efficient through additional space and technology enhancements. Phase 1 is scheduled to be completed by the end of the fourth quarter of 2021, and Phase 2 is scheduled to be completed by the end of the fourth quarter of 2024.

- Hotel and transit center:

  - – Size: 433,000 square feet.
  - – Hotel rooms: 519.
  - – Conference center: 37,500 square feet.
  - – Meeting rooms: One grand ballroom and one junior ballroom both divisible by 2 as well as 16 additional meeting/board rooms. There is also a 10,000 square foot pre-function area.
  - – Public plaza: 82,000 square feet.
  - – Commuter rail station: served by the University of Colorado A Line.

- Total number of concourse gates: 113 narrow-body contact gates and 24+ apron-load positions for commuter/regional aircraft. 39 new gates are being added to be completed in 2022.
- Peña Boulevard: Length: 12.3 miles long; average daily travel: 135,500 vehicles per day, depending time of year. This segment of the freeway between I-70 and E-470 is listed on the National Highway System (NHS), a system of roads that are important to the nation's economy, defense, and mobility. Expansion program is underway to widen the Boulevard to handle increased traffic.
- 37.5 asphalt lane miles of other roadways.
- 120.9 concrete lane miles of roadways.
- 54 bridges including box culvert/drainageway crossings.
- Parking spaces: 34,383 total spaces.
- A Concourse: 2640 feet long with an area of 1,266,000 square feet plus 524,000 square feet under construction due to be completed in 2022.
- B Concourse: 4244 feet long with a net area of 2,115,282 square feet plus 135,000 square feet under construction due to be completed in 2022.

- C Concourse: 1900 feet long with an area of 900,000 square feet plus 530,000 square feet under construction due to be completed in 2022.
- More than 170 concession locations.
- More than 190,000 square feet of concession space.
- Maintenance:

  - Number of assets in fleet: 1681
  - Number of alternative-fuel vehicles: 103 (CNG/electric/hybrid)
  - Number of snow-removal assets: airside 364; landside 324
  - Number of people trained for snow removal operations: airside 291; landside 270
  - Lane miles of roads cleared during snow operations: approximately 195 lane miles with 307 acres of parking lot facilities
  - Average runway occupancy time for snow removal: 12–15 minutes

- Environmental:

  - DEN features one of the largest commercial airport solar power arrays (Fig. 4.2) in the USA. Number of solar arrays: eight, including the new rooftop system on the Concourse B West expansion. Total solar generation design capacity: 25,500,000 kWh annually (enough to power 4150 homes). Two new solar projects will be built on the airport property and become operational in late 2022 with a generating capacity of 18.5 megawatts. The arrays will generate approximately 36 million kilowatt-hours of electricity each year, which is the equivalent to the electricity consumption of almost 6000 typical Denver residences.
  - Total electric vehicle charging stations: 52.

Fig. 4.2 DEN solar power array. (Courtesy of Denver International Airport)

- Water bottle filling stations: 29.
- Aircraft deicing fluid (ADF) collection: Collected 74% of the ADF sprayed during the 2019–2020 deicing season, preventing 1.7 million gallons of ADF from being released into the environment.
- Recycling: Almost 1300 tons in 2020.
- Composting: Over 30 tons in 2020.
- Food donation: Over 100 tons in 2020.

4.5 Critical Infrastructure of an Airport

4.5.1 Runways and Taxiways

Runways and taxiways today are mostly made of concrete, asphalt, or a combination of both, and the surface is usually grooved to increase friction when the surface is wet. Runways vary a great deal in length. There are no regulated lengths, but, obviously, they must be long enough to handle desired aircraft types. Aircraft should, as much as possible, take off and land into the wind which is why runway orientation is so important. Many airports maintain principal runways oriented according to prevailing winds at that location (this is more often east-west, but there are alternatives). DEN has four runways oriented north-south and two runways oriented east-west.

Runways and taxiways are evaluated regularly to assess pavement condition, including cracks, spalling, settling, and friction. Runways are installed in numerous concrete panels. The average panel size is about 25 square feet. DEN has a comprehensive airfield pavement management system that tracks the lifespan for each of these panels, resulting in the ability to replace individual panels rather than entire areas of the airfield at once. The pavement rehabilitation program uses a pavement scoring system requiring an evaluation of pavement every 3 years and requiring a pavement condition index (PCI) above 70. But even the simple task of performing the assessment requires the shutdown of a runway. Uniquely, DEN is the only airport to use laser scanning on a large scale for all six runways. This is to reduce the closure times for the runways (5–7 days of daytime closures vs. 1–2 nights per runway). The pavement condition assessment then provides the basis for planning the design and construction of rehabilitation projects, which in Denver's climate must occur in the warmer weather months. DEN typically shuts down only one runway per summer to perform rehabilitation work; therefore, the planning cycle for runway rehabilitation is currently on a 6-year rotation. If and when DEN builds out the remaining runways, this will become a longer cycle; however, it should be easier to shut down a runway when there is more capacity in the system. Imagine how much more difficult it is for smaller airports to perform runway rehabilitation when there are only one or two runways.

Taxiways are to runways like frontage roads are to a main highway – they allow for movement of aircraft out of the mainstream of traffic. Taxiways are just as important as runways in that they allow for movement from the gates out to the runway and they allow space for aircraft to queue before takeoff. Rehabilitation projects for taxiways can be just as challenging to schedule as runway work. It is more efficient for taxiways to act like one-way roads. However, there are times when traffic needs to flow both ways and can only do so one way at a time. This is similar to single lane traffic over a bridge during bridge work, using flaggers to control the one-way flow of traffic. Needless to say, it becomes complicated for the air traffic control tower to juggle the flow of aircraft traffic during the construction season.

4.5.2   Signage and Lighting

Runways, taxiways, and the apron area near the concourse gates use a comprehensive system of signs and markings to aid aircraft movement (Fig. 4.3). There are markings painted on the runway and taxiway surface as well as posted signs, lighted signs, navigation lights, and in-pavement lights. This is very similar to the painted crosswalks, stop bars, and lane markers on a roadway as well as street signs, traffic lights, and directional signs. The paint wears off and the lights need to be repaired.

Fig. 4.3 DEN taxiway lighting. (Courtesy of Denver International Airport)

Many recent projects at DEN have included airfield lighting and signage upgrades performed during runway and taxiway rehabilitation projects. With the increase in LED lighting efficiency and applicability to a wider range of applications, DEN has upgraded almost all of its airfield lighting to LED. In addition, underground wiring needed to be replaced as well, sometimes including replacement manholes, duct banks, and control vaults. A similar analogy would be to replace the street lighting, traffic lights, traffic light controllers, and all the associated underground wiring in a small city.

4.5.3 Fire Stations and Safety Equipment

There are specific safety guidelines for airports that require on-site firefighting capabilities as well as emergency response. The maximum aircraft size that an airport handles determines its categorization for fire and safety purposes. Requirements are then set for firefighting equipment and services that should be provided to respond to any incident possible by type of aircraft Sect. 4.4.

Airport fire stations must be provided as part of the airport infrastructure (Fig. 4.4). And they need to be large enough to accommodate the minimum equipment required and optimally located to achieve required response times. Each fire station building must be maintained and upgraded as systems reach their end of life. In

Fig. 4.4 DEN fire station. (Courtesy of Denver International Airport)

addition, firefighting equipment, fire trucks, and fire station equipment must be maintained on a regular basis while still providing round-the-clock readiness.

One of the most difficult aspects of firefighting at an airport is the use of aqueous film-forming foam (AFFF) which is very effective in extinguishing aircraft fuel fires. Aqueous film-forming foams (AFFF) are water-based and frequently contain hydrocarbon-based surfactant, such as sodium alkyl sulfate, and fluorosurfactant, such as fluorotelomers, perfluorooctanoic acid (PFOA), or perfluorooctanesulfonic acid (PFOS). Studies have shown that PFOS is a persistent, bioaccumulative, and toxic pollutant. Regulatory requirements of life safety systems require regular testing to prove systems will work when needed. Infrastructure upgrades for these systems are in dire need of a non-polluting alternative.

4.5.4 Snow Removal Rivaling Some Cities

In addition to fire stations, police, and paramedic support to keep everyone at the airport safe, there are other operations that enhance safety during the wintry season in Denver. And while snow removal is not often thought of as infrastructure, it is essential for keeping airport infrastructure functional. In addition, the task of snow removal (Fig. 4.5) must be kept in mind when designing and building airport infrastructure. Criteria such as lane width and ability for snow removal equipment to maneuver, where to push snow or pile snow for melting, what types of deicing materials can be used that minimize degradation of the pavement or indoor flooring, as well as the process for maintaining the snow removal fleet are all part of airport design and operation.

Fig. 4.5 DEN runway snow removal. (Courtesy of Denver International Airport)

DEN has earned the prestigious Balchen/Post Award many times, a national recognition for airports with outstanding snow and ice removal programs that maintain
airport operations during challenging winter conditions. It takes dedication, planning, and execution to maintain a safe and operating airfield to keep aircraft and
travelers moving in even the harshest of conditions. The award is based on a wide
range of criteria, including snow and ice control plans; equipment readiness; personnel training; overall safety awareness; timely communication with airlines, the
public, and other airport stakeholders; storm cleanup; and the effectiveness of snow
and ice control plans on runways and other surfaces.

A unique position to the airport is the "snowman," which is a role located in the
Federal Aviation Administration (FAA) control tower and provides a single point of
contact and coordination between the airport and controllers during a snow event.

DEN's airfield snow team averages 15 minutes to clear a runway, and teams have
managed over 80 inches of snow in a season. In 26 years of operation, DEN has only
closed the airfield six times due to snowy conditions.

4.5.5 Apron and Gate Areas

On the exterior of the concourse near the gate is an area called the apron (Fig. 4.6).
The interior of the concourse is a waiting room, often called the gate area or hold
room. The apron area is where the plane drives up and parks and waits for you to

Fig. 4.6 DEN apron and gate areas. (Courtesy of Denver International Airport)

board. While it is parked, the plane is refueled, water tanks are filled, sewage is removed, baggage is loaded/unloaded, food and beverages are restocked, cabin is cleaned, and perhaps the plane is deiced in cold weather prior to departure. And, if there is minor maintenance or repairs to be done on the aircraft, often that happens while the plane is parked at the gate as well. Imagine all the services and systems that are integral to these activities occurring at a gate. When one of those systems needs to be repaired or upgraded, it can have a significant impact on the operations for the airline, especially if the gate needs to be taken out of service for a long period of time. The planning and coordination of infrastructure upgrades in the apron area are significant and complex.

4.5.6   Jet Bridges

Jet bridges or passenger boarding bridges are a critical part of airport infrastructure today, connecting the concourse directly to the aircraft which obviously allows for much faster boarding and turnaround of a flight as well as convenience and comfort for passengers in inclement weather. We take them for granted these days and often complain when we have to board an aircraft without one.

The jet bridge was first seen in the late 1950s, with United Airlines installing them at New York JFK, Los Angeles, and San Francisco airports. Use soon expanded among the US airlines and then globally. Their basic design and operation, with a series of telescopic sections and wheels to guide alignment and move the bridge out of the way, has changed very little.

The next time you fly, pay particular attention that the jet bridges always connect to the left-hand side of an aircraft. This allows for better vision for the aircraft captain and simpler and faster loading of cargo, galley items, and fuel from the right-hand side.

Technology has improved over the years, with more sophisticated and automated controls being introduced to guide the jet bridge to meet the airplane. Jet bridges have also been made larger, and, sometimes, a second bridge has been added at the same gate to accommodate newer wide-body aircraft. Many airports have had to build new jet bridges and change gate areas to handle the A380 – the world's largest passenger wide-body aircraft.

Since 2018, automatic jet bridges have been introduced at a few airports. These use a range of sensors and cameras to align and dock with the aircraft automatically. Older jet bridges still function adequately, but without proper maintenance, they will last only about 20–30 years. Fixed portions of jet bridges require similar maintenance to taking care of a hallway in a busy high school – the carpets need to be replaced, the walls need to be painted, the lights need to be replaced, ceiling tiles need to be cleaned, and upgrades are required for fire alarms, security access, and signage. The moving portion that meets the aircraft requires significantly more maintenance and upgrades, much like the maintenance needed on a car. And lastly, the utilities needed for the aircraft while parked at the gate are routed on the underside of the jet bridge or alongside the bridge using a pantograph system which provides an economical and efficient method of carrying utilities across telescoping portions of passenger loading bridges. Anything that has moving parts is subject to increased maintenance and eventual failure and must have a planned replacement program.

4.5.7 Concourses

The concourses of an airport (Fig. 4.7) are the most significant area where passengers spend time while at an airport. While their first encounter is usually in the main terminal, most of the time spent at an airport is waiting for a flight on the concourses. And most passengers will immediately go to their departure gate to ensure they know where they need to be at boarding time, and then they may wander from there to get food, or perhaps shop, and use the restrooms prior to boarding. The concourse areas experience significant wear and tear, and the concourse infrastructure is difficult to maintain let alone replace.

As a visual analogy, imagine a major league sporting event occurring at one of the nation's largest venues, with tens of thousands of people walking from the entrance to their seat and then getting up and walking around again and again to get food, use the restroom, wander around, and eventually leave the venue. This activity happens several times a day at a large airport. Consider needing to repair, upgrade,

Fig. 4.7 DEN Concourse B West. (Courtesy of Denver International Airport)

or completely replace the flooring, hold room seating, ceiling tiles, lights, heating/ventilation/air-conditioning (HVAC), plumbing in restrooms, moving walkways, escalators, elevators, paging systems, or signage while the airport is active. For this reason, most of the work is performed during the short window of time overnight when there are very few flights, which then elongates schedules for completion. Passengers are the lifeblood of an airport, and customer service plus the overall experience is extremely important. In addition to the complexities of performing work in an active concourse, considerable attention is given to how the work will impact the passenger experience.

4.5.8   Terminal

The main terminal is usually the first building that a passenger will enter at an airport. The main terminal is where the airline ticket counters are located, as well as baggage handling, ground transportation services, and TSA security checkpoint screening.

But the basic layout of the main terminal building has remained much the same for many years. They are designed to offer a separated and well-organized flow of departing and arriving passengers as well as keeping screened passengers separated from non-secure areas. Often this is achieved using different levels of the building with departures usually at the top level and arrivals at lower levels. The baggage handling systems coming from the concourse gates into the main terminal building often are located underground or at ground level and are therefore better suited to connect at a lower level.

Security and clearance areas are much better designed and incorporated in modern terminals compared with those of the late twentieth century. Airport security (at least to the extent we see it today) is a relatively new concept, having changed significantly due to the attacks of 9/11/2001. Terminals designed in the 1960s and 1970s have had to be significantly modified to fit in the extensive security screening facilities now required as well as modern automated baggage handling systems. These systems have mostly unseen infrastructure that requires significant electrical, telecommunications, and systems controls wiring which is complicated to upgrade let alone troubleshoot within the confined limits of older buildings.

Probably the less complicated spaces within the main terminal building are the airline ticket counters; rental car and ground transportation counters; back of house office space for airlines, rental car companies, and airport personnel; as well as some concessions spaces. But these infrastructure areas can be just as complicated to maintain and upgrade as concourses, runways, and taxiways due to the operational impacts to the stakeholders and passengers.

4.5.9 Parking, Transportation, and Landside Support Services

The landside portion of an airport outside the secured fenced area includes many support services and associated infrastructure, most of which falls within the civil and utilities infrastructure realm, but also includes many different building types. Many of these infrastructure systems are discussed in other chapters in this book. These include:

- Access roadways from the surrounding area, such as Peña Boulevard at DEN – the main thoroughfare leading passengers into DEN and other minor roadways
- Rental car lots and associated customer, office, maintenance, and car wash areas
- Shuttle services from remote parking areas
- Parking structures and parking lots either remote or next to the main terminal building for passengers and employees
- Mass transit interface, such as the RTD commuter rail station and RTD bus service at DEN
- DEN passenger train between concourses and main terminal (Fig. 4.8)
- Utility infrastructure including water, sewer, natural gas, jet/diesel/gasoline fuel pipelines and storage tanks, stormwater, electric, solar, and telecommunications
- Secure access gates, guard houses, and miles of fencing
- Hangars for airlines, cargo carriers, and private (fly by owner) FBO
- Support facilities such as fleet maintenance buildings, materials storage warehouses, office buildings, and datacenters
- Construction contractor job trailers, batch plants, materials testing labs, and laydown yards

Fig. 4.8 DEN passenger train. (Courtesy of Denver International Airport)

4.5.10   Central Utility Plant

Since the opening of DEN in 1995, the single central utility plant (CUP) has continually evolved to keep up with growth and technology (Fig. 4.9). The heating and cooling systems provided by the CUP provide conditioning of the main terminal building and concourses as well as for systems that condition the jet bridges and provide pre-conditioned air for aircraft so that the aircraft doesn't have to run its engines to heat/cool the cabin. The history of improvements at the CUP includes responding to new code requirements, such as the phase out of R22 refrigerant, and changing utility rate structures.

When DEN was built, there were financial incentives to invest in natural gas-fired equipment over electric motors. But by the early 2000s, technology had changed. Major equipment efficiencies as well as utility rate structures and demand charges evolved making the original design parameters out of sync with current conditions. The original natural gas engine design became obsolete and was also prone to breakdowns. It also became increasingly difficult to find qualified natural gas engine technicians to maintain the equipment. Since utility rate demand charges were no longer a significant cost factor, the decision to replace the natural gas-fired equipment with electric motor-driven equipment became financially viable.

Replacement of major equipment in the CUP, let alone expansion, is very difficult at DEN because the CUP is completely land-locked bound by an active taxiway to the north, access roads to the south, incoming electrical service on the east, and an office building structure to the west and the cooling towers sit directly above it.

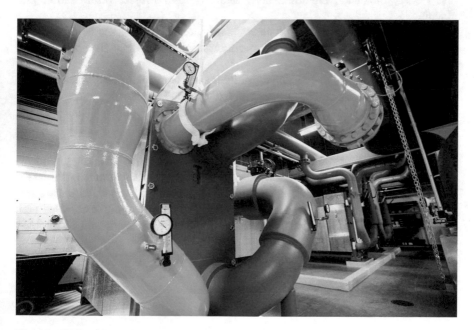

Fig. 4.9   DEN CUP piping. (Courtesy of Denver International Airport)

The main floor of the plant is located about 40 feet below grade. The CUP has faced many challenges over the years including the fact that the water treatment was not maintained well during the first 10 years of operation, and those issues are still haunting the equipment in place today. Maintenance and operations personnel are the most important factor in the longevity and efficiency of equipment. Heating and cooling systems don't run themselves, despite the sophisticated monitoring and controls systems in place even today. Rotating equipment and hydronic systems will fail at some point. The balance is to replace equipment before failure but not too soon as to waste precious infrastructure project dollars for more critical projects. Thankfully, equipment manufacturers have innovated their designs to fit within existing older equipment footprints, often being able to fit more equipment into the same footprint.

4.6   Asset Management for Airports

Asset management incorporates the optimized life-cycle management of physical assets (fleet, facilities, and infrastructure). Asset management supports robust decision analysis – being able to make the right decisions, at the right time, based on solid data and focus investments where they are most needed and have the strongest outcomes. The four key areas are:

Planning and Design: Capital project business case development, triple bottom line (social, financial, environmental) analysis, and project prioritization and planning.

Procure and Construct: Asset data standards and submittals, including building information management (BIM), asset commissioning and acceptance process, and warranty tracking.

Operate and Maintain: Preventive maintenance work plans and schedules and comprehensive asset-class, life-cycle maintenance, and monitoring plans.

Renew and Replace: Risk assessment, life-cycle cost and financial analysis, renewal and replacement planning, and funding analysis.

These combined initiatives can form a cohesive program that will create organizational alignment and incorporate business planning needs including required resources and investments. The plans should also incorporate recommendations for resource requirements (internal and external), progress tracking, performance monitoring, and benefits analysis, as well as retain flexibility to apply lessons learned, adapt the plan over time, and proactively respond to business and technology changes as appropriate.

And, most importantly, applying risk management tools and techniques to ensure that enterprise risks are proactively monitored and addressed and that investments are targeted toward the most critical infrastructure.

4.7   Conclusion

Airports incorporate almost every kind of infrastructure discussed in the other chapters in this book. An asset management program that ensures dollars are spent wisely and that assets are properly operated and maintained serves all the stakeholders at today's modern airports – often, like DEN, small cities themselves.

Airports are gateways to the world, and they can be an economic gateway for the surrounding community as well. Funding infrastructure the right way will inject money into local economies and help all of America keep up with our constantly changing, increasingly interconnected world.

References

1. "Building the Runway to Economic Growth, ACI-NA 2021 Infrastructure Report" March 2021, https://airportscouncil.org/wpcontent/uploads/2021/03/2021ACINAInfrastructureNeedsStudy.pdf
2. "2020 ASCE Report Card: Colorado Infrastructure Receives a C-" January 31, 2020, https://www.enr.com/articles/48599-asce-report-card-colorado-infrastructure-receives-a-c-
3. "US airports have over $115B in infrastructure needs, report finds" by Alex Gangitano, March 17, 2021, https://thehill.com/business-a-lobbying/543501-us-airports-have-over-115b-in-infrastructure-needs-report-finds
4. "Building the Runway to Economic Growth, ACI-NA 2021 Infrastructure Report" March 2021, https://airportscouncil.org/wp-content/uploads/2021/03/2021ACINAInfrastructureNeedsStudy.pdf
5. "Denver Airport Is a Model for Biden's Infrastructure Plan. How a "boondoggle" turned into an economic boon for Colorado." By Matthew A. Winkler, April 29, 2021. https://www.bloomberg.com/opinion/articles/2021-04-29/biden-infrastructure-plan-can-learn-from-denver-airport

Sandra Scanlon,   PE, LEED AP®, Senior Director of Airport Infrastructure Management Development at Denver International Airport (DEN), excels as a leader; a mother; an advocate for getting more women into science, engineering, and technology careers; and a community leader. Sandra began her career as an electrical engineer at Amoco Oil Company after graduating from Valparaiso University with a BS in Electrical Engineering and a Minor in Computer Science. She proceeded through increasingly responsible positions at Amoco as project engineer and electrical specialist. She is a successful entrepreneur founding Scanlon Consulting Services in 1997. In 2008, Sandra led her company through a successful merger, resulting in Scanlon Szynskie Group, Inc., and continued as President. In 2017, SSG MEP, Inc. was acquired by BCER Engineering, Inc., a four-office engineering firm providing mechanical, electrical, fire protection, technology, and sustainability consulting services. Sandra led BCER's electrical engineering department as an associate principal/owner and was chair of the board of directors until early 2020 when she joined DEN. At DEN, she leads four departments responsible for design, construction, and quality assurance of facilities and infrastructure projects throughout the airport property. She is a registered professional engineer in multiple states and a LEED-accredited professional.

She was a gubernatorial appointee to the Colorado State Board of Licensure for Architects, Professional Engineers, and Professional Land Surveyors for 8 years, where she served as Secretary, Vice Chair, and Chair of the Board. She is also an Emeritus Director and founding board member of the Denver School of Science and Technology (DSST) Public Schools and former board member of the DSST Foundation Board. Sandra has been an active member of the Society of Women Engineers (SWE) since college. She co-chaired the 2001 National Conference; spearheaded the establishment of the annual Girls Exploring Science, Technology, Engineering, and Math (GESTEM) event; and has held several significant positions within the Rocky Mountain Section including President. She also served in significant roles for SWE nationally including National Membership Committee Chair and is a Fellow Life Member and National Entrepreneur Award recipient. She has continued her community involvement and served on the board for Habitat for Humanity Metro Denver and was a coalition member of the Women's Foundation of Colorado (WFCO) STEM Coalition for the Colorado Education Initiative. Sandra is a senior member of the Institute of Electrical and Electronics Engineers (IEEE), member of the Denver IEEE Women in Engineering, and an IEEE Technical Expert.

Sandra's interest in engineering, project management, and all things STEM-related started when she was young. She thoroughly enjoyed her Tinkertoy Construction Set, Lincoln Logs, Lego's, Matchbox cars, and the Sunshine Family dolls as she created scenes of farm towns and small city layouts across the living room floor. She thought she wanted to be an architect and was encouraged in this pursuit by her mother and father. Sandra's mother, Colleen Wood, was consistently kept out of math and science classes as a young girl, and she wasn't about to let that happen to her daughter. Sandra's father, James Wood, brought home an antique drafting table from his manufacturing company for Sandra to use for her drafting classes she took in middle school and high school. She still has that table to this day. Once in college, one of only a handful of women in her classes at Valparaiso University, she quickly realized she loved engineering, especially electrical engineering and computer science. But along her career path, she rather enjoyed organizing tasks and projects, which led her to be a much better project manager and leader. Helping others to find their passion and enjoyment in what they love to do every day is what is most fulfilling, and that is where Sandra excels the most.

Chapter 5
Roadway Infrastructure

Lisa Brothers

Abstract Roads are an integral part of everyone's daily life and have a huge economic impact. This chapter discusses how roads are designed, shares best practices
for how to improve safety and reduce congestion, and reveals how engineers can
work with communities to design and construct roadway projects that improve
safety, protect or restore the environment, and meet the unique needs of the community. Using project case studies that range from rural municipalities to urban
environments, this chapter highlights how integrating green infrastructure solutions
into roadway projects allows communities to address resilience and environmental
concerns while simultaneously addressing roadway safety issues and capacity deficiencies. We'll then discuss the evolution of roadway infrastructure and what communities and engineers need to consider for the future.

Keywords Roads · Infrastructure · Green infrastructure · Traffic engineering ·
Transportation engineering · Resilience · Roadway safety · Traffic congestion ·
Roadway capacity · Multimodal roads

Roads take us where we need to go: to work and home, to visit family and friends,
and to see and connect with the world around us. Roads bring us the goods that we
need and want to live our lives, whether via trucks that stock the shelves of our local
stores or via overnight delivery vans. Roads, in short, are an integral part of everyone's daily life and have a huge economic impact.

Roads have always been important for people, but they became a bigger part of
the United States in the 1950s. With President Eisenhower's signing of the Federal
Aid Highway Act of 1956, the interstate highway system officially came into being.
In his 1963 memoir, *Mandate for Change 1953–1956*, Eisenhower reflected on the
roadway infrastructure system he fought for, saying, "More than any single action
by the government since the end of the war, this one would change the face of

L. Brothers (✉)
Nitsch Engineering, Boston, MA, USA
e-mail: LBrothers@nitscheng.com

P. Layne, J. S. Tietjen (eds.), *Women in Infrastructure*, Women in Engineering
and Science, https://doi.org/10.1007/978-3-030-92821-6_5

93

America. ... Its impact on the American economy – the jobs it would produce in manufacturing and construction, the rural areas it would open up – was beyond calculation" [1].

President Eisenhower was proven correct in assessing the economic impact of a connected roadway system. The thoughtful design and construction of roadway infrastructure has proven to be an important driver for the economy. In addition to providing jobs, roads are a critical method of transport for the goods and services that contribute to our economic vitality.

5.1 What Is Roadway Infrastructure?

Roadway infrastructure encompasses all different types of roads and their related components such as sidewalks, islands, bicycle lanes, etc. Our roadway infrastructure – which includes more than four million miles of public roads [2] in the United States – is a constantly changing network that relies on studying how people use the infrastructure, planning for changes in uses, and investing in solutions that make travel safe and efficient today while preparing for the future.

There are three main types of roadways, which are defined by the U.S. Department of Transportation's Federal Highway Administration (FHWA) [3] based on how the roadway functions with respect to *access* (opportunities for entry and exit) and *mobility* (level of travel friction), as well as trip length, speed limit, average daily traffic volumes, and number of travel lanes (among other categories):

- *Arterials:* Designed and constructed with high mobility for long-distance travel. Generally have directional travel lanes that are separated by some type of physical barrier, with high speed limits and limited access and egress (on- and off-ramps). Include interstates, freeways, expressways, and highways.
- *Collectors:* Designed and constructed to balance access with mobility in higher density areas of a community. These generally mid-sized roads connect arterials to local roads and often carry public transportation bus routes.
- *Local Roads:* Designed and constructed with high access for short-distance travel. Generally have many access points, fewer travel lanes, and low speed limits to provide direct access to specific destinations (e.g., houses). Local roads make up the vast majority of roads.

Over the last two decades, travelers have become more vocal about wanting their roadways to support all of the ways that they travel. This multimodal approach to roadway design means that it has become critical to provide safe roadway space for public transportation, bicycles, and pedestrians, as well as traditional motor vehicles. This more inclusive approach to roadway design reduces emissions by encouraging alternative transportation and – when done correctly – can improve traffic operation and flow.

Once built, roads require maintenance and reinvestment to continue to operate well. With more people driving further every year – the American Society of Civil

Engineers (ASCE) notes that vehicle miles traveled jumped to 3.2 trillion in 2019 (an 18% increase over 2000) – wear and tear on existing roads has increased, leaving 43% of our public roadways in poor or mediocre condition [2]. In addition, changing expectations of roadway users – such as increased interest in bicycling (predating but reinforced by the "bicycle boom" that doubled bicycle sales during the COVID-19 pandemic in 2020 [4]), walking, and public transportation – mean that many existing roadways need to be redesigned to accommodate all roadway users.

Roadway infrastructure is designed by transportation engineers – a specialty field within civil engineering that focuses on ensuring safety and efficiency for our roadway users. Transportation engineers work closely with transportation planners, who focus on balancing the sometimes competing needs of different modes of transportation with other land uses (e.g., site development), safety, and budgets. As our communities are impacted by climate change, with more severe weather events happening regularly, transportation engineers and planners also work to address the environmental impact of roadway designs and seek to integrate sustainability and resilience into projects through the use of low-impact development and green infrastructure solutions.

5.1.1   Key Issue 1: Focusing on Safety

The highest priority for all licensed professional engineers – including those who design roadway infrastructure – is to "hold paramount the safety, health, and welfare of the public" [5]. While of course safer vehicles are a key component of roadway user safety, the design of the infrastructure can play just as important a role. As more communities (particularly those in urban areas) adopt Vision Zero policies – which aim to eliminate all traffic fatalities and severe injuries – engineers design solutions that proactively help address traffic safety concerns through smart design that takes human behavior into account.

There are a number of roadway infrastructure improvements that can improve safety and reduce traffic fatalities by providing a better experience for all roadway users – vehicular drivers, bicyclists, and pedestrians – as shown in Table 5.1.

5.1.2   Key Issue 2: Managing Roadway Capacity and Mitigating Congestion

Traffic congestion keeps getting worse – just ask any driver. Since 2008, roadway congestion has increased annually by 1–3% [2]. In urban areas, a large portion of this is due to transportation network companies (TNCs) – aka ride-sharing services – while most regions have seen an impact from increased freight movement.

Table 5.1 Roadway infrastructure safety improvement options by road type

Improvement	Arterial	Collector	Local road
Wider shoulders, travel lanes, and clearances	X	X	X
Highly visible and well-maintained signage	X	X	X
Clear lane marking	X	X	X
Longitudinal (or center line) rumble strips	X	X	X
ADA-compliant sidewalks		X	X
Mid-block crosswalks		X	X
Elevated intersections			X
Curb extensions (also called neckdowns, bulbouts, etc.)		X	X
Driveway reconfiguration		X	X
Protected bicycle lanes		X	X

On top of trying drivers' patience, congestion costs motorists money. Extra time on the road and additional fuel costs combine for a total loss of $166 billion each year – that's over $1000 annually per auto driver. And that's on top of repair costs, which poor roadway conditions can contribute to [2].

Over time, the realization has been that more and/or wider roads aren't the answer, as more roads generally result in even more cars on the road and more sprawl. Instead, roads and highways need to be smart, looking to the future, and supporting more reliable and safe multimodal opportunities. Traffic signal timing and intersection design are critical components of addressing deficiencies, helping traffic flow more smoothly and efficiently. There are a variety of tools available that transportation planners and engineers can utilize to successfully manage traffic and mitigate congestion, with the goal of either adding more roadway capacity, designing existing roadways to be more efficient, or encouraging travel and land use patterns that lessen congestion [6] as shown in Table 5.2.

5.2   What Is Green Infrastructure?

While roadway infrastructure is critical for a wide range of reasons, it has traditionally resulted in an excess of impervious surfaces (e.g., sidewalks, driveways, alleys, and roadways). These surfaces generate rapid, large volumes of stormwater runoff that overwhelm storm sewer systems, compromise the health of water bodies, and interrupt the hydrologic cycle.

Impervious land cover and the historical and conventional stormwater management practices that focus on "end-of-pipe" solutions – gray infrastructure that is largely designed to move stormwater away from its origination point using large pipes – have the following negative consequences:

- Increased volume of runoff
- Decreased infiltration (groundwater recharge)
- Decreased evapotranspiration

Table 5.2 Roadway infrastructure congestion improvement options

Improvement	Description	Arterial	Collector	Local road
Adding more roadway capacity				
Removing physical bottlenecks	Redesigning roadways to improve physical capacity. Particularly important at highway interchanges and in areas where vehicles transition from large capacity roadways (e.g., arterials) to smaller capacity roadways (e.g., collectors)	X	X	X
Prioritizing high-occupancy vehicles (HOV)	Implementing HOV lanes provides a clear incentive for drivers to carpool	X		
Increasing transit system capacity	Providing more transit vehicles (including buses) or more frequent run times to allow more people to choose transit over individual vehicles. Could also include bus-only HOV lanes that make them more efficient than individual vehicles on the same road		X	
Designing existing roadways to be more efficient				
Implementing ramp metering	Creating regularly timed gaps between vehicles on busy on-ramps results in safer and more efficient merging conditions that can improve traffic flow for the entire corridor	X		
Optimizing traffic signal timing	Changing timing on traffic signals to support better flow and keep more vehicles moving		X	X
Improving work zone management	Scheduling and managing roadway construction to impact roadway users as little as possible	X	X	X
Integrating reversible commuter lanes	Designating a traffic lane as one on which the direction of travel can be changed based on traffic volume. Typically used on major commuter roads during peak/rush hour	X		
Restricting turns at key intersections	Prohibiting turns (typically left turns that cut across another lane of traffic) at an intersection in order to avoid disrupting traffic		X	X
Improving roadway design	Redesigning roadways with geometric improvements to better support traffic flow	X	X	X
Improving signage and lane markings	Implementing highly visible and well-maintained signage and lane markings	X	X	X
Encouraging travel and land use patterns that lessen congestion				

(continued)

Table 5.2 (continued)

Improvement	Description	Arterial	Collector	Local road
Creating programs that encourage non-vehicular transportation	Working to get vehicles off the road by promoting transit use, ridesharing, and non-motorized travel. This includes promoting land use options such as transit-oriented and high-density development that don't prioritize individual vehicles	X	X	X
Encouraging flexible work hours and telecommuting	Promoting work options that allow people to avoid traveling during peak/rush hour, thereby reducing the number of individual vehicles on the road	X	X	X
Implementing congestion pricing	Charging higher tolls during peak/rush hour incentivizes people to travel during different times, allowing vehicles to travel more efficiently	X		

- Increased peak flow of runoff
- Increased duration of discharge (detention)
- Increased pollutant loadings
- Increased temperature of runoff

These consequences have an overwhelmingly negative environmental impact that results in poor water quality in water bodies, an increased urban heat island effect, and climate change impacts. Roadway infrastructure design has been evolving to understand how these negative impacts can be prevented or mitigated within roadways; innovative design using low-impact development solutions and green infrastructure practices have emerged as best practices.

Low-Impact Development (LID) LID is a management approach and set of best management practices (BMPs) that can reduce runoff and pollutant loadings by managing runoff as close to its source as possible on a specific site. LID includes overall site design approaches and individual small-scale stormwater management practices that promote the use of natural systems for infiltration, evapotranspiration, and harvesting and reuse of rainwater. Within a roadway, this could include engineered-as-natural ecosystems such as porous pavement and curbside rain gardens that infiltrate, evapotranspirate, and/or harvest stormwater runoff, thereby reducing flows to closed drainage systems.

Green Infrastructure   Green infrastructure refers to an *integrated system* of natural elements and LID practices that provide broad environmental benefits across a larger area, such as a community or watershed. By managing water in a way that respects the natural hydrologic cycle through the use of vegetation, soils, and engineered-as-natural processes – as opposed to directing water into pipes and moving it away from the location – green infrastructure provides stormwater management while also providing flood mitigation, air quality management, climate change adaptation, habitat creation, and more.

Because traditional roadway infrastructure design uses large quantities of impervious materials, roadways and streetscapes traditionally have disrupted the hydrologic cycle and required stormwater to be directed to a closed drainage system consisting of underground pipes that discharge untreated water into water bodies. By implementing green infrastructure techniques that decrease imperviousness and slow, filter, absorb, retain, evaporate, and infiltrate stormwater runoff where it falls within a roadway profile, transportation engineers have the opportunity to positively impact the environment while also improving a roadway's appearance, the pedestrian experience, and sense of place.

Some of the key green infrastructure techniques that can be used within roadways include:

- *Bioretention:* Surface feature that compounds and treats the stormwater runoff, promotes evapotranspiration, and serves as visual amenities (native plantings); promotes groundwater recharge.

  - Designed to improve water quality and not to mitigate water quantity (i.e., flooding)
  - Functions similar to a sand filter to remove contaminants
  - Requires adequate pre-treatment, such as a sediment forebay, deep sump catch basin, or grass filter strip

- *Stormwater gardens:* Slows down and filters stormwater runoff, promotes evapotranspiration, and serves as visual amenities.
- *Constructed wetlands:* Replicates benefits of natural wetlands in managing water. Generally requires larger area than bioretention or stormwater gardens. Provides primary treatment and peak rate mitigation.
- *Tree box filters:* Creates a small bioretention system that can be used within a streetscape or other urban area as a planting area for a tree. Promotes groundwater recharge and evapotranspiration and serves as visual amenity.
- *Infiltration:* Directs water into the ground using drywells and leaching catch basins to provide groundwater recharge, some peak rate mitigation, and primary water quality treatment.
- *Permeable pavement:* Directs water into the ground by reducing impervious cover, promoting infiltration, and providing primary water quality treatment, groundwater recharge, and peak rate mitigation.

- *Green streets:* Increases plantings on roadways to provide pedestrian-friendly areas, creates natural shade to reduce heat-island effect, and adds areas for water quality treatment.
- *Rainwater harvesting:* Re-purposes rainwater for applications that do not require the use of potable water, such as irrigation. Rainwater harvesting reduces the volume of stormwater discharge and helps improve water quality.

5.3 How Do Roadways and Green Infrastructure Improve Communities?

Roads take us where we need to go, and green infrastructure helps restore a natural balance. When combined, roadway infrastructure and green infrastructure provide three key community benefits:

1. Creating space for people
2. Increasing resilience
3. Supporting environmental justice

5.3.1 Creating Space for People

Integrating green infrastructure solutions such as street trees (in tree box filters) and landscaping (that also serves as bioretention) into collector and local roads invites people to participate in the streetscape. Plants help create a sense of place and, when pedestrian amenities are included, make people feel comfortable walking and sitting. Foot traffic helps bring life to a road, revitalizing a community, and helps support local businesses.

5.3.2 Increasing Resilience

The impacts of climate change – more extreme weather events, shifts in timing of seasonal activities (e.g., spring flowering happening sooner), and rising sea levels, among others – are happening now. Communities are threatened by these impacts; rising temperatures are projected to add $19 billion each year to pavement costs by 2040 [7].

Green infrastructure techniques are an integral part of addressing climate change concerns and increasing community resilience. For example, an increase in vegetation lowers urban heat island effects and increases the natural evaporative cooling abilities of plants. Further, these softscapes act as natural "sponges" to absorb increased precipitation expected in humid climates, reducing the strain on aging infrastructure caused by everyday rainfall while buffering the impacts of damaging weather to protect development and investment.

Integrating green infrastructure solutions into roadway infrastructure provides communities with the opportunity to improve the environment and increase resilience, in land that is otherwise only contributing to the problem.

5.3.3 Supporting Environmental Justice

The US Environmental Protection Agency (EPA) defines environmental justice (EJ) as "the fair treatment and meaningful involvement of all people regardless of race, color, national origin, or income, with respect to the development, implementation, and enforcement of environmental laws, regulations, and policies" [8]. In practice, this means that each federal agency needs to pursue EJ by "identifying and addressing, as appropriate, disproportionately high and adverse human health or environmental effects of its programs, policies, and activities on minority populations and low-income populations" [9].

Integrating green infrastructure into roadway infrastructure provides a clear method for achieving environmental improvements within EJ communities. By helping improve water quality, air quality (via street trees), and climate/disaster resiliency particularly as it relates to flooding, green infrastructure can help build healthy and sustainable communities – something that is particularly important for communities that have seen historical under-investment.

5.4 Case Study: Peabody Square, Boston, MA

Located on Dorchester Avenue (a main artery to and from Boston), and adjacent to the Massachusetts Bay Transportation Authority (MBTA) Ashmont subway and bus station, Peabody Square is a principal crossroad with Talbot Avenue and Ashmont Street. Peabody Square functions as a vibrant center of the community, including a popular cluster of local businesses, public transportation access, and a public safety facility.

The revitalization of Peabody Square began in 2006 as part of the larger Dorchester Avenue improvement project that focused on improving pedestrian and vehicle safety, expanding multimodal transportation opportunities, enhancing green space, and addressing stormwater management. As the project entered the 75% design phase in 2007, the Massachusetts Department of Environmental Protection through the Charles River Watershed Association (CRWA) funded a grant to integrate LID techniques into the redesign of Peabody Square as a Green Street Pilot Demonstration Project.

The intent of the pilot project – the first of its kind in the City of Boston and early in the movement toward more "green streets" – was to examine how green infrastructure could be implemented into an urban street without sacrificing safety or creating long-term maintenance issues, with the goal of replicating successes throughout the City.

The key stakeholders for the project included the Boston Public Works Department (BPWD) as the owner, the Boston Transportation Department (BTD), and the St. Mark's Area Main Street non-profit group (who had been actively involved in initiating the project).

Key Project Milestones

Design start: 2006
Grant received to integrate green infrastructure: 2007
Construction start: Spring 2010
Peabody Square construction completion: Winter 2011
Dorchester Avenue construction completion: Summer 2012

5.4.1   Collaborating with the Community

Peabody Square was a highly trafficked area with an unnecessarily complex multi-legged configuration and all impervious hardscape that was uninviting to the pedestrian. Peabody Square had many channelizing islands and numerous signal phases, resulting in 13 crossings that created an unfriendly and unsafe environment for pedestrians, and congestion and long delays for motorists. At the initiation of the project, the accident rate was higher than the Massachusetts Department of Transportation's (MassDOT) Statewide and District averages. The safety improvement project was formulated with the goal of improving conditions by reducing the number of crossings significantly and improving the number of pedestrian walkways – with a seamless incorporation of green infrastructure.

The process began by dedicating time to collecting traffic data and other physical data from visiting the site and observing challenging areas. The design team recognized the excessive number of intersections and identified cut throughs used to access the many local businesses. A land survey resulted in base plans that the design team used to prepare concepts for five intersection redesign alternatives that would simplify and improve roadway layout, reduce points of conflict, create a safe environment for vehicles and pedestrians, provide for public plazas and area gateways, and revitalize the aesthetic appeal of the Square to promote commercial and community activity. The design team determined how to best incorporate sustainable design elements without compromising the safety and accessibility of the plaza.

After working with the BPWD and BTD to refine the concept designs, the design team managed a series of three public meetings where five alternative concepts for the Square were presented to the public. These meetings were approached with a goal of fact finding to pinpoint the issues that pedestrians had with the existing conditions. With this collected data, the design team phased out any options that immediately didn't meet the needs of the public and then modified the remaining designs to reflect the input from the community. The process of feedback and revisions continued, including on-site meetings with neighborhood associations, with input from each meeting used to better support the community desires for the area.

Through this in-depth community process, the five options were narrowed down to two options. The community and design team then unanimously agreed on one design alternative (Fig. 5.1) that eliminated the channelizing islands and long pedestrian crossings; discontinued Bushnell Street across the Square; realigned Talbot Avenue; reduced residential neighborhood cut-through traffic; added bicycle lanes; decreased traffic queuing (reducing air pollution); provided fire station signal preemption; addressed parking issues; created a socially inviting park and plaza that retained the area's historic clock tower and water trough; and added a variety of perennials, grasses, shrubs, and tree plantings.

5.4.2   Green Infrastructure Solutions

The design team collaborated with CRWA to implement sustainable design techniques to reduce stormwater runoff volume into the closed drainage system and remove pollutants from waterways. These LID techniques included integrating a bioretention basin, porous plaza pavers/pavement, and an infiltration trench within the planned plaza areas. The bioretention basin collects and treats stormwater runoff

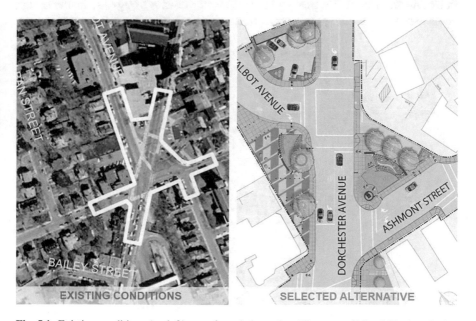

Fig. 5.1 Existing conditions (on left) vs. selected alternative. (Courtesy of Nitsch Engineering)

via engineered layers of mulch, soil, and plant root systems. The porous paver/pavement provides infiltration with an overflow protection connection to the storm drain system. The infiltration trench recharges and treats stormwater runoff from the adjacent parking lot.

Along with the goal of balancing safety improvements with sustainable design, the design team was charged by the City to make apparent the benefits that green infrastructure could have beyond its impact on the City's infrastructure. The vision of the Peabody Square pilot project was to create a socially inviting park and plaza that offered aesthetic benefits to the community all while managing the stormwater runoff using low-impact development designs. With this community-centric vision in mind, the design team maintained a line of open communication with the local public throughout the process. This was done through an interwoven community outreach approach throughout the design and construction administration processes.

On every project within the City of Boston, designs must be coordinated with the Boston Water and Sewer Commission (BWSC) which operates the drainage system. The Peabody Square project benefitted from the BWSC being open to establishing green infrastructure because of its benefits as an alternate water treatment method. The low-impact BMPs that were selected for the project provide numerous stormwater benefits, including runoff volume and rate reduction, groundwater recharge, natural treatment of stormwater runoff, and runoff temperature reduction. These benefits are particularly important because the stormwater runoff that discharges from the site into the City's storm drain system eventually makes its way to the Neponset River, which is on the Massachusetts list of impaired waters. The river is identified as impaired for organics, pathogens, and turbidity, all common pollutants in stormwater runoff. By treating the stormwater on-site using sustainable design components — including the bioretention basin shown in Fig. 5.2, porous pavers/

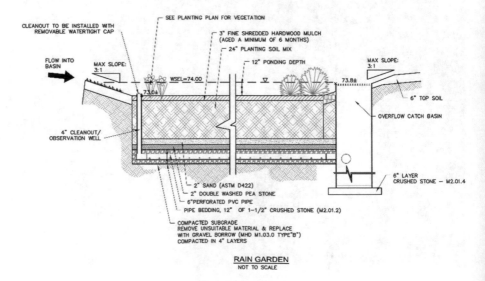

Fig. 5.2 Bioretention basin cross section. (Courtesy of Nitsch Engineering)

pavement, infiltration trench, and greening of the intersection — the project is doing its part to improve the water quality of the Neponset River.

Out of an abundance of caution on this pilot project, the BWSC requested that the design team also install a "traditional" closed drainage system as a secondary back-up in case the green infrastructure system failed.

The green infrastructure improvements are connected to the city system, so that any excess stormwater from heavy storms (e.g., 100-year, 50-year) can be sent into those connections to prevent area flooding.

5.4.3   Establishing Confidence in the New Peabody Square

A number of constraints and requirements contributed to the complexity of the Peabody Square project, including building over the subway tunnel cutting diagonally across the Square and within 3 feet of the surface in some areas; allowing for fire station operations including providing a location for the testing of the ladder truck at every shift change; working around the historic clock and water trough to lay out the new intersection; overcoming the reluctance to use sustainable design techniques; and establishing a public/private partnership for maintenance responsibilities.

5.4.3.1   Addressing Structural Support Concerns

The design team focused on maintaining the integrity of the subway tunnel. This included ensuring that the tunnel could withstand the weight of the fire station's equipment where the tunnel crossed under the station's parking lot that was regularly used for maintenance and cleaning of the station's equipment.

5.4.3.2   Maintaining Movement

During construction, the design team was continuously challenged by the many different movements that required continued access to Peabody Square, including coordinating with the fire station to avoid impacts to their services, accommodating neighborhood traffic that continued throughout the entirety of the project, and providing access to the surrounding businesses including curb cuts while still eliminating dangerous existing cut throughs. The design team was able to mitigate these challenges by working closely with members of the community and anticipating the needs of pedestrians through initial traffic studies and community outreach.

5.4.3.3 Leveraging the Landscape

To overcome hesitations the community had about changes to the plaza, the design team and the City focused on creating a stronger sense of place through creative landscape design that combined form with function.

The design team overcame some hesitations about green infrastructure by displaying the aesthetic benefits of the design. The selected green infrastructure components, such as the porous pavers, not only provided a cost-effective way to treat stormwater but also enhanced the beauty of Peabody Square by incorporating stormwater into landscape-based systems and aesthetic patterns. Bioretention basins with resilient perennials were used not only because of their ability to collect water but also because they provide a better aesthetic by creating additional green space. The design team incorporated the existing historic elements (i.e., the clock and water trough) that had been fenced off and inaccessible by installing the green infrastructure around these existing elements to create a cohesive landscape. The green infrastructure was used to re-imagine Peabody Square to provide a higher-quality environment that is accessible to the community, provides opportunities for pedestrian gathering, and is used for community events.

5.4.3.4 Planning for Operation and Maintenance

As one of the first implementations of green infrastructure elements owned by the City, developing an operation and maintenance (O&M) plan was critical to the success of the project. The design team conducted initial research and made recommendations, but the success of the project relied on a collaborative process between stakeholders and the design team to resolve construction and maintenance concerns. The public/private partnership between the BPWD, BTD, and the St. Mark's Area Main Street non-profit group helped address this challenge.

A key component of a successful O&M plan is the education of those who are responsible for long-term O&M. The design team educated the many project stakeholders (e.g., BPWD, BWSC, CRWA, the St. Mark's Area Main Street non-profit group, and the community) about how the innovative sustainable design techniques worked and should be maintained.

One key O&M challenge for the project involved how to care for porous pavers. While the technology existed, it had not been implemented much (if at all) in the City of Boston and presented a challenge for long-term maintenance. Porous pavers being applied to a project in the City required research into how to best care for them. The design team attended training at the University of New Hampshire, a national leader in pavement research, and used the information gained from the training to create a set of guidelines that established how to properly vacuum the porous pavers. Originally, the City anticipated having to contract out this work, but thanks to these guidelines, they instead discovered that they were able to adapt existing equipment to adequately perform this maintenance. This was a step in the direction of the City's long-term goal of integrating green infrastructure into future roadway infrastructure.

5.4.4   Impact

The Peabody Square project established a model for future green infrastructure projects in the City of Boston. The project highlighted how success can be achieved when there is investment from multiple entities (e.g., City of Boston, CRWA, BWSC, etc.). The vision and innovation of the project stakeholders and design team, combined with an iterative community engagement process, resulted in a project that could best meet the needs of its community, as shown in Fig. 5.3.

As the pilot project for the implementation of green infrastructure in the City, the project established a framework of integrating green infrastructure into complete streets that was able to be replicated repeatedly throughout the City and was used as a case study to illustrate the City of Boston's Complete Streets Guidelines.

Project Team

Transportation and Civil Engineer: Nitsch Engineering
Landscape Architect: IBI Placemaking
Structural Engineer: Lin Associates
Contractor: McCourt Construction

Fig. 5.3 Peabody Square landscaping includes bioretention, rain gardens, pervious pavers, and aesthetic improvements. (Courtesy of Nitsch Engineering)

5.5 Case Study: Kennedy Street Green Infrastructure Challenge, Washington, D.C.

The District of Columbia Water and Sewer Authority (DC Water) owns and operates a combined sewer system that serves more than 672,000 residents and 17.8 million annual visitors in the District of Columbia. As part of a 2005 consent decree from the EPA, DC Water began planning three storage tunnels under the DC Clean Rivers Project to minimize combined sewer overflows (CSOs) to District waterways, including the Anacostia River, the Potomac River, and Rock Creek (and ultimately the Chesapeake Bay watershed). By 2010, DC Water began investigating the application of green infrastructure as another tool for controlling CSOs, as they understood that the additional social and economic benefits associated with these techniques are much broader than the benefits associated with traditional "gray" infrastructure. After proposing to modify the consent decree to include green infrastructure in 2011, DC Water launched an international design competition in April 2013 that sought innovative green practices focused on capturing and absorbing stormwater to meet DC Water's goals of reducing CSOs.

The Kennedy Street Green Infrastructure Challenge Streetscape project began with this design competition. DC Water hoped to amend their consent decree obligations by accounting for the use of green infrastructure but had already begun building large sewer tunnels to store the overflow during large storm events. They aimed to target areas where tunnels had yet to be built and assess if widespread green infrastructure could reduce the size of the tunnels, or eliminate the need for them altogether, in order to fulfill their obligation to form a mitigation plan.

The Potomac and Rock Creek watersheds presented an opportunity for a hybrid approach incorporating smaller sewer tunnels with green infrastructure within an urban environment. DC Water also saw an opportunity to re-direct funds spent on the large infrastructure of the tunnels by reducing or eliminating the need to build the future tunnels that would otherwise store combined sewer overflow.

Key Project Milestones

Design competition: 2013
Design start: March 2015
Construction completion: September 2018

5.5.1 Competing for a More Sustainable City

As a design competition finalist, the design team proposed a streetscape design that integrated porous pavements, bioretention bump-outs and planters, infiltration opportunities, pedestrian boardwalks, and an engaging proposal for environmental art. The location for the improvements was in a commercial section of Kennedy Street NW, a 1.14-acre site located approximately four miles north of Capitol Hill in the Rock Creek watershed that had been pre-selected by the DC Water Clean Rivers staff. The street, which is located within an EJ community, had a redevelopment plan on the horizon and was a priority area in need of revitalization with large under-utilized sidewalks.

The design team began working with DC Water to advance the pilot project in March 2015, with the goal of designing improvements that could serve as a model for larger green infrastructure projects throughout the District. The pilot project was intended to allow DC Water and local permitting agencies to become more familiar with the intricacies of designing and building green infrastructure facilities in the District's urban environment.

DC Water and the EPA came to an agreement in 2015 to modify the 2005 consent decree to include green infrastructure strategies that could eliminate a large percentage of the CSOs in each of the three watersheds. If deemed practicable after the first large-scale projects, the green infrastructure facilities would reduce the size (and therefore the cost) of the tunnel needed in the Rock Creek and Potomac River watersheds.

5.5.2   Using Green Infrastructure as a Solution

The Kennedy Street pilot project was designed to provide clear and measurable environmental benefits that would ultimately reduce CSO discharges. The sustainable design included:

- 40 trees (5 existing; 35 new)
- 580 linear feet of infiltrative parking lanes
- 15 bioretention curb extensions
- 240 linear feet of landscape infiltration gaps
- 520 linear feet of recessed landscape infiltration
- 4 dry wells

By installing (and connecting) 5 technologies in 33 locations on 1 urban city block, the overall green infrastructure system design results in the reduction of 9000 square feet of impervious surface over the 1.14-acre site and the retention of 59,941 gallons of stormwater. The goal for the green infrastructure design was to retain the stormwater from a 1.2″ rainfall event over the project area. When it rained 1.2″ before, 28,000 gallons of stormwater drained to the combined sewer in 5 minutes. When it rains 1.2″ now, zero gallons drain to the combined sewer – in fact, the new green infrastructure facilities retain enough stormwater to mitigate a 2.1″ rainfall event. The travel time for water flowing from one end of the block to the other also slows to 20 minutes.

Although the Kennedy Street project was driven by the need to reduce CSO discharges, it became much more than a stormwater mitigation project. The seat walls, grates, and additional trees were intentionally included to activate the pedestrian streetscape and encourage people to socialize on the street while also providing education about stormwater management. The additional trees provide climate change adaptation benefits by reducing heat island impacts on the streetscape.

The team designed the facilities in a way that avoided the underground utilities (i.e., water, sewer, stormwater, gas, electric, and telecommunications) and preserved well-established street trees. This allowed DC Water to spend their money on green infrastructure interventions instead of utility relocations. However, it also required the team to design on the fly when unknown conditions were found underground – for example, when they found an electric vault that was much larger than anticipated and had to redesign a bioretention basin and seat wall to accommodate it.

5.5.3   Establishing Lines of Defense

As a pilot project for DC Water, the Kennedy Street green infrastructure project was primarily focused on developing unique and innovative green infrastructure applications that could serve the District. To capture the largest quantity of stormwater – and therefore provide the most benefit to the District – the engineers designed a unique interconnected system that provides multiple lines of defense.

The green infrastructure BMPs used on site provide three lines of defense: above-ground rainfall capture through the enhanced tree canopy, street-level capture through a combination of landscape-based strategies and permeable parking, and below-grade infiltration using drywells for stormwater traveling down the existing alleys between the buildings. This detailed design of multiple lines of defense allowed for flexibility when challenges arose.

By designing the 33 green infrastructure BMPs to connect in a series, as shown in Fig. 5.4, the system provides enhanced treatment and infiltration of stormwater. In this system, any water that cannot be infiltrated in a green infrastructure BMP will flow into the next BMP, with water flowing from east to west. Along the sidewalk, a trench drain is innovatively used to capture surface water and convey it to a series of recessed planters – which, in turn, overflow to the bioretention curb extensions. The goal was both to provide volume and to slow down the travel of water.

Another unique application of existing technologies can be found along the northern side of the streetscape: landscape infiltration gap facilities (LIGs). These facilities include the first known implementation of LIGs in public space in the District (and possibly the United States, as the design team could only find prior information on the practice from European installations). LIGs are small strips of grass that break up a paved area, allowing smaller quantities of stormwater to directly infiltrate into the ground. The LIGs created a perception of more green and open space than a standard permeable paver that has only sand. By breaking up pavement areas, they also help mitigate the heat island effect. The design team installed six LIG facilities on the north side of the street to test out this practice in the District, as shown in Fig. 5.5.

Each system was designed to be fully dedicated to collecting rainwater from only a small area. Rainwater from the backside of the sidewalk was captured with LIGs

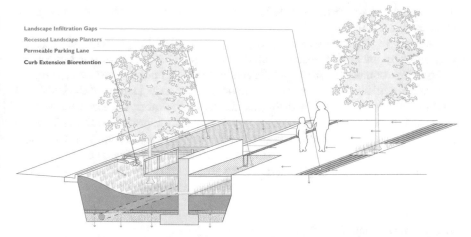

Landscape Infiltration Gaps
Recessed Landscape Planters
Permeable Parking Lane
Curb Extension Bioretention

Fig. 5.4 Cross section showing how four BMPs connect on Kennedy Street. (Courtesy of Nitsch Engineering)

Fig. 5.5 Landscape infiltration gaps on Kennedy Street. (Courtesy of Nitsch Engineering)

on one side of the street and a sidewalk planter on the other side of the street. Rainwater from the parking lane gets absorbed by permeable paving. Water from the roadway runs into the gutter and into the bioretention area. The design team broke the streetscape down to interconnected, micro-managed pieces rather than allowing everything to run to the end of the street and one large piece of infrastructure. As a result, the design optimizes the performance of every green infrastructure facility.

Multiple regulatory agencies had jurisdiction over the project, including the District Department of Transportation (DDOT), the District Department of Energy and the Environment (DOEE), DC Water, and the local utilities. Permitting the project was a significant challenge, as the unique interconnected nature of the BMPs did not fit within the authorities' existing standards. Furthermore, because LIGs had not been used in the District, there was no specific permitting process in DOEE's online permitting database. To address these complexities, the design team worked closely with DOEE staff to show how the interconnected system met the functionality requirements and to develop an approach to receiving permits for these facilities.

5.5.4 Creating Place

As climate change reshapes communities, those in the engineering profession continue to work on practical solutions to mitigate or prevent damage. Green infrastructure is an integral tool. The unique way that green infrastructure BMPs are connected on Kennedy Street, and the first application of LIGs in the District, serves as implementable inspiration to other engineers as they also seek to reduce the strain on aging infrastructure while buffering the impacts of damaging weather to protect development and investment. In addition, the success of these BMPs provides a critical example of how these practices can be successful in highly urbanized areas.

Engineering success is reliant on public acceptance: the very best idea would exist only on paper if the community didn't support it. It was important not only to gain acceptance of the project from the Kennedy Street neighborhood but also to demonstrate to ratepayers located throughout the watershed and DC area the importance of the project for its long-term impacts all over. The goal was to display how this demonstration project could be replicated in the future to positively benefit the watershed in its entirety. Through two public engagement meetings, residents learned about the design and how stormwater moves through an urban environment. The process allowed the design team the opportunity to educate the public about the existing issue of CSOs and high-cost infrastructure and to bring awareness about the ability of green infrastructure to store water and address the CSO problem while benefiting the neighborhood with beautification, heat island mitigation, and solutions to other urban environmental issues. Many residents left the meetings excited about the positive impact these improvements would have on their community.

5.5.5 Telling the Story of Water

Although the project was conceived as part of a larger strategy to use green infrastructure techniques to address CSOs, it became much more than a stormwater management project. Along Kennedy Street, the design team sought opportunities to include education and placemaking elements that also met other community needs.

To highlight the movement of stormwater, while still providing wide sidewalks, the team designed steel grating above some bioretention basins. This design retains an ADA-accessible walking area – a particular concern of the DDOT – while also using the space to provide more green infrastructure facilities. Although not part of the stormwater functionality of the project, the grates over the bioretention areas allow pedestrians to experience the movement of stormwater as they walk above it.

A seat wall integrated a public art element: an engraving of a map of Washington, D.C., highlighting the water bodies of the watershed the project was designed to protect. This was incorporated to enhance the streetscape, draw people to the sidewalk, and provide public education through the engravings. The design also intentionally sought to improve pedestrian safety on the block; the bioretention curb extensions narrowed the roadway, which reduced traffic speeds and provided shorter crosswalks. Thirty-five additional street trees were incorporated to improve stormwater functionality but also provide shading to reduce heat island impacts.

Great care was taken to intentionally incorporate design elements into the project that would enhance the streetscape and create a more sustainable, resilient, and walkable place. The public can then begin to understand why money is being spent on green infrastructure and how it can provide ample benefits to their local communities. The aim with including additional education and placemaking elements was to create an experience that told the story of the water as people utilized the roadway. These educational and experiential details – and the care taken to ensure their incorporation into the project – are unique when compared with other CSO mitigation projects.

5.5.6   Planning for the Future

The success of the Kennedy Street pilot project was critical for the future of DC Water's green infrastructure program and for compliance with the consent decree: if the project had not been successful, DC Water's plans for large-scale green infrastructure implementation would have needed to be revised. The success of this project has also paved the way with other regulatory authorities for future green infrastructure implementation in the District, providing direct value to engineers working within the District.

The Kennedy Street pilot project has provided DC Water with a test run for the design, permitting, and maintenance of green infrastructure facilities in a densely urbanized environment. DC Water had signed an agreement with DDOT to do the long-term O&M. Still, DC Water was challenged with training their staff in green infrastructure maintenance, so they signed a contract with a maintenance company for coverage while working on a framework for creating new "green" jobs.

DC Water has been demonstrably happy with the project process, speaking at many conferences about the project, and sending out regular updates via social media channels.

5.5.7 Impact

The Kennedy Street Improvement Project was an opportunity to experiment with communicating the value of sustainable design in the face of concerns over spending and investment. Project Manager Nicole Holmes, licensed professional engineer (PE), noted, "It is challenging to validate the cost of green infrastructure for stormwater mitigation if you're comparing it to gray infrastructure alone. Green infrastructure will cost more to manage the same amount of water, so it's extremely important to validate the many other benefits of green infrastructure through a long-term life cycle cost benefit analysis. You're saving so much in all of these other ways: energy, property values, standard of living, and other environmental benefits. It requires a close partnership to let everyone account for and contribute to the benefits that everyone could receive long-term from choosing green infrastructure."

In the case of the Kennedy Street project, shown in Fig. 5.6, there was a challenge when interacting with public agencies who were tasked with working with ratepayers and stakeholders and explaining the benefits of spending more money for what at first was seen as the "same outcome." DC Water successfully established a

Fig. 5.6 Kennedy Street. (Courtesy of Nitsch Engineering)

partnership of trust and understanding with all of the agencies that would benefit from this project, making it possible to explain the many additional benefits of choosing green infrastructure.

Project Team

Civil Engineer: Nitsch Engineering
Landscape Architect: Urban Rain|Design and Warner Larson Inc.
Land Surveyor and Geotechnical Engineer: EBA Engineering, Inc.
Permitting: McKissack & McKissack
Community Engagement: Tina Boyd & Associates
Contractor: Capitol Paving

5.6   Case Study: Roadway Improvements, Buckland, MA

Hurricane Irene brought devastation to western Massachusetts in the summer of 2011 – including the Town of Buckland, which borders the Deerfield River. The river and its connecting streams flow through Buckland and into the neighboring tourist town of Shelburne Falls. In Shelburne Falls, the river passes beneath the Bridge of Flowers, which attracts visitors in the spring who are delighted to walk across admiring the variety and abundance of blooms, and highlights the Glacial Potholes, which usually show smoothed rock surfaces with the anomalies of deep eroded craters and a backdrop of water piling over the man-made dam. Hurricane Irene raised water levels to the bottom of the Bridge of Flowers, submerged the glacial potholes, and caused the roadways of Conway Street, Summer Street, South Street, and Shelburne Falls Road to flood over. This storm event created a debt of damage that left elements of the roadways in disrepair, disconnecting the community.

Covering 1.7 +/− miles of roadway, the Buckland Roadway Improvements project includes the reconstruction and widening of Conway Street, South Street, and Conway Road from Bridge Street to the Conway Town Line. These roadways were in dire need of improvement due to the damage and devastation from Hurricane Irene. A MassDOT Transportation Improvement Program (TIP) funding grant made improvements possible. Otherwise, the rural nature and low population of the Town and a corresponding lack of budget for repairs would have precluded the project from moving forward. The improvements were designed to meet MassDOT's Complete Streets standards, which focus on encouraging safer multimodal transportation while integrating LID elements that result in a greener street.

Key Project Milestones

Design start: January 2016
Construction start: Fall 2021 (estimated)
Construction completion: September 2023 (estimated)

5.6.1 Re-building and Growing the Town Center

The main economic driver in the Town of Buckland is the tens of thousands of visitors travelling to Shelburne Falls each year. The downtown area, collectively called Shelburne Falls with part of Shelburne, is a major tourist and shopping area that includes the historic manufacturing plant and shop of Lamson and Goodnow. The Town's economic development and local employment base depend on its ability to present a vibrant and attractive gateway to the "Shelburne Falls" village community.

The Buckland Roadway Improvements project is focused on supporting the economic growth of the center of Buckland through three interconnected goals: repairing damage done to existing infrastructure by Hurricane Irene; improving access to the business district through reconstruction of the roadway and sidewalk; and encouraging safer multimodal travel through Complete Streets design.

5.6.1.1 Repairing After Hurricane Irene

After Hurricane Irene, the essential infrastructure of Conway Street, South Street, and Conway Road was in need of repair. The many culverts that supply passages for the Deerfield River tributaries along the roadways overtopped, the roadway was washed away, and 1104 feet of sidewalk were damaged by the flooding of the road. There were tripping hazards where the roadway connected to the existing sections of sidewalk, and road edges were rough, uneven, undefined, and dangerous for both pedestrians and bicyclists, as shown in Fig. 5.7. This lack of defined road edge, sidewalks, and curbing made for inconsistent and dangerous parking for local homeowners and businesses, as well as for pedestrians.

The repair and improvement project prioritizes creating a smoother, more consistent riding surface on the roadway and establishing defined shoulders. The project design includes 1104 feet of sidewalks in highly trafficked areas along portions of Conway, Summer, and South Streets and makes the existing sidewalks ADA-compliant. The improvements benefit motorists, pedestrians, and bicyclists alike by encouraging safer multimodal transportation and improving pedestrian connectivity and accessibility through the Town of Buckland to Shelburne Falls.

The project also replaces two existing culverts with larger culverts to help prevent future roadway overtopping and redesigns the roadway infrastructure by replacing drainage lines, sewer lines, and water piping to bring everything up to MassDOT standards. One concrete culvert with scour will be repaired, as identified by the bridge inspection team.

5.6.1.2 Improving Pedestrian Access

Within the Town of Buckland's center are community amenities that drive both pedestrian and vehicle use. A community ball field is accessed from South Street, and the back of the field runs along Summer Street; bicyclists and joggers use this

Fig. 5.7   Edge of road and sidewalk connectivity issues. (Courtesy of Nitsch Engineering)

route daily. A school bus route for the elementary school, high school, and Franklin County Technical School means that students walk along the poor sidewalks or road edge where the sidewalk is either poor or missing completely. The Police Station and the Highway Department are located on Conway Street, and the Wastewater Treatment Facility is located just off Summer Street; all these Town departments would benefit from improved road and drainage conditions improving their response times. This route is also a major connection from Routes 2 and 112 to Route 116 in Conway, and many Conway residents and Southern Ashfield residents use this route daily.

The existing sidewalk – where it exists – offers little safety as there is no definition from the road edge on much of it and it is flush with the lane of travel. Much of the sidewalk is severely cracked and heaving and contains trip hazards, and there are sections that are impassable to wheelchairs. Wheelchair ramps are nonexistent or in poor shape along most of the route. These issues had resulted in concern for bicyclist and pedestrian safety in this area for some time.

The roadway improvement project will either improve, upgrade, or install new roadside appurtenances including signs, curbs, sidewalks, pavement markings, drainage facilities, barriers, and guardrails. Outdated, inappropriate, and missing signage will be removed, replaced, and installed in accordance with the FHWA's Manual on Uniform Traffic Control Devices (MUTCD) standards. Damaged and non-functional guardrails will be replaced along Conway Road. Drainage will be updated and improved along the entire project, including replacement of some "fabricated" structures.

Pedestrian and bicyclist safety were a priority for the design team. Road edge and pavement markings will be installed, allowing for lane markings. Wheelchair ramps will be installed to meet ADA standards. New curbing and sidewalks where there currently are none will tie the area to the downtown business district.

Several businesses along the corridor will benefit from enhanced definition of curb cuts from both safety and aesthetic perspectives. Sidewalk improvements and expansion create a more walkable community, business, and tourist district and are expected to help revitalize this edge of Shelburne Falls where there are several shops and offices as well as a small park and observation deck over the Deerfield River to view Salmon Falls (a traditional Native American fishing site that the town was built around). Increased pedestrian and bicycle activities are also expected from the improvements to road edges.

5.6.1.3 Creating a Complete Street

The priority of the project was to design both a Complete Street and a Green Street, focusing on safe multimodal travel, including bikeways and sidewalks; drainage improvements including new catch basins and stormwater design to meet MassDOT standards; and culvert outfall improvement with stone riprap to help prevent erosion.

At project initiation, traffic count data from the Franklin Regional Council of Governments showed that from 2005 to 2010 there was an increase from 1750 cars to 1980 cars tracked by their Average Annual Daily Traffic Count data on Conway Street. The data also showed that there were seven crashes on this route, with the majority having taken place on Conway Road. Most accidents were reported to have happened when the road was described as "wet, icy, sandy, or slushy" due to poor drainage that caused pooling and freezing.

The repairs and improvements meet MassDOT Complete Streets design criteria by implementing roadside stormwater improvements. In the more rural portion of the project, the design team maintained and enhanced roadside vegetated swales – one of the first LID solutions – along Conway Road. These swales enter drop inlets and then are piped to the other (down grade) side of the road. They are designed with vegetation and check dams to slow water, reduce pollutants, and convey runoff from the road to periodically spaced drain inlets. In other parts of the project – more residential areas with limited right-of-way space, presence of ledge, and steep topography – the design team updated the closed drainage system with deep sump and hooded catch basins that captures runoff from about 3450 feet of roadway.

The upgrades to the closed drainage system and road runoff along the entire route improve the quality of the stormwater discharged into the Deerfield River and

Salmon Falls. This has a lasting impact of reducing damage done to the road and will extend the life of the infrastructure. The project design improves drainage and lessens the probability of pooling water and icing, thereby reducing the frequency and potential for accidents.

5.6.2 Redesigning Culverts to Prevent Flooding and Support Coldwater Fish

One of the streams carried by the roadway's culverts is classified as a Coldwater Fish Resource (CFR). The Massachusetts Division of Fisheries and Wildlife notes, "A CFR is a waterbody (stream, river, or tributary thereto) used by reproducing coldwater fish to meet one or more of their life history requirements. CFRs are particularly sensitive habitats. Changes in land and water use can reduce the ability of these waters to support trout and other kinds of coldwater fish" [14].

The goals of the culvert replacement were to alleviate the flooding condition by meeting MassDOT criteria for hydraulic design while also improving fish and wildlife passage in a sensitive habitat. To achieve these goals, two culverts were redesigned as a part of this project.

The original two culverts, shown in Fig. 5.8, were 4 to 4.5 feet wide, and the replacement culverts are more than 10 feet wide, which classifies them as bridges. Analysis of hydraulics, performed in accordance with the MassDOT LRFD (Load and Resistance Factor Design) Bridge Manual, showed that the dimensions of both culverts needed to be larger to increase the flow capacity, reduce flood risk over the roadway, and align with the Massachusetts Stream Crossing Standards (which provide guidelines for specific culvert dimensional requirements to enhance fish and wildlife passage). The increased size of the culverts met both design goals of reducing the risk of overtopping the roadway and improving fish passage.

Designing for fish habitat in CFR also requires considering the materials within the culvert. The design team sought to re-create the riverbed within the replacement culvert to encourage the transient wildlife to utilize the pass-through in a way that the previous culvert hadn't done. Additionally, the culvert along the CFR was originally in a perched condition – meaning the culvert was 4 feet higher at its outlet – and so fish were unable to move back upstream. The design team redesigned the replacement culvert to remove the perch and create a stream bed within the culvert, thereby reconnecting the stream to the downstream Deerfield River. Because of the velocity of the water through the culvert, multiple large boulders are included as eddies to act as resting areas for fish traveling upstream.

In order to build this new connection, the design team proposed a three-sided culvert, shown in Fig. 5.9, to allow for the most accurate imitation of the riverbed by allowing adjustments and observation before the top is then enclosed. A full culvert would have limited this re-creation of the natural stream bed.

Fig. 5.8 Existing roadside culverts. (Courtesy of Nitsch Engineering)

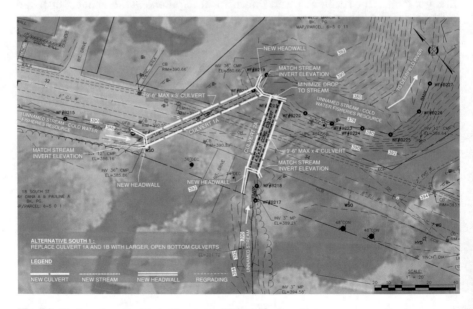

Fig. 5.9 Design plan of culvert. (Courtesy of Nitsch Engineering)

This project is the first example of a stream reconnection of a culvert on a MassDOT project. This project results in a new connection between a regularly stocked upstream pond and the Deerfield River, allowing the fish that travel downstream at maturity to return upstream to again spawn, creating a more abundant cycle of wildlife in the area.

5.6.3 Impact

The Buckland Roadway Improvements project highlighted the importance of and benefits of collaboration. The design team worked closely with the Town and MassDOT to improve the resilience through Complete Streets design goals.

The Buckland Roadway Improvements project improves accessibility while protecting environmental sensitivities. The project allowed a Town that lacked funding due to a small population and limited tax base to bring a roadway that was in disrepair to MassDOT Complete Streets standards, improving the standard of living in the neighborhood. New benefits such as bike lanes, improved geometry with widening of the road, and improved drainage encourage community activity and growth.

Project Team

Civil, Transportation, and Structural Engineer; Land Surveyor: Nitsch Engineering
Environmental Engineer: LEC Environmental Consultants
Geotechnical Engineer: Lahlaf Geotechnical Consulting

5.7 Conclusion

5.7.1 Keys to Successful Implementation

Selecting appropriate green infrastructure solutions for integration into roadway infrastructure projects requires engineers and owners to consider a few things:

1. *Designing flexible, context-sensitive solutions:* Each community that is threatened by climate change faces a unique situation that requires tailored solutions: one-size-fits-all solutions are not an option.
2. *Choosing the right tool from the toolkit:* Many communities have benefited from working with an engineer to develop a toolkit of solutions that could work for their specific issues and needs and then choosing the right BMP for the specific project.
3. *Considering long-term operations and maintenance:* Like anything else, appropriate operations and maintenance is key to making green infrastructure last for the long term. Roadway infrastructure owners need to have a clear asset management plan that helps them effectively operate, maintain, and improve their assets. Generally speaking, a structured schedule helps accomplish this.

5.7.2 Future of Roadway Infrastructure

Roadway infrastructure is a long-term commitment, with roads generally designed to last around 20 years [10]. Because of that long lifespan, every roadway design project needs to anticipate and consider the future, so that roadway profiles and layouts, as well as how they impact and protect the environment, are ready to support the altering climate future. Some of the key trends that will impact roadway infrastructure design in the future are as follows.

5.7.2.1 Emphasizing Sustainability and Resilience

The rise in severe weather events that result from climate change have impacted roads and added increased costs. Today, FHWA requires state DOTs to consider resilience in their roadway design. Because what is required and gets funded gets built, public policy is trending toward regulations that support green infrastructure at the local, state, and federal levels. Similarly, there are now many different grant programs that provide states and municipalities with money dedicated to making roadway improvements that prioritize sustainability and resilience.

The Institute for Sustainable Infrastructure (ISI) developed the Envision rating system: "a consistent, consensus-based framework for assessing sustainability, resiliency, and equity in civil infrastructure" [11]. Created in 2010 by the American Public Works Association (APWA), ASCE, the American Council of Engineering Companies (ACEC), and the Zofnass Program for Sustainable Infrastructure at the Harvard University Graduate School of Design, ISI's Envision system pull the best practices for green infrastructure into one cohesive system. Like the LEED certification system for buildings, Envision seeks to create a clear standard for what sustainability and resilience mean within the roadway. With the focus on sustainability and resilience on the rise, rating systems like Envision will continue to grow in prominence as long as costs for registration can be managed.

Innovative materials and technologies could make a huge impact on the sustainability and resilience of roadway infrastructure. Increased use of permeable paving materials will reduce storm runoff by recharging water to the ground below and could help prevent flooding. Advanced pavement monitoring systems, which embed sensors into the pavement to measure moisture and temperature, allow owners to understand pavement conditions more efficiently and help prioritize maintenance and improvements that help roads last longer. More research and development of innovative paving materials could provide even greater impacts, and moving toward sustainable pavement solutions that achieve engineering goals while using resources effectively and preserving the environment should be the goal [12].

5.7.2.2   Designing for Vehicles of the Future

Transportation remains the largest contributor to greenhouse gas (GHG) emissions in the United States. As costs drop, electric vehicles will become a larger economic driver. Automobile manufacturers, looking to the future, are ramping up the production of electric vehicles. For example, in January 2021, General Motors announced that they would exclusively produce electric vehicles by 2035. The US federal government has begun to replace their fleet of vehicles with electric ones, including Postal Service vehicles.

As electric cars become more popular, there will be a necessary shift in infrastructure support. The United States will require a much more reliable electric grid and more charging stations. Some companies are looking into ways to charge electric vehicles as they are driven, which would require integrating wireless charging capabilities into roadway surfaces.

As electric vehicles move more into the mainstream, so too do autonomous vehicles. If these vehicles are implemented correctly, they have the potential to have hugely positive impacts on roadway safety and mobility, reducing congestion and improving the environment [2]. From a roadway infrastructure perspective, designing roads for autonomous vehicles involves integrating new materials (e.g., special paint for lane stripes to help the vehicle's computer determine location) and/or technologies (e.g., wireless signals in overhead lights and traffic signals to connect cars). There may also be maintenance implications from autonomous vehicles, which may require roads to be in better condition than many owners currently deliver to operate safely, thereby requiring more maintenance.

5.7.2.3   Addressing Congestion

Traffic congestion is everyone's least favorite thing. While it's too early to understand whether the rise in remote work (working from home) during the COVID-19 pandemic will permanently impact traffic patterns, the importance of mitigating congestion will remain high. Engineers and roadway infrastructure owners will continue to drive best practices for managing traffic forward. For example, ASCE reports that decentralized traffic lights promote traffic flow [2], and studies by FHWA show that practices like congestion pricing can help address peak hour traffic congestion.

In addition, non-vehicular travel continues to rise in popularity, as communities find that if safe and comfortable non-car options are provided, more people will use them. More roadway infrastructure will need to be dedicated to bicycle lanes, sidewalks, and bus lanes.

5.7.2.4 Leveraging Technology

Technology advances at an ever-increasing rate, and transportation engineers will continue to leverage these advancements to drive progress for the future. New technologies can help address each of the trends for the future already identified: new climate models and interfaces, such as the Massachusetts Climate Resilience Design Standards Tool, make it easier for owners to see how climate change may impact a specific site; smart roads technologies that support electric and/or autonomous vehicles will continue to grow in importance; and traffic simulation modeling tools such as PTV Vissim make complex traffic simulations more realistic and intuitive to understand for lay audiences.

5.7.2.5 Funding Maintenance and Improvements

Roadway infrastructure, in spite of bipartisan support from both the public and the largest political parties, is consistently underfunded, which has resulted in 40% of the system being in poor or mediocre condition [2]. Right now, spending is focused on system preservation (roadway repairs and maintenance), which has been underfunded to the point that ASCE estimates a $786 billion backlog of road and bridge capital needs [2]. ASCE further estimates that current spending levels must be increased by 29% to address the current and anticipated backlogs [2].

The Highway Trust Fund (HTF) funds federal roadway investment using user fees from the federal gas tax, which has not been raised since 1993, even as inflation has cut its purchasing power by 40%. At the same time, increased vehicle fuel efficiency has resulted in drivers buying less gas. The Congressional Budget Office estimates that the HTF will have a $15B deficit by 2022 as current spending levels exceed user-fee revenues [2]. While some states have worked to increase their portion of funding through raising and/or reforming gas taxes, and exploring new revenue sources such as mileage-based user fees, not enough has been done to even maintain the US current roadway infrastructure – let alone make improvements.

Underfunding infrastructure (including roadway infrastructure) has a negative impact on both the US economy and its citizens' lives. From a purely economic perspective, ASCE projects that if the funding gap is not addressed, the US economy will lose more than $10.3 trillion in GDP by 2039 – and each household will lose more than $3300 per year in disposable income [13].

There are many ideas for how to address the funding gap. No matter which solutions are chosen, the end result must be the development of public policy – at the municipal, state, and federal levels – that provides the required funding to maintain existing roadway infrastructure, make improvements that better support environmental goals, and prepare for the future.

References

1. Weingroff R (1996) Federal-Aid Highway Act of 1956: Creating the Interstate System. United States Department of Transportation (USDOT) Federal Highway Administration (FHWA). https://www.fhwa.dot.gov/publications/publicroads/96summer/p96su10.cfm. Accessed 9 June 2021.
2. American Society of Civil Engineering (ASCE) (2021) 2021 Report Card for American's Infrastructure: A Comprehensive Assessment of America's Infrastructure. Available via ASCE. https://infrastructurereportcard.org/wp-content/uploads/2020/12/National_IRC_2021-report.pdf. Accessed 9 June 2021.
3. United States Department of Transportation (USDOT) Federal Highway Administration (FHWA) (2013) Highway Functional Classification Concepts, Criteria, and Procedures. Available via FHWA. https://www.fhwa.dot.gov/planning/processes/statewide/related/highway_functional_classifications/fcauab.pdf. Accessed 9 June 2021.
4. Zipper D (2020) Can the Bike Boom Keep Going? Bloomberg CityLab. https://www.bloomberg.com/news/articles/2020-10-29/how-the-feds-could-keep-the-bike-boom-rolling. Accessed 9 June 2021.
5. National Society of Professional Engineers (NSPE) (2019) Code of Ethics for Engineers. Available via NSPE. https://www.nspe.org/resources/ethics/code-ethics. Accessed 9 June 2021.
6. United States Department of Transportation (USDOT) Federal Highway Administration (FHWA) (2004) Traffic Congestion and Reliability: Linking Solutions to Problems, Executive Summary. Available via FHWA. https://ops.fhwa.dot.gov/congestion_report_04/executive_summary.htm. Accessed 9 June 2021.
7. Underwood B, Guido Z, Gudipudi P, and Feinberg Y (2017) Increased costs to US pavement infrastructure from future temperature rise. Nature Climate Change 7, 704–707. https://doi.org/https://doi.org/10.1038/nclimate3390.
8. Environmental Protection Agency (2021) Learn About Environmental Justice. https://www.epa.gov/environmentaljustice/learn-about-environmental-justice. Accessed 9 June 2021.
9. Executive Order 12898 of February 11, 1994 (1994): Federal Actions to Address Environmental Justice in Minority Populations and Low-Income Populations. Available via National Archives and Records Administration. https://www.archives.gov/files/federal-register/executive-orders/pdf/12898.pdf. Accessed 9 June 2021.
10. United States Department of Transportation (USDOT) Federal Highway Administration (FHWA) (1998) Developing Long-Lasting, Lower Maintenance Highway Pavement By The Research And Technology Coordinating Committee. Available via FHWA. https://www.fhwa.dot.gov/publications/publicroads/98julaug/developing.cfm. Accessed 9 June 2021.
11. Institute for Sustainable Infrastructure (ISI) About Envision. https://sustainableinfrastructure.org/envision/overview-of-envision/. Accessed 9 June 2021.
12. Ozer H, Al-Qadi I, and Harvey J (2016), Strategies for Improving the Sustainability of Asphalt Pavements. Available via FHWA's Sustainable Pavements Program. https://www.fhwa.dot.gov/pavement/sustainability/hif16012.pdf. Accessed 9 June 2021.
13. ASCE and EBP (2021) Failure to Act: Economic Impacts of Status Quo Investment Across Infrastructure Systems. Available via ASCE. https://infrastructurereportcard.org/wp-content/uploads/2021/03/FTA_Econ_Impacts_Status_Quo.pdf. Accessed 9 June 2021.
14. Mass.gov (2021) Coldwater Fish Resources. Division of Fisheries of Wildlife. https://www.mass.gov/info-details/coldwater-fish-resources. Accessed 9 June 2021.

Lisa Brothers, PE, ENV SP, LEED AP BD+C, was first introduced to the idea of a career in engineering by her high school
business teacher. While she was not yet familiar with what a profession in the engineering industry entailed, Lisa chose to apply
to the University of Massachusetts Lowell's College of
Engineering. She ultimately gravitated toward the field of civil
engineering based on her interest in being outside and observing
the building process. She graduated from UMass Lowell with a
BS in Civil Engineering in 1984.

Upon graduating, Lisa began working for the Massachusetts
Department of Public Works (now MassDOT) as an Assistant
Roadway Engineer and Bridge Engineer for a $30-million high-
way construction project. In this role she was the first female
engineer assigned to construction at the District. After 3 years
working in the public sector, Lisa chose to move into the private sector where she could better
pursue her entrepreneurial interests. In the spring of 1986, Lisa enrolled in the part-time MBA
program at Northeastern University, earning that degree in 1991 while continuing to work full-time.

After working as a design engineer for a couple different firms, Lisa found her calling when her
colleague announced that she was going to start her own firm in 1989. She has long been an advocate for women to create their own opportunities and knew immediately that this was hers. Lisa
followed her colleague out of the room saying, "Not without me!" Her MBA definitely helped
position the new company for success.

Lisa now has more than 35 years of experience in the design, construction, and management of
roadway, site development, sustainable design, and infrastructure-related projects. As President
and CEO of Nitsch Engineering since 2011, Lisa is responsible for the vision, growth strategy,
strategic direction, and overall performance of the firm. She also serves as Principal-in-Charge for
many of the firm's design projects.

The love for the outdoors that initially drew Lisa to civil engineering extends beyond her work
life. She enjoys staying active as much as possible through a variety of outdoor activities such as
hiking, biking, and kayaking – particularly alongside her husband and two adult children.

Lisa is dedicated to supporting UMass Lowell; she recognizes that she would not be in the
position she is today without the affordable, exemplary engineering education she received and
continues to give back to the University. She currently sits on the Chancellor's Advisory Council,
is a founding member of the Center for Women and Work Advisory Board, and is a past Chair of
the College of Engineering/Industrial Advisory Board.

Raising awareness about and making progress on issues of equity, diversity, and inclusion
within the engineering industry is a passion of Lisa's. She serves on the Boston Women's Workforce
Council, which is working to close the gender and racial wage gap, and is Chair of the American
Council of Engineering Companies' (ACEC's) National Diversity, Equity, Inclusion, and
Belonging Committee.

A registered professional engineer in Massachusetts and 11 other states, Lisa is involved in a
wide range of professional activities. She has been actively involved in the American Council of
Engineering Companies/Massachusetts (ACEC/MA) for more than 25 years; Lisa currently serves
on the ACEC/MA Board as National Director, as the PAC Committee Chair, and on the Government
Affairs Committee, and she is a past President of the Member Organization. She is a member of
the Environmental Business Council of New England Board of Directors. Lisa also served as
President of the Women's Transportation Seminar-Boston (WTS-Boston) Chapter.

Lisa's contributions to her community have been well recognized by a range of organizations.
She was named a 2015 Woman of Influence by the Boston Business Journal and received the EY
Entrepreneur of the Year™ 2014 Award in the New England region's services category. She also
received the 2017 Leadership Award and was named the 2008 Woman of the Year by WTS-

Boston; received the 2004 Citizen Engineer Award from the Boston Society of Civil Engineers; received the 2018 University Alumni Award and 2003 Francis Academy Distinguished Engineering Alumni Award from UMass-Lowell; was honored with a 2002 Pinnacle Award as an Emerging Executive from the Greater Boston Chamber of Commerce; and received the BSCES Lester Gaynor Award in recognition of her exemplary service as a public official in Wilmington, Massachusetts, in 2001.

Chapter 6
Roadway Lighting and "Smart Poles"

Sandra Scanlon

Abstract It is estimated that street-level lighting by municipalities accounts for 30% of all the energy used to generate electricity for outdoor lighting. If we converted our roadway lighting to light-emitting diode (LED), we could save an enormous amount of energy usage, which would help our already stressed electric utility infrastructure. In addition to energy savings, the maintenance savings are also significant. Now imagine if we used the light pole as another backbone for technology infrastructure – "smart poles" capable of supporting a diverse set of Internet of Things (IoT) devices.

Keywords LED · Lighting · Pole · Roadway · Haitz · Color · Daylight · Circadian · Incandescent · Fluorescent

6.1 Introduction

It is estimated that street-level lighting by municipalities accounts for 30% of all the energy used to generate electricity for outdoor lighting [1]. Another 60% goes toward lighting parking lots and parking garages. Think about the little amount of time you spend in a parking lot or parking garage versus at your desk or at home. If we converted our roadway and parking area lighting to light-emitting diode (LED), we could save an enormous amount of energy usage, which would help our already stressed electric utility infrastructure. In addition to energy savings, the maintenance savings are also significant. Now imagine if we used the light pole as another backbone for technology infrastructure – "smart poles" capable of supporting a diverse set of Internet of Things (IoT) devices.

S. Scanlon (✉)
Denver International Airport, Denver, CO, USA
e-mail: sandra@cooljest.com

© The Author(s), under exclusive license to Springer Nature   129
Switzerland AG 2022
P. Layne, J. S. Tietjen (eds.), *Women in Infrastructure*, Women in Engineering
and Science, https://doi.org/10.1007/978-3-030-92821-6_6

Walk into any given Starbucks, and you'll probably notice that the majority of patrons have their faces buried in an electronic device. Whether it is a cellular phone, laptop, or tablet, we are a 24/7 connected society through myriad forms of technology all around us. As we start to design infrastructure that supports our connectedness, new trends are developing beyond handheld devices to address innovation and sustainability in smart feature-filled streetscapes.

Cities are looking to companies, utilities, and engineering professionals that are leading the charge to meet the needs of smart city design. Innovation and sustainability go hand in hand. Mass adoption of LED lighting in households and interior spaces is well documented, but LED usage in exterior and roadway lighting applications has been slow, with very recent trends showing a significant increase in adoption.

Companies such as Panasonic, Siemens, Philips, American Tower, and Cisco are developing products and solutions to improve infrastructure features while maximizing benefits to both corporate and environmental triple bottom line of people, planet, and profitability.

6.2   What Is an LED? [2]

A light-emitting diode (LED) is a semiconductor light source that emits light when current flows through it. Electrons in the semiconductor recombine with electron holes, releasing energy in the form of photons. The color of the light (corresponding to the energy of the photons) is determined by the energy required for electrons to cross the band gap of the semiconductor. White light is obtained by using multiple semiconductors or a layer of light-emitting phosphor on the semiconductor device.

Appearing as practical electronic components in 1962, the earliest LEDs emitted low-intensity infrared (IR) light. Infrared LEDs are used in remote-control circuits, such as those used with a wide variety of consumer electronics. The first visible-light LEDs were of low intensity and limited to red. Modern LEDs are available in visible, ultraviolet (UV), and infrared wavelengths, with high light output.

Early LEDs were often used as indicator lamps, replacing small incandescent bulbs, and in seven-segment displays. (A seven-segment display is a form of electronic display of numerals that is an alternative to the more complex dot matrix displays.) Seven-segment displays are widely used in digital clocks, electronic meters, basic calculators, and other electronic devices that display numbers. Recent developments have produced high-output white-light LEDs suitable for room and outdoor area lighting. LEDs have led to new displays and sensors, while their high switching rates are useful in advanced communications technology.

LEDs have many advantages over incandescent light sources, including lower energy consumption, longer lifetime, improved physical robustness, smaller size, and faster switching. The first white LEDs were expensive and inefficient. However, the light output of LEDs has increased exponentially. The latest research and development has been propagated by Japanese manufacturers such as Panasonic and

Nichia and by Korean and Chinese manufacturers such as Samsung, Kingsun, and others. This trend in increased output has been called Haitz's law after Roland Haitz. Haitz's law, the metric he formulated, is the equivalent for LEDs and lighting to Moore's law for transistors and integrated circuits. Haitz's law correctly predicted the timescale and degree to which LEDs would triumph over all other lighting technologies in efficiency and cost. Light output and efficiency of blue and near-ultraviolet LEDs rose, and the cost of reliable devices fell. This led to relatively high-power white-light LEDs for illumination, which are replacing incandescent and fluorescent lighting, as well as other light sources used in highway, street, and parking lot lighting.

6.3   Quality of LED Light Versus Other Types of Light

If you have ever walked into a Home Depot or Lowe's to buy a light bulb, you may have noticed the consumer displays to show the different color temperatures of LED lamps available. Some people dislike the seemingly blueish bright light of cool LED lamps versus the more yellow appearance of warm LED lamps. There are two ways designers describe LED light – color rendering index (CRI) and color temperature. The visible light spectrum contains red, orange, yellow, green, blue, indigo, and violet. Some artificial light sources illuminate objects better in some parts of the visible light spectrum than others. The higher the light source's CRI, the more of the spectrum it will illuminate, and the truer the color of the object that will be seen. For example, daylight, incandescent, and halogen bulbs all illuminate the entire spectrum and have a CRI of 100. Low-pressure sodium lights (found in many older streetlamps) illuminate with only yellow light. But just because a light source has the maximum CRI of 100 doesn't mean you'll have the true colors of what you're looking at.

Color temperature refers to its color balance in terms of more blues and greens versus more reds and yellows. Daylight has a bluer quality, and incandescent lights glow with a cozy yellow light. When we look at the objects in our homes under low-CRI light (like a fluorescent lamp), we notice that they look flat and lifeless. A color temperature that is too cool (like cool-white LEDs) may seem uninviting, even sterile. The color temperature of the light source as well as the CRI affects how we see the color of objects in our surroundings (Fig. 6.1).

Figure 6.2 compares light types by spectral response in different kinds of light sources including daylight, incandescent bulbs, LED, and fluorescent bulbs. You can see how different even full-spectrum light sources like daylight, halogen, and incandescent light can be in terms of how much of each color you get: daylight gives more intense but cooler light.

Fig. 6.1 Color temperature equated to time of day from ambient light [3]

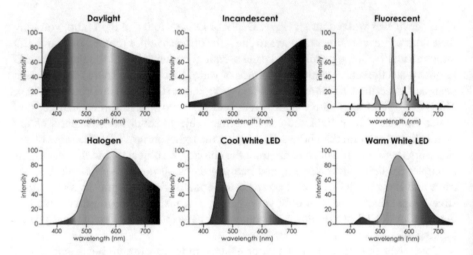

Fig. 6.2 Comparison of light source spectral response [4]

6.3.1 Blue-Rich Light

Certain blue LEDs and cool-white LEDs can exceed safe limits of the so-called blue-light hazard as defined in eye safety specifications such as "ANSI/IESNA RP-27.1–05: Recommended Practice for Photobiological Safety for Lamp and Lamp Systems." In 2006, the International Electrotechnical Commission published IEC 62471 Photobiological safety of lamps and lamp systems, replacing the application of early laser-oriented standards for classification of LED sources.

Blue-rich light suppresses melatonin secretion which controls our circadian rhythm (Fig. 6.3). A circadian rhythm or circadian cycle is a natural internal process that regulates the sleep-wake cycle and repeats roughly every 24 hours.

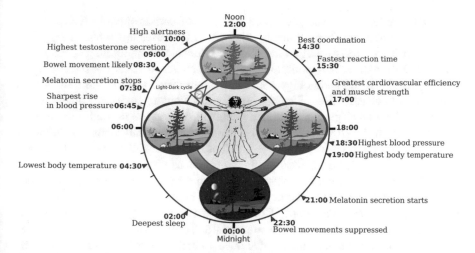

Fig. 6.3 Circadian rhythm light-dark/sleep-wake cycle [5]

In 2016, the American Medical Association (AMA) released a report discussing how blue-rich light affects individuals and the environment. It expressed concern over the possible adverse influence of blueish street lighting on the sleep-wake cycle of city-dwellers. Many industry professionals claim exposure levels are not high enough to have a noticeable effect. Others believe it affects not only humans but wildlife as well.

While the report focused on streetlights, all blue-rich light sources, including cell phones, televisions, laptops, tablets, and computer screens, can potentially have an effect. The AMA report encourages attention to optimal design and engineering features when converting streetlights to LED. These design features include shielding to control glare, lighting controls for dimming and scheduling, and limiting the color temperature of the LED light source. The Board of Directors of the Illuminating Engineering Society of North America (IESNA) published a response to the AMA's report, acknowledging the need to reduce glare of all streetlights and dimming of lights in off-peak hours. IESNA also addresses the impact of melanopic content, a factor in all lighting and its effects on individuals, saying it contributes to sleep disruption. Fortunately, good design, lower melanopic content, and replacing streetlights based on lumen output rather than wattage can address the adverse side effects of LEDs.

6.4   LEDs and Sustainability

LEDs use a lot less energy compared to other light sources such as incandescent and halogen lamps. In the United States, we predominantly burn fossil fuels to produce our electricity, and therefore there is a direct link between the reduced amount of electricity we use to power LEDs and the level of carbon dioxide (CO_2) that is entered into our atmosphere. Since lighting with LEDs uses much less energy, less carbon dioxide is produced. According to The Climate Group, lighting accounts for nearly 5% of global CO_2 emissions. A global switch to energy-efficient light-emitting diode (LED) technology could save more than 1400 million tons of CO_2 and avoid the construction of approximately 1250 power stations. Following the historic global Paris Agreement, effective climate actions are needed. With energy savings of up to 50–70%, LED lighting has been recognized as one of the most actionable and ready-to-implement technologies for cities to transition to a low-carbon economy and peak emissions in the next decade [6].

While LEDs do not contain mercury like fluorescent lamps, they contain other hazardous metals such as lead and arsenic. LED bulbs will last three times longer than compact fluorescent lamps (CFLs) and ten times longer than halogen and incandescent light bulbs. It takes more material and energy to manufacture a replacement for a non-LED bulb. Since LEDs have a much longer life cycle when compared to CFs, halogen, or incandescent bulbs, they help to conserve our natural resources as well as reduce the amount of CO_2 emissions produced during their manufacturing process. LEDs are environmentally friendly only if they are also recycled. More than 95% of an LED bulb is recyclable, and there are waste management companies that will collect and recycle LEDs, some for a small fee.

The recycling process generally involves the LED bulbs being crushed and separated into constituent components. Glass, aluminum, and lead that are present in LED lights, as well as non-ferrous metals, are separated. The glass can then be used in other products, as can the aluminum.

Though overall LED lights are more environmentally friendly than conventional, incandescent bulbs, it is still important that LEDs are not simply dumped into landfills, as this could have long-term environmental effects. Given the huge rise in demand in recent years for LED lights and the scarcity of natural resources that are used to produce them, it is important that advocacy for recycling of LEDs expands so that LEDs can continue to be used well into the future.

6.5   Standards

LED Street Light Standards – Because the quality of LED lighting allows you to see true colors and see more clearly, the perception of safety improves mobility – not just of cars but also of pedestrians. Despite industry design standards for illumination levels and pole spacing, the easiest retrofit approach is to replace, one for one, each high-pressure sodium (HPS) or other light source with LED (Figs. 6.4 and 6.5). However, the major advantage to LED lighting design is the more efficient and targeted application of light. While a one-for-one replacement is certainly an installation cost advantage, an optimal design is one where the pole spacing of LED street lighting is applied at the maximum efficiency – which is almost always a farther pole spacing than traditional street lighting design. Many cities approach their

Fig. 6.4 Before and after replacing one-for-one HPS with LED [7]

Fig. 6.5 Parking lot comparison of HPS and LED [8]

retrofit projects, however, with a one-to-one approach to save on the retrofit/replacement of the underground infrastructure of conduit and wiring to each pole location. In order to provide the necessary infrastructure backbone for a smart city streetscape, it may make sense to design for maximum efficiency – putting in new conduit infrastructure for both line-voltage power conduits and communications conduits.

There has been a general lack of standards in the LED exterior application field. There are various standards relating to LEDs for automotive use and traffic signals, but not much has been written specifically for outdoor applications. Efforts are increasing by code authorities to address this. There are plenty of guidelines on what not to do, but not per se on good design practices for what to do. Code authorities such as the International Commission on Illumination (also known as CIE from its French title, the Commission Internationale de Photométrie), National Electrical

Manufacturers Association (NEMA), and Illumination Engineering Society (IES) are working diligently to expand the breadth of focus areas for exterior LED lighting applications.

6.5.1 Roadway Lighting LED Adoption

Ann Arbor, MI, was the first city to change out streetlights citywide to LED in 2007. Perhaps a smaller city is more nimble or able to adopt change than a larger corporate entity or city? Denver was one of utility company Xcel Energy's pilot cities that helped pave the way for their new national standard. However, LED streetlights were not even an option in Xcel Energy's approved design standards until 2016.

6.6 Smart Poles

Streetlights are mounted on poles. Gone are the days when a pole only held up a streetlight. Now they are smart – smarter than ever when paired with the myriad of IoT devices and systems to make our streetscapes and cities interconnected. Devices to control lighting, provide Wi-Fi access, monitor intersections for traffic control or police surveillance, sense the weather, and so on can be mounted to or integrated within the pole itself (Fig. 6.6). The height of the pole can be an advantage for device placement as well as hiding necessary wiring within the pole for the items attached to it. Feeding the base of the pole is an underground network of conduits for power and communications. The challenge for existing streetlight infrastructure is that often there isn't a separate communications conduit feeding the pole, which by code is required to be kept separate from line-voltage power wiring. Another aspect of this infrastructure is the option for solar-powered devices, again, using the pole to mount a small solar panel used to store energy in small batteries for the devices mounted on the pole.

6.6.1 The Internet of Things (IoT)

The smart light pole is an important source of not only information collection but interconnectedness for people. It is the next significant infrastructure network in the future IoT field and an important part of any smart city. Smart light poles collect information through integrated sensors. These sensors collect data that can interact with city traffic management systems, police surveillance systems, even financial management, and procurement systems to provide multiple data support for smart city big data applications.

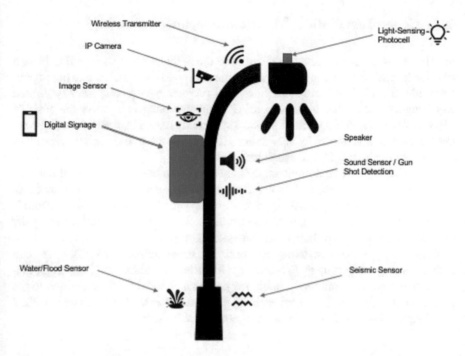

Fig. 6.6 Smart street light pole. (Courtesy of the author)

Smart streetlight poles have applications that include not only the obvious street lighting control for on/off, dimming, scheduling, and even LED color changing but also Wi-Fi antenna base station, advertising through digital signage, emergency call system, weather monitoring, electric vehicle charging, and parking space management.

Spying – But is this new infrastructure becoming too much like "Big Brother"? Back in the fall of 2019, Hong Kong protestors targeted dismantling smart lampposts being tested by the government that were equipped with Bluetooth connectivity, sensors, and cameras that officials said collected weather and traffic data. But protestors felt the lampposts stood as beacons for increasing government surveillance, fueling fears that the poles actually hid technology to snoop on passersby monitoring their movements and communications. The secretary of Hong Kong's Innovation and Technology Bureau described the idea as "a conspiracy theory." The company that developed the smart lampposts intended the radio-frequency identification packs to transmit location information to applications, such as one linked to specially designed canes for the visually impaired to provide them with directions [9].

6.7   Costs: Installation, Maintenance, and Operating

Installation – Design professionals can help cities make good choices that have a positive impact on their community and achieve desired savings. Despite quick payback, initial capital cost can sometimes stall projects by stressing already limited city budgets. Most cities have been using their own budgets to cover the costs of LED retrofits. While LED retrofit projects seemingly have a high initial capital cost, the payback period is usually very short. Regardless, the initial capital investment can exceed the limits of small municipal budgets.

Maintenance – When lighting controls are added to the LED retrofit of conventional streetlight systems, maintenance costs can be reduced by as much as 80%. Inherent in many lighting controls systems are the added features of real-time monitoring for outages, improving lighting reliability, service delivery, and reducing the need for premature lamp changes in conventional lighting.

Operating – When converting conventional street lighting to LED, cities can achieve energy savings up to 70% and provide brighter and more uniform light distribution to enhance community safety for pedestrians and drivers. According to the Department of Energy (DOE), widespread use of LEDs in the United States by 2027 could equate to more than $30 billion in energy savings.

6.8   Feature-Filled Future

With the rise in popularity of electric cars comes the need to charge. Companies like Ubitricity, PlugShare, and Ovo have developed solutions to turn existing city streetlights into charging stations, providing electric car owners with a way to locate charging stations across North America and even sell energy back to the grid.

As a way to offset the upfront costs associated with implementing smart features, cities are opting for digital kiosks to not only provide public transportation information, tourist activities, and city services but also allow companies to pay into city budgets for advertising opportunities. Digital kiosks can also provide Wi-Fi connectivity, weather information, and news.

6.9   Conclusion

Municipalities are quickly realizing the advantages of converting roadway lighting to light-emitting diode (LED), which will reduce energy usage (and budgets) and help put less stress on aging electric utility infrastructure. Some experts think that though many cities are interested in LED streetlights, the greatest barrier to widespread adoption of the technology may simply be education. The more the "good story can be told" to explain that proper LED lighting application is really a human-centric approach to creating a world around us that meets our needs and is environmentally friendly, the more adoption of outdoor LED lighting will grow. Some cities will be faster to adopt smart city features too, and each city can pursue a variety of features to meet their needs. Designers and planners have the opportunity to not only meet current municipal needs but also look to evolving interconnectivity to meet the demands of the future. LED lighting, smart poles, digital kiosks, and lighting control could all play a part that might someday be as common as today's handheld digital devices.

References

1. "LED streetlights save cities millions, but raise health issues – and urban residents hate them" by Kelly Hodgkins, September 29, 2016, accessed 9/12/2021 https://www.digitaltrends.com/cool-tech/people-dislike-bluc-light-led/
2. Light-emitting diode, https://en.wikipedia.org/wiki/Light-emitting_diode, accessed 9/12/2021
3. https://niligo.com/wp-content/uploads/2019/03/color-of-sun-sunrise-and-sunset.jpg accessed 9/12/2021
4. Jennifer Priest, The DIY Decoratos Eco-Friendly Lighting, Dilemma, http://housecraft.ca/eco-friendly-lighting-colour-rendering-index-and-colour-temperature/, September 30, 2012.
5. Overview of biological circadian clock in humans, Yassine Mrabet. https://commons.wikimedia.org/w/index.php?curid=3017148 accessed 9/12/2021
6. LED, Climate Group, https://www.theclimategroup.org/led, accessed 9/12/2021
7. News + Media, City of Los Angeles Public Works, Bureau of Street Lighting, LED Conversion Program - Before and After Photographs - 6/7/12, http://bsl.lacity.org/led-news-media.html#prettyPhoto, accessed 9/13/2021.
8. "Advantages of LED vs. High Pressure Sodium, Area Lights" by Courtney Patrick, March 6, 2018. https://www.homelectrical.com/advantages-led-vs-high-pressure-sodium-area-lights.6.html accessed 9/12/2021
9. "Protesters Spy an Enemy: Lampposts," by Eli Bender, The Wall Street Journal, 8/31-9/1/2019

Sandra Scanlon, PE, LEED AP®, Senior Director of Airport Infrastructure Management Development at Denver International Airport (DEN), excels as a leader; a mother; an advocate for getting more women into science, engineering, and technology careers; and a community leader. Sandra began her career as an electrical engineer at Amoco Oil Company after graduating from Valparaiso University with a BS in Electrical Engineering and a Minor in Computer Science. She proceeded through increasingly responsible positions at Amoco as project engineer and electrical specialist. She is a successful entrepreneur founding Scanlon Consulting Services in 1997. In 2008, Sandra led her company through a successful merger, resulting in Scanlon Szynskie Group, Inc., and continued as President. In 2017, SSG MEP, Inc. was acquired by BCER Engineering, Inc., a four-office engineering firm providing mechanical, electrical, fire protection, technology, and sustainability consulting services. Sandra led BCER's electrical engineering department as an associate principal/owner and was chair of the board of directors until early 2020 when she joined DEN. At DEN, she leads four departments responsible for design, construction, and quality assurance of facilities and infrastructure projects throughout the airport property. She is a registered professional engineer in multiple states and a LEED-accredited professional.

She was a gubernatorial appointee to the Colorado State Board of Licensure for Architects, Professional Engineers, and Professional Land Surveyors for 8 years, where she served as Secretary, Vice Chair, and Chair of the Board. She is also an Emeritus Director and founding board member of the Denver School of Science and Technology (DSST) Public Schools and former board member of the DSST Foundation Board. Sandra has been an active member of the Society of Women Engineers (SWE) since college. She co-chaired the 2001 National Conference; spearheaded the establishment of the annual Girls Exploring Science, Technology, Engineering, and Math (GESTEM) event; and has held several significant positions within the Rocky Mountain Section including President. She also served in significant roles for SWE nationally including National Membership Committee Chair and is a Fellow Life Member and National Entrepreneur Award recipient. She has continued her community involvement and served on the board for Habitat for Humanity Metro Denver and was a coalition member of the Women's Foundation of Colorado (WFCO) STEM Coalition for the Colorado Education Initiative. Sandra is a senior member of the Institute of Electrical and Electronics Engineers (IEEE), member of the Denver IEEE Women in Engineering, and an IEEE Technical Expert.

Sandra's interest in engineering, project management, and all things STEM-related started when she was young. She thoroughly enjoyed her Tinkertoy Construction Set, Lincoln Logs, Lego's, Matchbox cars, and the Sunshine Family dolls as she created scenes of farm towns and small city layouts across the living room floor. She thought she wanted to be an architect and was encouraged in this pursuit by her mother and father. Sandra's mother, Colleen Wood, was consistently kept out of math and science classes as a young girl, and she wasn't about to let that happen to her daughter. Sandra's father, James Wood, brought home an antique drafting table from his manufacturing company for Sandra to use for her drafting classes she took in middle school and high school. She still has that table to this day. Once in college, one of only a handful of women in her classes at Valparaiso University, she quickly realized she loved engineering, especially electrical engineering and computer science. But along her career path, she rather enjoyed organizing tasks and projects, which led her to be a much better project manager and leader. Helping others to find their passion and enjoyment in what they love to do every day is what is most fulfilling, and that is where Sandra excels the most.

Chapter 7
Community Engagement + Community Partnerships = Community Projects: Implementing Successful Rail Transit Projects

Kimberly Slaughter

Abstract The implementation of a rail program in our communities is complex and affects people physically, emotionally, and financially. It requires not only professionals who bring technical expertise to a project, but ones who also recognize the full range of impacts that come from implementing new infrastructure. The most successful professionals in the rail industry are program/project managers who can acknowledge the impacts, communicate the potential impacts to a broad audience, and lead a team to implement a program that reflects the voice of the affected community. This chapter highlights the benefits of partnerships with the residents and businesses of communities to develop better infrastructure projects that support and sustain the communities they impact.

Keywords Partnerships · Community voice · Rail infrastructure · Successful project managers

The implementation of a rail program in our communities is complex and affects people physically, emotionally, and financially. Successful rail programs are led by professionals who not only apply their technical expertise but also recognize the full range of impacts to the communities served by the project. The most successful professionals in the rail industry are program/project managers that can identify and acknowledge the impacts, communicate them clearly to a broad audience, lead a team to implement a program that reflects the voice of the affected community, and recognize that community partners are the key to bringing those voices.

This chapter will explore the value of community champions as partners to implementing agencies. These partnerships are key to the successful development of transit programs that include rail. We will also provide case studies of several

K. Slaughter (✉)
SYSTRA, New York, NY, USA
e-mail: kslaughter@systra.com

© The Author(s), under exclusive license to Springer Nature Switzerland AG 2022
P. Layne, J. S. Tietjen (eds.), *Women in Infrastructure*, Women in Engineering and Science, https://doi.org/10.1007/978-3-030-92821-6_7

projects from across the nation and explore how they partnered with the community to shape their final projects.

7.1   Background

The first American-built passenger rail example was powered by a steam engine and tested on the Baltimore and Ohio Railroad in 1829. It is now known famously as the Tom Thumb locomotive, built by Peter Cooper. It lost the race against a horse [just barely!], but it more than proved its ability as a reliable source of mechanical transportation. As the industry greatly expanded during the 1850s, railroad tycoons with endless bank accounts were more interested in earning more money for themselves than in public safety (this lack of safety foresight helped bring stiff, arguably overbearing government regulation, which later in the 1960s and 1970s resulted in the near collapse of the industry) [1]. Early passenger trains faced additional challenges that included Indian sabotage and attack, particularly in Western states where Native Americans fought for control of their land. The Native Americans were fighting to keep their land, homes, and communities that were being usurped by the government to do "what is best" for them without obtaining their input.

The value that rail infrastructure has brought to our communities is unquestionable. This development creates jobs and brings goods and materials that we all rely on everyday. The State of New York Department of Transportation documented a history of passenger railroads in their state. "Besides the tremendous impact of construction and opening of the Erie Canal, it would be difficult to discuss the incredible growth and development of New York (a State that increased in population from 1820 to 1900, from 1.4 million people to almost 11 million people) without highlighting the role played by railroads. Starting in the 1830s, throughout the length and breadth of the Empire State, railroads large and small tied together city and farm (later suburbs), bringing foodstuffs and raw materials toward the cities, and in turn, bringing manufactured goods and summer vacationers out to the country" [2]. While these changes improved most people's quality of life through those jobs and access to goods and materials, this infrastructure also displaced people from homes, took family lands, and isolated neighborhoods.

It is not fair to place all of the burden of negative impacts on rail projects. Roadway infrastructure has also contributed to the bifurcation of neighborhoods through displacement, taking of land through eminent domain, and isolating neighborhoods on a larger and more local scale. But, the well-known reference of "being from the wrong side of the tracks" has never been modified to being from the wrong side of the freeway. In a social impact assessment by the Global Journal of Commerce and Management Perspective, the social impacts of infrastructure projects to communities were defined as the cost of the human population by any public and private actions that change the way people live, work, play, relate to one another, organize to meet their needs, etc. The major types of social impacts relate to lifestyle, cultural, community, quality of life, and health-related impacts. The report concluded that the early consideration of social impacts, the alignment of activities with regional and community planning objectives, and meaningful participation of communities in decision-making are key features of a policy regime that will demonstrate best practices and support the sustainable development of resources and communities [3].

Another report by the United States Department of Transportation also found that engaging stakeholders early and continuously to gather feedback and gain support for implementation is an essential strategy for implementing successful transportation projects. Further, the project teams should work with the community to seek solutions that implement placemaking and economic development strategies while protecting affordability, equity, and character [4]. In my experience, the members of the community have valuable direct knowledge of what is currently missing in their communities and what may be needed to create a sense of place.

In 2021, Secretary of Transportation Pete Buttigieg, under the Biden Administration, is working to right some of these historical wrongs. Every decision about transportation is not necessarily a decision about justice. It is a decision about our future. And to understand our future, we must learn from our past. It was 65 years ago, on June 26, 1956, that President Eisenhower signed a bill creating the Interstate Highway system.

And it was an extraordinary achievement. But we know that the planners behind it also made choices that often routed new highways directly through Black and Brown neighborhoods, doing lasting damage to those communities. For example, I-81 in Syracuse, New York, was constructed in three stages, opening between 1959 and 1969, and, as Senate Democratic Leader Chuck Schumer described, right over and through the 15th Ward. It displaced nearly 1300 residents from what had been a close-knit, middle-class, Black neighborhood. Those who remained were cut off, in many ways, from opportunity. And over the next few decades, much like my own hometown of South Bend, Indiana, Syracuse lost about 30% of its population.

Robust public participation is vital to the rulemaking process. By providing opportunities for public input and dialogue, agencies can obtain more comprehensive information, enhance the legitimacy and accountability of their decisions, and increase public support for their rules. Agencies, however, often face challenges involving a variety of affected interests and interested persons in the rulemaking process. In this discussion, I am paralleling rulemaking to public decision-making.

Today, communities of every size grapple with rapidly changing mobility options, a recognition of systemic inequities, and the increasing effects of climate change. Transit and development investments shape communities for generations, defining possibilities for existing and new residents. Too often, these investments have led to displacement of existing residents. Consequently, those involved with transit, mobility, land use and development policies, plans, projects, implementation, and operation must be equipped to make decisions with and on behalf of those who live there now, as well as those who will be there in the future.

Let us explore the benefits of the forum created by Rail~Volution to a concerted space to include the public voice in rail infrastructure. Let us also explore a couple of recent rail projects and how they successfully engaged the public's voice to create the final rail infrastructure project.

7.2 Rail~Volution

The pattern of designing and implementing rail infrastructure without the consideration of what the average citizen wanted or even asking their opinions has happened for many years. Some would contend that it continues today. To address this, in 1989, Rail~Volution, a network of leaders, practitioners, and advocates inspired by the potential for major transit investments to shape more vibrant and equitable communities, was formed. Believing that the focus of any new investment is not so much about the project – whether rail, bus, trail, or development, the Rail~Volution network believes it is more about the people and what they want their communities to become. To that end, Rail~Volution began to host a series of outreach and advocacy events geared toward developing real advocates for the Portland, Oregon, metropolitan region's MAX Light Rail System [5].

At the conference in 1994, Congressman Earl Blumenauer (District 3, Oregon) announced that in 1995, Rail~Volution would become a national transport conference. From this point, Rail~Volution acted as a loose federation of sponsoring partners, united by common interests and dedication. In 2000, the National Steering Committee realized the need for a more formal organization and developed it into a 501(c)(3) non-profit charitable organization. Since then, Rail~Volution has hosted more than 20 transportation conferences throughout the United States. From Seattle, Washington, to Miami, Florida, Rail~Volution has showcased the innovations and transportation projects as evidence that investing in transit systems creates jobs, increases health, and creates vibrant livable cities.

Rail~Volution gathers and supports a large and diverse network of leaders, professionals, and advocates who see the potential for transit, mobility, land use, and development to create great places to live, especially for those excluded from access to opportunity.

Through its annual conference and programs, Rail~Volution engages diverse, cross-sector stakeholders, providing ways to share, learn, connect, and recharge. Throughout its 25-year history, Rail~Volution has shaped the vision for transit-oriented communities, influencing policies, planning, and the way projects are delivered. Today, its focus is to equip individuals, organizations, and regions with the tools to integrate transit, land use, and community development decisions with what is important for each community, keeping transit and mobility options at the center.

7.2.1 Vision

The organization envisions America's cities and regions transformed into livable places – healthy, economically vibrant, socially equitable, and environmentally sustainable – where people have transportation choices.

7.2.2 Mission

Serving as a catalyst for the movement to build livable communities with transit, Rail~Volution inspires people in communities and regions to make better transit and land use decisions. This is done by partnering, equipping, and connecting people and institutions at all levels.

7.2.3 Values

7.2.3.1 Impact and Results

Rail~Volution is committed to making a measurable difference toward building livable cities and regions with transit. The focus is on sensible, real-world-tested, effective approaches to development that improves the way people live, work, and travel.

7.2.3.2 Inclusivity

Diverse stakeholders are welcome to the Rail~Volution table, honoring what people of different races, nationalities, professional disciplines, and geographic and economic backgrounds bring to the mission. The network believes diversity promotes strength and embraces varied opinions and perspectives, recognizing that lively discussion fosters greater understanding, energy, creativity, and momentum.

7.2.3.3 Collaboration and Partnership

The Rail~Volution community believes that more can be accomplished through collective action than through individual efforts. The network of people and institutions – connected by common values and goals – is the organization's greatest asset. Rail~Volution aspires to form strong partnerships built on respectful, open, and honest relationships.

7.2.3.4 Innovation

Rail~Volution strives to be visionary and cutting-edge by promoting cross-sector and interdisciplinary discussions and solutions. The organization encourages fun and energizing exchanges and creative problem-solving, grounded in best practices. Rail~Volution pushes traditional boundaries to invent new, effective approaches to multimodal transportation planning and community development.

7.2.3.5 Quality

Rail~Volution is committed to achieving and maintaining the highest standards of quality in everything it does and continuously evaluates its effectiveness and seeks improvement.

7.3 Rail Infrastructure Case Studies

7.3.1 Northern Indiana Commuter Transportation District (NICTD) West Lake Corridor Project Case Study

The Northern Indiana Commuter Transportation District (NICTD) proposes to expand its commuter rail service through an approximate 8-mile extension of the South Shore Line (SSL), known as the West Lake Corridor Project (Project). The line would be extended to the south to provide passenger rail service to three municipalities in Lake County, Indiana: Hammond, Munster, and Dyer (Fig. 7.1) Trains on the new branch line would connect with the existing SSL and ultimately with the Metra Electric District (MED) line to the north to Millennium Station in downtown Chicago – a total distance of approximately 30 miles [6].

This project would provide a vital transportation link connecting northwest Indiana with Chicago and Cook County, Illinois. It would also expand NICTD's service coverage, improve mobility and accessibility, and stimulate local job creation and economic development opportunities for Lake County.

As the project area includes a wide range of ages, income levels, backgrounds, and native and non-native language speakers, as well as varying access to digital information, the agency used a wide variety of tools to reach community stakeholders. NICTD developed a project website, issued press releases, mailed postcards, held in-person open houses, and attended a variety of community group meetings, as well as engaged elected officials early and often as the project progressed. Information was made available in hard copy format and digitally at local libraries.

According to Nicole Barker, NICTD Director of Capital Investment and Implementation, the most successful project tool was the in-person open house format for communication with the public. These events attracted a large number of

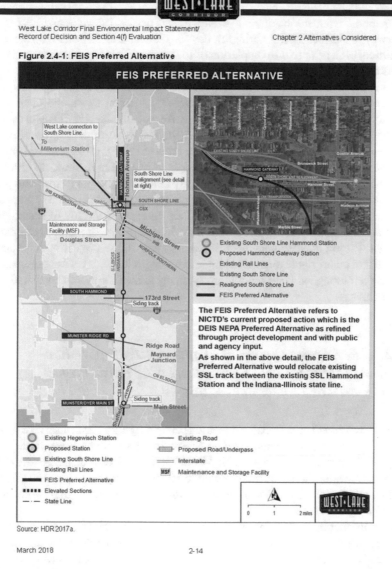

Fig. 7.1 West Lake Corridor Project preferred alternative [7]

people to learn more about project details. These meetings were not always easy, as various project opponents at times voiced their concerns, sometimes quite loudly. It was important to also approach those individuals who were less vocal with their opinions to ensure everyone's concerns were understood. The staff knows that no project is universally supported, but they believe stakeholders appreciated the NICTD team's accessibility at these sessions.

They are seeing that a large portion of the community embraces and supports the project. Elected officials, business leaders, and real estate professionals have lent their support from the beginning. Many transit-oriented development efforts are underway, which are actively bringing new businesses and real estate opportunities to the area, even prior to the completion of the project. This is generating excitement and momentum for the project itself.

Extensive public outreach occurred, resulting in creative ideas to redesign stations and parking lots to minimize real estate acquisition needs and open up key areas for future transit-oriented development. As a result, the NICTD team modified various plans in response to stakeholder input. The final product truly reflects the communities' wishes in station and parking lot design, as well as reserved areas for transit-oriented development. The biggest lesson learned is to show up and listen. Once stakeholders know they have been truly heard, they begin to trust the project leadership. NICTD repeatedly stated at community meetings and when speaking with elected officials that the designs definitely changed as a result of stakeholders' views and concerns. There was a noted shift in acceptance of the project once this occurred. Project stakeholders, even those opposed to the project as a whole, felt and still feel comfortable today reaching out to the project team to ask questions and express their thoughts. Their input literally changed the footprint of each station and parking area.

7.3.2   Los Angeles County Metropolitan Transportation Authority (Metro) Crenshaw/LAX Transit Project Corridor Project Case Study

Metro's Crenshaw/LAX Transit Project will extend light rail transit service from the existing Metro E Line (Expo) at Crenshaw and Exposition Boulevards in Los Angeles, California, and merge with Metro C Line (Green) at the Aviation/LAX Station on Aviation Boulevard and Interstate 105 in the City of El Segundo (Fig. 7.2). The line will travel 8.5 miles and serve the cities of Los Angeles, Inglewood, and El Segundo, California [8].

Once in operation, this line will offer additional transportation options to congested roadways and provide significant environmental benefits, economic development, and employment opportunities throughout LA County. Riders will have easier connections within the Metro Rail system, as well as to municipal bus lines and other regional transportation services. The project includes the Southwestern Yard Maintenance Facility, a facility with the capacity to service and store up to 70 light rail vehicles (LRV). This facility will serve the new Crenshaw/LAX Project and the existing Metro C Line (Green), housing general administration, operation, and support services. The facility will be equipped to perform inspections, body and heavy repairs, and cleaning and washing of Metro's growing LRV fleet.

Fig. 7.2 Crenshaw/LAX Transit Project Corridor [9]

The Project is routed through two historically underserved communities, which have the highest per capita mass transit ridership in the Greater Los Angeles Area. This project's goal is to revitalize the community by allowing visitors to Los Angeles and Angelenos from the Greater Los Angeles Area to experience the rich culture and history of these areas while also giving greater access to opportunity to the residents for job and educational opportunities.

As part of the environmental clearance process, a number of alternatives were considered and vetted. This included consideration of different alternative alignments as well as modes of transportation (bus vs. rail). At the onset during the environmental clearance, the selected alignment and mode of transportation were influenced by the community. There was a desire for rail instead of bus…this was changed to rail even though the analysis indicated that a bus system would be adequate. There were two stations added to iconic areas along the corridor, which did not necessarily enhance ridership or may have created less than optimal station spacing, but were important community points of interest with significant historical and cultural importance.

The first step toward opening up communication was to first build trust. A group of well-established and trusted community members were engaged and made a part of the Community Leadership Council (CLC) and were advisers to the project on messaging. They were also part of monthly workshops on topics of interest, such as transit-oriented development, which the CLC would present to the overall community during regularly scheduled meetings with the community. In addition, construction that would impact businesses and residents, such as street closures and their durations, were influenced by the community monthly through the CLC.

Metro focused on proactive tactics such as the CLC and two other steps such as:

(i) Business Interruption Fund. This was a program to ensure that businesses impacted by the construction activities would survive. Grants were provided to eligible businesses. In the end, if the businesses and the community did not survive the construction process, it would defeat the purpose of building the rail line for that community.
(ii) Business Solution Center. This was a program to ensure that the businesses were being managed as efficiently as possible and taking advantage of opportunities such as the internet to augment their client base.

There were several phases of deployment. The communities which were impacted by the project include Crenshaw and Westchester (both in the City of Los Angeles) and the City of Inglewood. Crenshaw and Inglewood both have historically been underrepresented, which created an environment of distrust that had to be overcome to be successful. Because this area includes those locales impacted by the Watts riots and the 1992 riots associated with Rodney King, this project needs to be handled with extra sensitivity. After the 1992 riot, local and state politicians pushed to improve the conditions in these areas through the addition of an effective mass transit system. These communities have the highest per capita mass transit ridership for the Los Angeles Area and yet only have access to standard buses.

The path took some effort because it took time to build that trust. For example, there was a section of the line, approximately 1 mile long, that was to be constructed at-grade. The community felt that they were being given a less effective product and wanted to have the section be placed underground in a tunnel, as was done in more affluent areas on other projects. The at-grade alternative was the preferred alternative and approved by the Metro board, but the community resisted it well into the construction phase. That resistance included a lawsuit. The slogan for the community was "it ain't over till it's under." However, once the communities saw the possibilities, they took the next steps to use the rail line to include other programs, adding a linear museum along the corridor. The museum is along the street and experienced from the sidewalks or train and captures the culture of the community (which the community felt was at risk of being lost due to gentrification). The museum also attracts outsiders to the area while increasing the interest and pride of the community members.

In addition, there was the tree issue along Crenshaw Blvd. The Space Shuttle had been brought through Crenshaw on its way to the museum in Exposition Park. Limited advisory time was given to the community, and hundreds of mature trees were lost as a result of the Shuttle move. Two years later, Metro comes along the same corridor stating that some trees will need to be cut down. This was not accepted well by the community even though Metro was inclusive in the process and brought up the issue a couple of years before the trees would be cut down. Some community members claimed that some of the trees were planted by the Reverend Martin Luther King; therefore, they should be considered monuments that could not be cut down. However, this was not practical as these trees were in the very section where the rail was to run at-grade. In the end, Metro worked with the city council and the community to compromise. All trees were repurposed and donated to a non-profit group that made them into African drums and taught the area youths to play the drums.

In the end, once the community started to understand the visibility of the community and the inherent benefits of this rail corridor being the gateway to Los Angeles from Los Angeles World Airports, they started to embrace the project and created opportunities along that same alignment to draw those riding the train to the businesses along that corridor. Trust is difficult to earn, but if one has the best interest of the community at heart, it will be seen as such by the community.

The project had different managers for Metro at different points in time. The first project manager's tenure was during the environmental and bidding phase. Budget and political challenges cut that tenure short. Of course, when that person left, there was a need to rebuild trust, as there was the perception that there were promises that were not being kept.

The second project manager (also referenced as executive officer by Metro) came on board for construction. The desire was to be proactive and transparent. Multiple programs were started as the result of challenges that were being seen in the community including two of the programs mentioned above: the Business Interruption Fund and the Business Solution Center. Multiple community meetings were held along the corridor each month to inform and stay in touch with the community.

7.4   Conclusion

Rail infrastructure enhances the quality of life in our communities. It connects us to jobs, education, healthcare, and entertainment to improve our personal and communal economies. However, these types of infrastructure have a history of being implemented in communities without recognizing the value of the affected community's input. Despite the good that rail infrastructure provides, there have been negative impacts also. In order to secure the right-of-way needed to implement the projects, some people have been displaced from the only homes they have known for generations, and communities are left bifurcated and isolated from the resources of the greater community.

Professionals that manage the planning, design, and implementation of these projects are recognizing that their technical expertise is only one element in delivering a successful project, but it is not enough on its own. We need to partner with members of the affected communities and/or self-selected advocates for those communities to identify their priorities and concerns. This partnership leads to a final project that is supported by and reflects the communities served. Examples of these partnerships with communities can be seen in the case studies shown for rail projects in Northwest Indiana and in Los Angeles, California. While different approaches were used, the results in both cases were better projects based on community input.

The need to sustain a place for the community voice has also manifested itself in an organization called Rail~Volution. Rail~Volution began to create a space to have those community voices heard and has evolved into an advocacy and technical forum for exploring lessons learned on implementing transportation projects, transit-oriented development, training, podcasts, and more. This need has reached the White House administration through Transportation Secretary Pete Buttigieg. It is under his leadership that we have seen the administration recognize both the positive and negative impacts of these projects and drive policy to assure that our historical mistakes are not repeated. These actions give us hope that we will continue to build all infrastructure with the communities as our partners and not collateral damage.

References

1. www.american-rails.com, "Early Passenger Trains: Rail Travel In The 19th Century", (07/05/2021)
2. New York State Department of Transportation, Passenger Rail Service in New York State. (2019/2020)
3. Ms Nirali Shukla & Dr. H.J.Jani, Global Institute for Research and Education, Global Journal of Commerce & Management Perspective Social Impact Assessment of Road Infrastructure Projects, Volume 7(1), January-February 2018, ISSN: 2319-7285.
4. USDOT Ladders of Opportunity, Every Place Counts Design Challenge Summary Report, 2016.
5. www.railvolution.org/about (07/10/2021).
6. Nicole Barker, Director of Capital Investment and Implementation, NICTD; May 24, 2021.

7. Final Environmental Impact Statement/Record of Decision and Section 4(f) Evaluation (FEIS/ROD), Northern Indiana Commuter Transportation District, West Lake Corridor, http://www.nictdwestlake.com/resources/, accessed October 1, 2021.
8. Charles Beauvoir, former Executive Officer Crenshaw/LAX Transit Project (through September 2019), Los Angeles County Metropolitan Transportation Authority (07/08/2021).
9. Metro, Crenshaw/LAX Transit Project, https://www.metro.net/projects/crenshaw_corridor/,a accessed October 1, 2021.

Kimberly Slaughter with a background in planning, policy development, and project funding and over 30 years in the rail and transit industry, Kimberly has helped shape public transportation across the United States. A transit industry leader and executive, she has worked with public transit agencies and private transportation consulting firms. Kimberly was born and raised in Houston, Texas. She is a proud graduate of the University of Texas at Austin for both her Bachelor of Arts in Government/Pre-Law and her Master of Science in Community and Regional Planning.

Her career path has not been a straight line. It has been a winding path that has evolved and materialized over time. She discovered community and regional planning through taking a course in her senior year of undergraduate school in social psychology. This class allowed her to explore how communities are designed and what resources are made available to us and how they affect our mental well-being and, subsequently, our quality of life. She secured an internship with a consulting firm 1 month before starting graduate school where 80 percent of that firm's work was in transportation. Once she realized how transportation and access to transit had the potential to change people's lives, she knew that she had found her passion. She wanted to influence improving the mobility of our communities for all people regardless of their geography or socioeconomic station in life. Kimberly took the initiative to round out her graduate curriculum with courses from the civil engineering and public affairs departments to better prepare her for her chosen career path.

She initially leveraged her computer programming skills to become a travel demand forecaster and to differentiate herself in the industry. She was a bit of a unicorn in this aspect of the industry because she was both a woman and black. It was rare that she encountered another female colleague that was a modeler and they were never women of color. She knew that she had to be excellent at her craft because she was always being judged. The fact that she was also young and walking into working environments in a leadership or teaching role added yet another layer of complexity. She was 22 years old when she started this career and only spent a short tenure as a part-time intern. Her employer was very impressed with her skills and very quickly engaged her on a full-time basis, sending her to work directly with clients. At first, when going to work with clients, they thought she was a receptionist or providing some administrative support. They were always astonished to learn that she was the technical staff who ran their forecasting models or taught their staff how to run the models. Her true differentiator lay in the fact that not only could she write the source code, run the model, and analyze the results, but she also clearly communicated the results of the model to various audiences. This included community members, peers, board of directors, and elected officials.

Kimberly is currently SYSTRA USA's chief executive officer. She is focused on promoting the company's vision of being the signature team for transportation solutions. SYSTRA's focuses on delivering planning, design, architecture, engineering, and program and construction management solutions to rail and transit agencies throughout the United States. The company's mission is to contribute to the cities and regions in which we work and live by creating, improving, and modernizing their transportation and infrastructure systems while furthering sustainable development. Prior to joining SYSTRA, Kimberly was the national rail and transit market sector leader at HNTB and before that was vice president/central region transit market director at HDR. Earlier, she worked for the Metropolitan Transit Authority of Harris County in Houston, where she held several positions, including senior vice president of service design and development and associate vice president of planning.

Kimberly is a passionate force behind helping to provide access to safe, reliable, affordable, and convenient public transit to all communities. Putting her passion into practice, she serves on several professional and community-based association boards and in committee leadership positions. Kimberly serves on the board of directors of the Houston Equity Fund, the Mineta Transportation Institute and the American Public Transportation Association (APTA). She previously served on APTA's Business Members Board of Governors and the Business Council of the African American Mayors Association. She holds membership in Women's Transportation Seminar (WTS) and the Conference of Minority Transportation Officials (COMTO). In 2020, she had the honor of co-chairing the APTA Post COVID Mobility Restoration & Recovery Task Force. In 2019, she served on Chicago Mayor Lori Lightfoot's transportation transition team.

Kimberly is a proud member of Alpha Kappa Alpha Sorority, Inc. She is most proud of her role as a mother to two intelligent, tenacious, resilient, and beautiful daughters who are starting their journeys as future role models.

Chapter 8
Public Transportation Ridership Patterns: Past, Present, and Possible Future Trends

Jill Hough and Susan Handy

Abstract Public transportation has played a key role in mobility since its inception in the late 1600s. Urban and rural areas have utilized various modes of transportation to provide public mobility. Ridership patterns have fluctuated through time, and this chapter addresses some of the factors, such as demographic shifts, work-style shifts, new competition, and underinvestment, that influence ridership trends. The authors also address social and environmental impacts of public transit and what it will take to increase ridership.

Keywords Mobility · Demographics · Ridership trends · Social · Economic · Environmental importance · Equity · Transportation network companies (TNCs)

8.1 What Is Public Transit and Why Is It Important?

Simply, public transportation is a system, typically made up of buses or trains, that moves people from one location to another and is available to the public for a fee. Merriam-Webster dictionary defines public transportation as a system of trains, buses, etc., that is paid for or run by the government [23].

Government support for public transportation is relatively recent, however. Reports of services that resemble public transportation date back to 1662 when Blaise Pascal invented the omnibus in Paris. The system charged a fare for users of horse-drawn buses which followed fixed routes and schedules [16]. After 15 years,

J. Hough (✉)
Upper Great Plains Transportation Institute, North Dakota State University, Fargo, ND, USA
e-mail: jill.hough@ndsu.edu

S. Handy
Department of Environmental Science and Policy, University of California, Davis, Davis, CA, USA

P. Layne, J. S. Tietjen (eds.), *Women in Infrastructure*, Women in Engineering and Science, https://doi.org/10.1007/978-3-030-92821-6_8

the fares became too high for most people, and public transportation service ceased until the nineteenth century. As cities grew in response to the Industrial Revolution, the need for faster ways of getting around cities led to a series of innovations in public transportation, starting with the omnibus, followed by horse-drawn street-cars, steam-powered cable cars, and eventually electric streetcars by the last decade of the century. In the twentieth century, cities like New York, Boston, and Chicago invested in subways and elevated trains, but in most cities, privately owned and operated streetcar systems were the dominant means of transport for the public. Following World War II, private companies replaced streetcars with buses and eventually sold their systems to cities in the face of growing competition from private cars.

Modern public transit systems date back to the 1960s, when the federal government invested in public transit for the first time. During the 1960s, the federal government realized that urban areas were having trouble with transportation [30]. Out of the need for greater equity, among other problems, the Urban Mass Transportation Act of 1964 created the federally sponsored Urban Mass Transportation Administration (UMPTA) to oversee public transportation. Although urban transit existed prior to the Urban Mass Transportation Act, rural transit systems did not, and it was this act that brought transit service to rural communities and cities with less than 50,000 people. In 1991, UMPTA was renamed the Federal Transit Administration (FTA). In the decades since, federal funding, as administered by FTA, has been critical to the viability of public transit systems and has been especially important for capital investments such as the construction of light rail transit systems and the purchase of new buses.

Public transport has undergone many changes since its inception, but it has always stayed true to its mission of moving people from one location to another. Today, 6800 nonprofit organizations provide rides to individuals for various purposes. Of those nonprofits, 2207 transit agencies receive funding from the FTA for the specific purpose of providing rides [1]. Of those FTA-funded agencies, 1279 are categorized as rural, serving cities with fewer than 50,000 people, and 928 are urban, providing service to cities with more than 50,000 people. Public transportation plays an important role in the cities and communities where it exists, generating social, economic, and environmental benefits as described in the next section.

8.2 Social, Economic, and Environmental Importance of Public Transportation

8.2.1 Social and Economic Importance

Public transportation is not just a mobility option; for some people, it is the only option. It provides travel opportunities for historically disadvantaged groups including those with low income, people with disabilities, and older adults, as well as people without access to a vehicle. According to data from the 2019 American Community Survey (1-year estimates), 8.7% of households in the United States do not have access to a vehicle [33].

Without access to a vehicle, it is difficult to get to work and healthcare appointments, purchase necessities such as food, attend worship services, and participate in social functions. Without public transportation, individuals without access to a vehicle may live socially isolated lives. Studies show that the benefits of public transit are numerous [10]. Benefits of public transportation can be classified into four categories: mobility, efficiency, land use, and economic development (Table 8.1). Mobility benefits occur from travel that would not otherwise take place, particularly for people who do not have a driver's license, have a disability, or are unable to drive for other reasons. Efficiency benefits occur when people use transit rather than driving, particularly at peak hours when roads are most congested; this benefit is more pronounced in larger cities than rural areas. Another benefit of transit is that it can help to shape land use patterns by increasing the value of the land served by the system. This impact is most pronounced with rail rather than bus systems and is thus more applicable to large urban areas rather than rural areas. The economic development benefits of transit stem from increases in employment which in turn lead to increases in economic activity. Transit agencies employ many people, including drivers and administrative staff. Often transit workers live within the community where the service is provided meaning that they spend their income within that area, creating more economic activity. The many specific benefits of transit within each of these broad categories are outlined in Table 8.1.

The economic benefit to a community of an investment in transit is often assessed using cost-benefit analysis. Researchers have estimated cost-benefit ratios for transit ranging from 0.47 to 9.70, meaning that a community receives as much as $9.70 in benefits for every dollar spent on transit [2, 10, 11, 29]. Although the estimated benefits vary depending on the context, the cost-benefit analysis is often positive for transit. Some might argue that funds spent on public transportation could be diverted to another public good that provides a larger return. However, the improved social and economic equity that transit provides for disadvantaged populations is essential.

8.2.2   Environmental Importance

Public transportation benefits go beyond social and economic; they also relate to the environment. Use of public transportation instead of personal vehicles can reduce greenhouse gas emissions (carbon dioxide (CO_2), methane, nitrous oxide, and fluorinated gases with carbon dioxide being the largest contributor) and reduce energy usage. Transportation is among the largest contributors to greenhouse gas production, emitting 29% of the greenhouse gases as shown in Fig. 8.1.

The Environmental Protection Agency [8] data in Fig. 8.2 illustrates that transportation emissions have increased from 1990 to 2019 primarily due to greater

Table 8.1 Categories of transit benefits and descriptions

Benefit category	Description
Mobility benefits	*Benefits from increased travel that would not otherwise occur*
Direct user benefits	Direct benefits to users from increased mobility
Public services	Support for public services and cost savings for government agencies
Productivity	Increased productivity from improved access to education and jobs
Equity	Improved mobility that makes people who are also economically, socially, or physically disadvantaged relatively better off
Option value/ emergency response	Value of having mobility options available in case they are ever needed, including the ability to evacuate and deliver resources during emergencies
Efficiency benefits	*Benefits from reduced motor vehicle traffic*
Vehicle costs	Changes in vehicle ownership, operating, and residential parking costs
Chauffeuring	Reduced chauffeuring responsibilities by drivers for non-drivers
Vehicle delays	Reduced motor vehicle traffic congestion
Pedestrian delays	Reduced traffic delay to pedestrians
Parking costs	Reduced parking problems and non-residential parking facility costs
Safety, security, and health	Changes in crash costs, personal security, and improved health and fitness due to increased walking and cycling
Roadway costs	Changes in roadway construction, maintenance, and traffic service costs
Energy and emissions	Changes in energy consumption, air, noise, and water pollution
Travel time impacts	Changes in transit users' travel time costs
Land use	*Benefits from changes in land use patterns*
Transportation land	Changes in the amount of land needed for roads and parking facilities
Land use objectives	Supports land use objectives such as infill, efficient public services, clustering, accessibility, land use mix, and preservation of ecological and social resources
Economic development	*Benefits from increased economic productivity and employment*
Direct	Jobs and business activity created by transit expenditures
Shifted expenditures	Increased regional economic activity due to shifts in consumer expenditures to goods with greater regional employment multipliers
Agglomeration economics	Productivity gains due to more clustered, accessible land use patterns
Transportation efficiencies	More efficient transport system due to economies of scale in transit service, more accessible land use patterns, and reduced automobile dependency
Land value impacts	Higher property values in areas served by public transit

Source: Litman 2018, as in Mattson et al. [21]

demand for travel. In total, passenger cars and light- and heavy-duty trucks have long been major contributors to greenhouse gases, while buses emit lower amounts of greenhouse gases in total.

But transit ridership is an important factor to consider when looking to reduce greenhouse gas emissions. Although a bus produces more emissions than a car, the more passengers that ride that bus, the lower the emissions per passenger mile. When enough people choose public transportation, the emissions per passenger

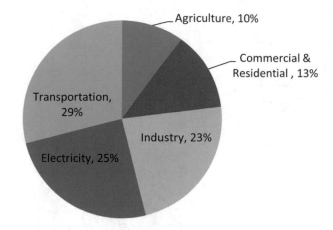

Fig. 8.1 Sources of US greenhouse gas emissions in 2019. (Source: U.S. Environmental Protection Agency [8]. Inventory of U.S. Greenhouse Gas Emissions and Sinks: 1990–2019)

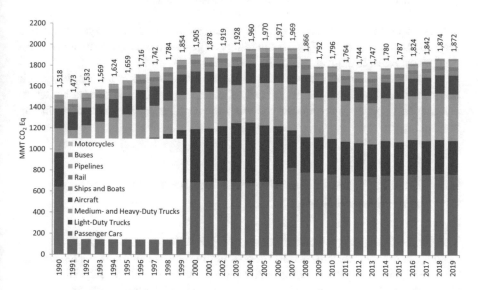

Fig. 8.2 Trends in transportation-related greenhouse gas emissions (In 2011, FHWA changed its methods for estimating VMT and related data. These methodological changes included how vehicles are classified, moving from a system based on body type to one that is based on wheelbase. These changes were first incorporated for the 1990 through 2008 Inventory and apply to the 2007 to 2019 time period. This resulted in large changes in VMT data by vehicle class, leading to a shift in emissions among on-road vehicle classes. This change in vehicle classification has moved some smaller trucks and sport utility vehicles from the light truck category to the passenger vehicle category in this inventory). (Source: EPA [8]. Inventory of U.S. Greenhouse Gas Emissions and Sinks: 1990–2019, p. 2–38)

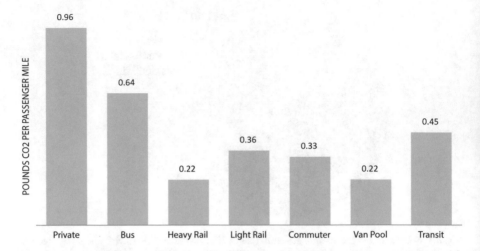

Fig. 8.3 Estimated CO_2 emissions per passenger mile for transit and private autos. (Source: U.S. Department of Transportation, Federal Transit Administration [34])

mile is lower than for driving alone. Heavy rail, subway and metro systems, which serve urban areas, produce 76% less greenhouse gas per passenger mile than single-occupancy vehicles at given ridership levels (Fig. 8.3).

Emissions from buses are expected to decline even more as cities transition their fleets to alternative energy sources such as electricity. Although several strategies are needed to reduce greenhouse gas emissions, riding public transportation can play an integral role.

8.3 Transit Needs in Rural Versus Urban Areas in the United States

The United States has a total population of 328 million people with 264 million residing in urban areas and 64 million residing in rural areas (Table 8.2). Although rural residents are in the minority, their transit needs are still substantial, and the challenges of providing adequate transit service in rural areas are greater.

Rural communities are home to a disproportionate share of vulnerable populations who depend on transit. For the United States, 16.5% of the population is 65 or older; in rural areas, the share is 19.8% compared to urban areas with 15.7% (Table 8.2). Rural areas also have a greater percentage of people with disabilities, 15% compared to 12.2% in urban areas. Other demographic differences suggest higher demand for transit in urban areas. In urban areas, 12.7% of individuals are below the poverty line versus 10.8% in rural areas. Though rural areas have a lower median household income in dollars at $64,314 compared to $66,047 in urban areas, the cost of living is also lower in rural areas. Urban areas are home to more

Table 8.2 Characteristics of US urban and rural populations

	United States	Urban	Rural
Total population (million people)	328	264	64
Average household size	2.6	2.6	2.6
Gender (%)			
Male	49.2	48.9	50.5
Female	50.8	51.1	49.5
Age			
Median age	38.5	37.4	43.6
65 or older (%)	16.5	15.7	19.8
85 or older (%)	1.9	2.0	1.8
Population with a disability (%)	12.7	12.2	15.0
Race (%) [a]			
White	75.0	71.6	89.3
Black or African American	14.2	16.0	6.8
American Indian and Alaska native	1.7	1.5	2.6
Asian	6.8	8.0	1.8
Hispanic or Latino	18.4	21.1	7.2
Foreign born (%)	13.7	16.1	3.9
Highest education level completed (%) [b]			
Did not complete high school	11.4	11.6	10.9
High school	26.9	25.2	33.7
Some college, no degree	20.0	19.7	20.9
Associate degree	8.6	8.3	9.8
Bachelor's degree	20.3	21.5	15.7
Graduate or professional degree	12.8	13.7	9.0
Economic characteristics			
Individuals below the poverty line (%)	12.3	12.7	10.8
Median household income (dollars)	65,712	66,047	64,314

Source: American Community Survey, 2019; 1-year estimates as in Mattson and Mistry [22]
[a] Alone or in combination with another race
[b] Population 25 years or older

foreign-born residents, 16.1% compared to 3.9% in rural areas, who are more likely to use public transit than US-born residents [5]. They are also home to more non-white residents, who are also more likely to use public transit [5]. Though their demographic characteristics differ, rural and urban communities both need public transit.

Urban areas throughout the United States have a variety of types of public transportation available to users such as buses, subways, light rail, and commuter rail. In some places, monorail, passenger ferry boats, trolleys, inclined railways, and people movers are also an option. Because rural areas have low population densities with people spread across large geographic spaces, public transportation in rural areas is primarily made up of buses and vans which offer flexibility and require minimal capital investment compared to rail systems. In rural areas, the buses may run on a fixed route or deviated fixed-route schedule, but often are dispatched through a demand-response process. In a demand-response process, a passenger reserves a ride with the transit agency (usually 24 hours in advance) to go to a specific destination. The driver picks up the passenger at a designated location to provide the ride to the specified destination. Rural transit agencies are typically county based, whereas urban transit agencies are usually city based but may also be county based.

The Rural-Urban Continuum Codes (RUCCs) define a rural-to-urban spectrum of places. As presented in the *2021 Rural Transit Fact Book*, the RUCCs classify counties on a scale of 1–9, in which the higher the number, the more rural the county. For example, counties in a metro area of one million people or more receive a RUCC of 1, while counties with populations of less than 2500 urban residents which are not adjacent to a metro area are completely rural and have a RUCC of 9 (Table 8.3). The continuum within the United States, illustrated in Fig. 8.4, shows the Midwest has the most rural counties, while the East and West Coasts and pockets throughout the country have urban cities.

8.4 Transit Ridership Trends and Factors Influencing Those Trends

Transit ridership has fluctuated over time, and various factors have influenced the trends. Ridership grew steadily through the 1920s, but declined with the introduction of the automobile. Although people liked their personal automobiles, monetary difficulties during World War II prompted increased ridership which reached an

Table 8.3 Rural-urban continuum codes

Code	Description
1	Counties in metro areas of one million population or more
2	Counties in metro areas of 250,000 to one million population
3	Counties in metro areas of fewer than 250,000 population
4	Urban population of 20,000 or more, adjacent to a metro area
5	Urban population of 20,000 or more, not adjacent to a metro area
6	Urban population of 2500 to 19,999, adjacent to a metro area
7	Urban population of 2500 to 19,999, not adjacent to a metro area
8	Completely rural or less than 2500 urban population, adjacent to a metro area
9	Completely rural or less than 2500 urban population, not adjacent to a metro area

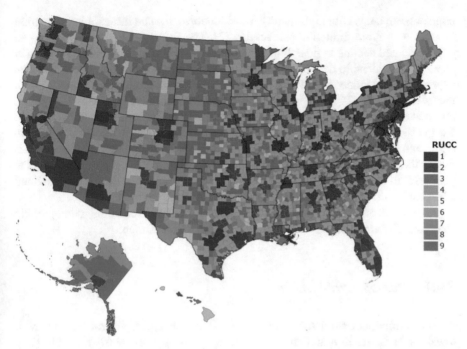

Fig. 8.4 County-level 2013 Rural-Urban Continuum Codes. (Source: 2021 Rural Transit Fact Book. Fargo, North Dakota)

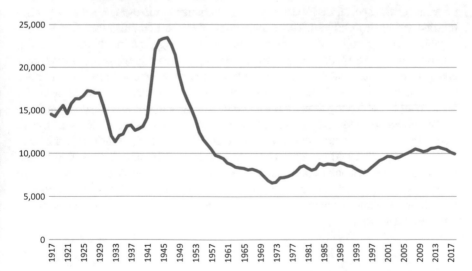

Fig. 8.5 Total transit ridership by all modes, in millions, 1917 to 2018. (Source: Adapted from [1])

all-time high of 24 billion one-way trips (Fig. 8.5). Following World War II, transit ridership declined but peaked again in 2011 with 10.4 billion passenger trips. Since 2012, public transit ridership has continued to decline. One factor is a decline in population in areas with high-quality transit coupled with an increase in population in areas with more limited transit service [7]. The COVID-19 pandemic led to an unprecedented decline in ridership in 2020 that has rebounded far more slowly than driving. The ridership decline varied by agency, but both urban and rural agencies experienced a decline. The Washington, D.C., ridership declined by about 90% on the Metrorail and 75% on bus [36], whereas ridership in San Antonio, Texas, reported a decline of 30% by the end of March 2020 [35]. Rural transit agencies in North Dakota experienced an estimated average decline of 68% ridership in March, April, and May 2020 during the pandemic [24]. It is estimated that ridership trends during the first full year of COVID-19 resulted in about 65% less ridership than the year prior to the beginning of COVID-19 [27]. Yet, public transit service remained a lifeline for the essential workers who could not work remotely.

Transit ridership may be down for several reasons, including demographic changes, lifestyle shifts, new competition, and underinvestment.

8.4.1   Demographic Shifts

Age is an important predictor of transit use [5], but changing generational behavior toward suburbanization and automobile usage also impacts transit ridership [28]. Prime years of transit use are between ages 16 and 30 as individuals in this age range may not be as reliant on owning an automobile. However, around age 30, reliance on a car for transportation increases owing to changes in life-stage, such as getting married and having children. The older portion of the Millennial Generation is moving into the age group with the lowest public transit usage, a trend that may account for some of the overall decline in transit ridership [5]. Riders who had substantially changed their transit use – increasing it or decreasing it – were more likely to have moved within the past 2 years, to have increased their income, and to be under 40 years old [14]. Young riders represent a declining share of all transit riders [7].

The effects of demographic factors can be seen in changing patterns of transit ridership across neighborhoods. Between 2000 and 2015, Southern California neighborhoods that had been relatively transit dependent saw decreases in transit ridership as their populations became "less poor," and "significantly less foreign born," and the number of households with a vehicle increased [19]. The number of households without a vehicle is one of the strongest predictors of transit use in Southern California.

8.4.2   Work-Style Shifts

Another important trend has been an increase in the number of workers telecommuting and working from home that has reduced the number of work trips taken regardless of the mode. Data from the National Telecommunications and Information

Administration (NTIA) showed that nearly one third of American employees worked remotely on some occasion in 2019. Of the employed urban workers, 32% teleworked, while 22% of rural workers teleworked in 2019 [12]. In early 2020 when the COVID-19 pandemic began to impact the world, offices abruptly closed, and the number of workers telecommuting or working remotely increased, leading to a precipitous decline in transportation ridership. However, not all employed adults were able to work from home. The Pew Research Center released data showing that lower-income workers were less likely to have the teleworking option and needed to report to work [26]. The service that transit agencies still offered through the worst of the pandemic was a vital option for these essential workers.

8.4.3 New Competition (TNCs)

Public transportation has faced increasing competition in recent years from various forms of privately operated shared-use mobility options including bike share, scooter share, and ride-hailing companies (also known as transportation network companies (TNCs) such as Uber and Lyft). These alternative modes are both complements to and substitutes for traditional public transportation. Researchers are examining the impact of the various non-transit shared-use mobility options on public transportation. Data analyzed from San Francisco between 2010 and 2015 found that TNCs were responsible for a net ridership decline of nearly 10% [9]. TNCs are taking away transit trips primarily in dense cities such as Boston [14]. The study also found that transit riders tend to take TNCs when transit is not reliable (a substitution effect) or when it stops before the rider's destination (a complementary effect).

8.4.4 Underinvestment

Despite billions of federal dollars going to public transportation agencies each year, transit agencies are struggling to modernize and expand their systems. The FTA has identified a $90 billion backlog for public transit assets and infrastructure. Federal funding for transit, as substantial as it is, is dwarfed by federal funding for highways [4]. Transit investments thus depend on local sources of funding far more than highway investments do [1]. Local funding sources include transit fares but also dedicated local sales taxes, among other sources. With the declines in ridership, transit agencies are seeing a decline in fare-box revenues that only exacerbates the financial challenges for transit agencies. Declines in ridership fuel a vicious cycle by leading to declines in revenues that lead to declines in service that lead to further declines in ridership.

8.5   How Do We Increase Transit Ridership?

Even before the pandemic, transit agencies were struggling to win back riders. To compete with driving, transit systems need to minimize travel time from a rider's trip origin to their destination, but the systems also need to provide frequent, direct, reliable, comfortable, and safe travel at a reasonable cost [31]. The frequency of service is important because it helps to reduce uncertainty about how long a rider will have to wait for the next bus or train. Direct trips are important because every time a rider needs to make a transfer from one bus or train to another as a part of their trip, their uncertainty goes up [15]. Riders also need to feel confident that the service will reliably get them to their destination at the scheduled time [25] Comfort, another important consideration for transit riders, depends on the design of the transit vehicle as well as the smoothness of the ride. Safety is also paramount, including safety getting to and from the transit vehicle as well as safety while on the transit vehicle [3]. Although transit has a far better safety record than driving, passengers, especially women, may feel vulnerable to other passengers [18]. A growing number of US transit agencies have implemented programs to reduce sexual harassment on their systems, while some systems elsewhere in the world have designated train cars for women riders. The price of the ride also matters, especially to low-income riders.

   Transit riders differ from one another, and people are unique in how they decide what trips to make and how to make them [14]. As noted earlier, many people do not have the choice to drive, making them dependent on transit. Such riders have sometimes been called "captive riders." Many transit agencies offer discounts to those most dependent on transit, such as older people, students, and people with disabilities [6]. The Americans with Disabilities Act (ADA) requires transit agencies to provide service that is accessible to this last group, whether on regularly scheduled service or through demand-response service known as "paratransit." ADA requirements have led to changes in vehicle design as well as greater attention to the design of bus stops and rail stations to ensure universal access [32]. By making their service work better for captive riders, transit agencies can encourage more use of the system and thereby increase ridership.

   To substantially increase ridership, however, transit agencies must attract more "choice riders," that is, people who have the option to drive [19]. Attracting these riders requires attention to all of the qualities noted above: speed, frequency, directness, reliability, comfort, safety, and cost [37]. An individual may opt for transit if transit outperforms driving on the sum total of all these qualities. Rail service has a better chance of competing with driving than does bus service, and transit agencies may be tempted to shift funding from bus service, which tends to serve captive riders, to rail service, which has a better chance of attracting choice riders. In 1994, the Bus Riders Union sued the Los Angeles MTA over this issue [13]. Bus Rapid Transit has become an increasingly popular investment for transit agencies in the United States (and elsewhere in the world), as it combines many of the benefits of rail service with the flexibility of bus service at costs closer to the latter than the former [17].

Following the height of the pandemic, transit agencies are working to recover from the decline in ridership by retaining and regaining riders. To win back riders, transit systems in large cities including Washington, D.C.; Boston; Cleveland; Las Vegas; and San Francisco are offering reduced fares or even free rides [20]. Other cities, such as Los Angeles, are exploring pilot projects that would give certain groups such as students and low-income riders free rides. In March 2020, during the pandemic, the Kansas City Area Transportation Authority removed fares and does not plan to bring them back at this time. To further encourage transit use, cities can look for ways to discourage driving by increasing parking prices, reducing the number of parking spots, and increasing the overall cost of driving. San Francisco and New York are considering congestion pricing programs that would both discourage driving and raise revenues that would be invested in their transit systems.

8.6   What Will It Take?

What will it take to lure people back to public transportation? In short, it will take a myriad of strategies involving collaborations between transit agencies, local governments, and others. It will require transit systems to understand the needs of their riders and offer fast, frequent, direct, reliable, comfortable, safe, and affordable service. It will also require the political will of residents, elected officials, local businesses, and other interest groups, along with funding from local, state, and federal sources. Public transportation generates important social, economic, and environmental benefits and is an essential, if sometimes underappreciated, component of the transportation system.

References

1. American Public Transportation Association. (2020). *2020 Public Transportation Fact Book.* Washington, D.C.
2. Burkhardt, J. (1999). "Economic Impact of Rural Transit Services." Transportation Research Board, Transportation Research Record 1666: 55-64.
3. Ceccato, V., & Newton, A. (Eds.). (2015). *Safety and security in transit environments: An interdisciplinary approach.* Springer.
4. Congressional Research Service. (2021). "Highway and Public Transit Funding Issues." *In Focus.* March 1. https://crsreports.congress.gov/product/pdf/IF/IF10495
5. Coogan, M., G. Spritz, T. Adler, N. McGuckin, R. Kuzmyak, and K. Karash. (2018). *TCRP Report 201: Understanding Changes in Demographics, Preferences, and Markets for Public Transportation.* Transportation Research Board, Washington, D.C. https://doi.org/10.17226/25160
6. Darling, W., Carpenter, E., Johnson-Praino, T., Brakewood, C., & Voulgaris, C. T. (2021). Comparison of Reduced-Fare Programs for Low-Income Transit Riders. *Transportation Research Record*, 03611981211017900.

7. Driscoll, Richard, Kurt Lehmann, Steven Polzin, and Jodi Godfrey. (2018). "The Effect of Demographic Changes on Transit Ridership Trends." *Transportation Research Record*. Vol. 2672(8): 870-878. https://doi.org/10.1177/0361198118777605

8. Environmental Protection Agency. (2021). *Inventory of U.S. Greenhouse Gas Emissions and Sinks 1990-2019*. https://www.epa.gov/ghgemissions/inventory-us-greenhouse-gas-emissions-and-sinks

9. Erhardt, G., R. Mucci, D. Cooper, B. Sana, M. Chen, J. Castiglione. (2021). "Do Transportation Networks Companies Increase or Decrease Transit Ridership? Empirical Evidence from San Francisco. *Transportation*. https://doi.org/10.1007/s11116-021-10178-4

10. Ferrell, Christopher. (2015). *The Benefits of Transit in the United States: A Review and Analysis of Benefit-Cost Studies*. Mineta Transportation Institute, San Jose State University. https://transweb.sjsu.edu/sites/default/files/1425-US-transit-benefit-cost-analysis-study.pdf

11. Godavarthy, R., J. Mattson, and E. Ndembe. (2014). Cost-Benefit Analysis of Rural and Small Urban Transit. North Dakota State University Upper Great Plains Transportation Institute Small Urban and Rural Transit Center, http://www.nctr.usf.edu/wp-content/uploads/2014/07/77060-NCTR-NDSU03.pdf.

12. Goldberg, Rafi. (2020). "Nearly a Third of American Employees Worked Remotely in 2019, NTIA Data Show." Blog post. (https://www.ntia.gov/blog/2020/nearly-third-american-employees-worked-remotely-2019-ntia-data-show).

13. Grengs, J. (2002). Community-based planning as a source of political change: The transit equity movement of Los Angeles' Bus Riders Union. *Journal of the American Planning Association*, 68(2), 165-178.

14. Higashide, Steven and Mary Buchanan. (2019). *Who's not on Board 2019: How to Win Back America's Transit Riders*. Transit Center, New York. www.transitcenter.org.

15. Horowitz, A. J., and Zlosel, D. J. (1981). Transfer penalties: Another look at transit riders' reluctance to transfer. *Transportation*, 10(3), 279-282

16. LANTA, The History of the Bus, https://www.lantabus.com/2012/05/16/the-history-of-the-bus-2, May 16, 2012.

17. Levinson, H. S., Zimmerman, S., Clinger, J., & Rutherford, G. S. (2002). Bus rapid transit: An overview. *Journal of Public Transportation*, 5(2), 1.

18. Loukaitou-Sideris, A. (2014). Fear and safety in transit environments from the women's perspective. *Security journal*, 27(2), 242-256.

19. Manville, Michael, Brian Taylor, and Evelyn Blumenberg. (2018). *Falling Transit Ridership: California and Southern California*. Institute of Transportation Studies, University of California Las Angeles.

20. Marshall, Arrian. "Transit Agencies Are Trying Everything to Lure You Back." *WIRED*. 06.21.2021. https://www.wired.com/story/transit-agencies-trying-everything-lure-you-back

21. Mattson, Jeremy, Del Peterson, Jill Hough, Ranjit Godavarthy, and David Kack. (2020). *Measuring the Economic Benefits of Rural and Small Urban Transit Services in Greater Minnesota*. Minnesota Department of Transportation Report No. MN 2020-10. St. Paul, Minnesota. http://www.dot.state.mn.us/research/reports/2020/202010.pdf

22. Mattson, Jeremy and Dilip Mistry. (2021). *Rural Transit Fact Book 2021*. SURTCOM 21-07, Small Urban and Rural Center on Mobility, North Dakota State University, Fargo.

23. Merriam-Webster. (n.d.). Public transportation. in *Merriam-Webster.com dictionary*. Accessed August 28, 2021. from https://www.merriam-webster.com/dictionary/public%20transportation

24. Molina, Antonio, Narendra Malalgoda, Ali Rahim Taleqani, Jeremy Mattson, Kenechukwu Ezekwem, Taraneh Askarzadeh, and Jill Hough. (2021). Surveys of Transit Riders and Agencies During the COVID-19 Pandemic. UGPTI Staff Paper 188.

25. National Association of City Transportation Officials (NACTO). (2016). *Transit Street Design Guide*. (Washington, DC: Island Press). ISBN: 9781610917476

26. Parker, Kim, Juliana Horowitz, and Rachel Minkin. (2020) *How the Coronavirus Outbreak has – and Hasn't – Changed the Way American Work."* Pew Research Center.

27. Polzin, Steven. (2021). *Public Transportation Must Change After COVID-19*. Reason Foundation Policy Brief. https://reason.org/wp-content/uploads/public-transportation-must-change-after-covid-19.pdf
28. Shaheen, S., and A. Cohen. (2018). "Is It Time for a Public Transit Renaissance?: Navigating Travel Behavior, Technology, and Business Model Shifts in a Brave New World." *Journal of Public Transportation*. 21(1): 67–81.
29. Skolnik, J., and R. Schreiner. (1998). "Benefits of Transit in Small Urban Areas: A Case Study." Transportation Research Record 1623: 47–56.
30. Smerk, George. (1991). *The Federal Role in Urban Mass Transportation*. Indiana University Press, Bloomington and Indianapolis.
31. Taylor, B. D., & Fink, C. N. Y. (2013). Explaining transit ridership: What has the evidence shown?. *Transportation Letters*, 5(1), 15-26.
32. Transit Cooperative Research Program (2003). "Impact of the Americans with Disabilities Act on Transit Operations," Legal Research Digest No. 19. http://onlinepubs.trb.org/onlinepubs/tcrp/tcrp_lrd_19.pdf
33. United States Census Bureau. (2020). Selected Housing Characteristics. Accessed July 28, 2021. https://data.census.gov/cedsci/table?d=ACS%201-Year%20Estimates%20Data%20Profiles&tid=ACSDP1Y2019.DP04&hidePreview=false
34. United States Department of Transportation. (2010). *Public Transportation's Role in Responding to Climate Change*. Federal Transit Administration, Washington, D.C.
35. VIA Metropolitan Transit. (2020). *VIA Continues to Run Essential Service in a Safe Environment*. 3/24/2020. https://www.viainfo.net/covid-19/
36. WMATA. (2020). Metro and COVID-19: Steps We've Taken. March 30-April 3, 2020. https://www.wmata.com/service/covid19/COVID-19.cfm
37. Zhao, J., Webb, V., & Shah, P. (2014). Customer loyalty differences between captive and choice transit riders. *Transportation Research Record*, *2415*(1), 80-88.

Dr. Jill Hough directs the Small Urban and Rural Center on Mobility (SURCOM), an arm of the Upper Great Plains Transportation Institute at North Dakota State University (NDSU). Hough also serves on the graduate faculty at NDSU where she has advised masters and Ph.D. students, served on thesis and dissertation committees, and taught courses on Public Transportation and on Leadership, Ethics, and Academic Conduct. She also serves as the deputy director of the University Transportation Center, Small Urban, Rural and Tribal Center on Mobility led by Montana State University.

After earning her B.S. and working on her M.S. in Agricultural Economics at NDSU and writing her thesis on "Transportation and Economic Development," she was hired as a research assistant at the Upper Great Plains Transportation Institute at NDSU. Her career aspiration changed with this decision when she began working in the field of transportation rather than in commodity marketing as she had planned. She joked with other graduate students that because she grew up on a farm in the country and was always trying to figure out how to get to town and the only bus that passed by was the school bus that she was destined to work in transportation! Hough also attributes her passion for transportation to her father, who owned and operated a transport company while she was young. Early in her research career, Hough worked on projects relating to logistics of the grain industry, transportation, and economic development and on low-volume roads projects. One day, she read about public transportation and the impact it has on people's lives, and she was instantly hooked. She felt that she wanted to focus her career on moving people to where they needed to go as well as providing them access to necessities and social functions.

While researching public transportation, Hough was inspired to apply intelligent transportation system technologies to help aid in the movement of people. This new interest in transit and technology led to a short-term on-site work experience at the Federal Transit Administration Headquarters in Washington, D.C. During this time, she took on the role of the Transit Intelligent Vehicle Initiative Platform Leader until returning to NDSU. To further her understanding and education, Hough sought a Ph.D. from the University of California, Davis, in Transportation Technology and Policy.

While at the University of California, Davis, pursuing her Ph.D., Hough collaborated with her colleague at NDSU to conceptualize and write a proposal and pursue the development of a rural and small urban transit research center. In 2002, Hough became the first director of the Small Urban and Rural Transit Center which changed to the Small Urban and Rural Center on Mobility in 2020 to reflect the expanding center focus to include shared-use mobility. While directing the Center, Hough has served as a participant and as a leader on numerous national and international committees.

Most recently, Hough has become interested in applying her skills and energy to address how transportation can play a role in combating food insecurity. She looks forward to continuing to focus on transportation and its role in making people's lives better.

Dr. Susan Handy is a Professor in the Department of Environmental Science and Policy and the Director of the National Center for Sustainable Transportation at the University of California, Davis, where she also chairs the graduate program in Transportation Technology and Policy. After earning a B.S.E. in Civil Engineering from Princeton University in 1984 and an M.S. in Civil Engineering from Stanford University in 1987, she decided to broaden her perspective with a Ph.D. in City and Regional Planning from the University of California at Berkeley in 1992. Her first academic position was in the Community and Regional Planning program at the University of Texas at Austin, where she taught from 1993 to 2002 before moving to UC Davis.

Handy attributes her interest in transportation to countless hours staring out of the window of the family station wagon during trips around the Western United States. Although she missed the early years of the construction of the Interstate Highway System, she experienced its completion first hand and observed with interest and dismay the impacts of highways on the sprawling development of California. Because she was good at math in high school, teachers encouraged her to consider majoring in engineering, and civil engineering seemed like a good fit, but she came to realize that an engineering perspective is too narrow for understanding – and improving – the transportation system.

Her research focuses on the intersection between transportation and land use, particularly the impact of neighborhood design on travel behavior. Recent projects examine bicycling as a mode of transportation and strategies for reducing automobile dependence. In her 28 years as a professor, she has taught hundreds of undergraduate and graduate students about urban planning, transportation planning, and the connections between the two. Her students, of whom she is very proud, have gone on to important positions in local, regional, and state agencies, as well as consulting firms and universities.

Part II
Making Connections

Chapter 9
Creating Bridges as Art

Linda Figg

Abstract This chapter on creating bridges as art will explore how infrastructure is both structure and symbol, creating functional bridge sculptures that embrace community connections. Multiple bridge case studies will be described, highlighting various bridge styles and the fundamental principles that arrive at sustainable solutions. These principles include respect for the natural and built environment, sensitivities to communities and cultures, innovation and technology, economy and value-based investments, and community involvement in the bridge design. Topic areas on bridge creation and choices will lead the discussion, showing what is needed to put a bridge project together that achieves significant benefits to quality of life. Every bridge has a story. This chapter will explain how that story comes together in remarkable ways.

Keywords Bridges · Bridge design · Functional bridge structures · Bridge creation · Bridge project

9.1 The Bridge Story Begins

Every bridge has a story (Fig. 9.1). It begins with a vision, a vision that reflects the beauty of the community's sense of place. A bridge tells the story of the technology of its time and respects the natural and built environment in a holistic and context-sensitive design. Bridges are born weaving together an orchestra of voices that inspire and challenge a design team to create a solution that honors both function and aesthetics equally.

L. Figg (✉)
FIGG Group, Tallahassee, FL, USA
e-mail: lfigg@figgbridge.com

© The Author(s), under exclusive license to Springer Nature
Switzerland AG 2022
P. Layne, J. S. Tietjen (eds.), *Women in Infrastructure*, Women in Engineering
and Science, https://doi.org/10.1007/978-3-030-92821-6_9

Fig. 9.1 Winners of three Presidential Design Awards through the National Endowment for the Arts. Only five of these awards for bridges were ever given by the presidents of the United States (left to right): I-275 Bob Graham Sunshine Skyway Bridge (Florida), Natchez Trace Parkway Arches (Tennessee), and Blue Ridge Parkway Viaduct (North Carolina). (Courtesy of FIGG)

Functionality, typically the primary focus when a new crossing is needed, is fundamental to a bridge's identity. Although a bridge facilitates mobility, its more profound purpose is enabling and deepening a community's sense of identity and human connection. It does this by telling the story of a very particular place and time. The storyteller is the bridge itself. It speaks through its aesthetics, which emerge from three distinct yet interdependent conditions: context, design, and connection. A bridge's beauty, functionality, and identity arise from the ongoing, dynamic relationship between these conditions and are expressed by the structure and those who use it.

Designing a bridge that inspires at the microscale, macroscale, and human scale relies on understanding the laws of nature as well as human nature. Since a bridge is always experienced in its totality, its design is inextricable from its location, which encompasses the site's climate, geology, existing infrastructure, and traffic patterns. Its design also cannot be separated from the region's history, culture, landmarks, and socioeconomic conditions. Most evidently, a bridge engages those who use it; considered in this way, the public is an active participant in a bridge's aesthetics as well. In a successful bridge, the ongoing interplay of design, location, and end user resonates daily.

Given the sizable investment of funds in a landmark bridge, it must stand the test of time physically as well as visually. The impression the structure leaves on those who live in the community and on those who visit remains for as long as the bridge does. Hence, bridge designers owe it to the public to craft reliable, beautiful, and memorable bridges.

9.1.1 Context

At one time, America was composed of distinctive regions with unique characteristics and cultures. Today, there is a perception that we have become a largely homogenous society and, given technology's interconnective power, that "place" doesn't matter anymore. Nothing could be further from the truth! Communities across the nation are beginning to say, "Wait a minute. Let's dig deep and find out what's special about us." As the urbanist Jane Jacobs pointed out, successful communities—places where people like to live and work—are multidimensional and diverse. Those who live in them have a sense of their unique identity and want to enhance and protect it.

Communities want bridges that are visually pleasing, reflective of their place, and sensitive to the environment. Additionally, the public wants construction to be completed quickly with minimal disturbance to traffic, expectations that parallel the owner's desire for swift and economical construction. By demanding more of a voice in what is built in their communities, the public is driving higher aesthetic and durability standards for bridges.

Each bridge site is unique. Although similar design approaches may be utilized, creating a signature bridge requires exploring and embracing what makes a place like no other. In identifying solutions that will speak to a community's needs, we take into consideration the area's particular character, seasonal nuances, salient landmarks, and geotechnical demands—all of which make the genius loci, or spirit of the place, visible.

Structural engineer Buckminster Fuller once said, "When I am working on a problem I never think about beauty. I only think about how to solve the problem. But when I have finished, if the solution is not beautiful, I know it is wrong." We start each project by deciding how the bridge is going to be built, which drives much of its economy and lays its aesthetic foundations. This same premise is evident in the Gothic cathedral, where the technology of the flying buttress made possible the great height that gives the church its spiritual meaning. The very process of finding a solution—how to build the tallest cathedral—became the basis of its ability to inspire wonder.

Bridge alignment has the first and most profound effect on the overall design approach. Whether a bridge is a replacement structure, parallel structure, or new structure, it must conform to existing transportation networks. The alignment considers grades and elevations of existing and proposed roads and the terrain traversed, which may include bodies of water, roadways, railroads, and other existing or proposed site constraints. The length of the structure, especially the length of the visually dominant main span, will dictate the most suitable type of bridge and how it will be constructed.

Determining the construction method and bridge type must take existing site constraints—environmental conditions, traffic patterns, bodies of water, limited right-of-way—into consideration. The owner's schedule, contractor's equipment, and the project size are also critical factors. The basic bridge type has a great deal to do with the distance that needs to be spanned and the functional purpose of the crossing, such as passage for vehicles, trains, and pedestrians. These types can be grouped into general configuration categories of long bridges over water, long-span girder bridges, urban bridges, train/rail bridges, environmentally sensitive bridges (natural landscapes with pristine conditions), arch bridges, and pedestrian bridges (Fig. 9.2).

For particularly large projects, both span-by-span and balanced cantilever methods may be utilized, speeding construction by allowing the simultaneous erection of spans at different locations. In the instance of New Jersey's Victory Bridge (Fig. 9.3), construction was expedited by using span-by-span erection of the approach spans and, simultaneously, erection of the main span using balanced cantilever construction.

In restricted urban corridors, some of the most challenging expansions can be solved by building upward within the existing right-of-way, eliminating the expense and complexity of obtaining new rights-of-way from commercial and residential owners. One example is the Selmon Expressway (Fig. 9.4), an elevated toll road in Tampa, Florida, that was constructed in the existing median and provides six vehicular lanes of capacity in only six feet of space. The bridge roadway features

Fig. 9.2 Collage of bridge types. (Courtesy of FIGG)

Fig. 9.3 New Jersey's Victory Bridge. The Victory Bridge features a record-setting 440′-long precast concrete segmental main span over the Raritan River in New Jersey and pays tribute to WWI Veterans with a series of bronze plaques recognizing US Marines, US Navy, US Army, Air Corps, and Red Cross. (Courtesy of FIGG)

Fig. 9.4 The Selmon Expressway in Tampa, Florida, consists of 5.13 miles of elevated highway built in the median of the existing expressway using sculpturally shaped piers that are 6′ wide. (Courtesy of FIGG)

reversible express lanes that alleviate congestion and provide additional capacity when and where it is needed most. Similarly, a 2.3-mile section of AirTrain JFK (Fig. 9.5), the transit system serving New York's JFK Airport, was built in the existing median of the Van Wyck Expressway, an extremely congested highway that is densely bordered with residential and commercial properties. The erection method allowed construction to proceed with minimal traffic disruption, while the bases of the piers creatively utilize the space within the 10′-wide median.

Dense urban contexts typically drive aesthetics. In certain cases, however, aesthetic choices can physically reshape a region's topography and re-establish a community's identity. The South Norfolk Jordan Bridge (Fig. 9.6) is located in the City of Chesapeake in Virginia's Hampton Roads region. It crosses the Elizabeth River, part of the Intracoastal Waterway and one of the many bodies of water that define the region's geography and long colonial history. Designed and constructed as a

Fig. 9.5 AirTrain JFK consists of nine miles of elevated rail transit that carries passengers into JFK International Airport in New York. (Courtesy of FIGG)

Fig. 9.6 The South Norfolk Jordan Bridge in Virginia is a 5375′-long high-level fixed bridge that keeps rail, vehicular, pedestrian, and maritime traffic moving at all times. (Courtesy of FIGG)

private bridge, the new crossing alleviates the area's burgeoning congestion problem, reducing the typical commuter's drive time. In this instance, the context encompassed historic neighborhoods, industrial businesses on the riverfront, and a significant US military presence.

To tap into the existing right-of-way and road networks, the South Norfolk Jordan Bridge follows and extends the curved path of the original roadway alignment. To accommodate shipping traffic, the bridge has a 145′ vertical and 270′ horizontal navigational clearance. To achieve this height and provide a roadway and pedestrian walkway that meet ADA-compliant safety standards, the new bridge is longer. Rising dramatically from sea level, the bridge is a visual landmark that also offers breathtaking views in every direction, allowing renewed appreciation of the region's many historical and natural features.

9.1.2 *Design*

Grounded in and subordinate to the laws of the physical universe, aesthetic form always follows well-designed function. Consequently, iconographic power is the happy by-product of technical prowess and contextual awareness. Generated from these considerations, a bridge's aesthetics are evident in the structure's efficiency; clarity of form; sustainability; durability; and harmonious shapes, colors, textures, and lighting. In a successful project, the parts come together in a whole that satisfies the senses as well as the need for reliable transportation. When this happens, the bridge enjoys a lasting, memorable presence in the community and gradually accrues symbolic meaning as well.

During the conceptual design phase, span lengths and the depth and dimensions of the superstructure and substructure are determined based on an analysis of the site and understanding how to optimize construction in ways that balance economy, functionality, and lasting visual quality.

Span lengths are determined after a bridge's overall length and alignment have been established and existing site constraints identified. In establishing span length, the goal is to determine the optimal span length and consistently use this length throughout the project, a repetition that streamlines construction and also provides pleasing visual continuity. This repetition can become a unifying strategy for the overall design. Span length, in concert with structure depth, establishes a bridge's aesthetics and how well it flows into its surroundings.

Pleasing, efficient aesthetics are achieved through consistent design elements and shapes. Harmonious forms, lines, and patterns draw the viewer's eye from one element to the next, creating visual continuity and sensory satisfaction. Achieving this requires utilizing the same superstructure cross section over the bridge's full length and identifying, and also consistently using, the optimal span length.

Structural depth, in concert with span length, sets the stage for the structure's overall appearance and determines how well it harmonizes with its surroundings. Many variables determine structural depth. Typically, maintaining a constant box girder depth and constant cross section will greatly simplify casting and erection operations. For longer spans, however, it is often more economical to vary the depth of the superstructure instead of maintaining a constant depth. A deeper box girder section is required to resist the higher forces close to the piers, while a shallower section at midspan is adequate to resist lower forces. In these instances, gradually decreasing the structural depth over the length of the span will minimize materials used and, by reducing the structure's visual mass, will result in a more graceful, slender structure.

Determining an ideal span-to-depth ratio is another essential consideration. We have found that span-to-depth ratios ranging from 20 to 30 will result in superior aesthetics. On uniform spans, a span-to-depth ratio of 15 is also visually attractive, but as a rule, less than 15 is not.

Long spans may be accomplished with a cable-stayed design. In this type of construction, the preferred length for back spans is half the length of the main span.

When this is not possible, unequal span lengths can be balanced by providing additional superstructure weight or other compensations, yielding a pleasing, asymmetrical appearance.

In a successful bridge, the overall form, superstructure, and substructure connect seamlessly to each other and the landscape. The visual experience of a bridge depends on these larger geometries, which gently tie the structure to the land and lead a path through the landscape while respecting and enhancing what existed before. Light and shadow, colors and textures, materials, and the various vantage points from which the structure will be seen also shape a bridge's visual impact.

Crafting a design that meets the needs of a given site relies on creating a workable sequence of shapes and a versatile structural program. The proper shape can create openness in the structure; allow longer spans; and, when going over land, enable the use of vertical space – multiple uses of the space underneath the bridge are possible. At the South Norfolk Jordan Bridge, parks at either approach also extend beneath the bridge, providing residents with new swaths of green space. In other instances, the vertical space can be used to avoid impacting right-of-way or to expedite construction. The Selmon Expressway and AirTrain JFK, for example, were built upward from the center median to make use of existing transportation corridors and avoid impacting right-of-way. Sometimes, the bridge must be built from the top down, as was the case with the Blue Ridge Parkway Viaduct around North Carolina's Grandfather Mountain, to preserve a pristine natural environment of extraordinary beauty. Similarly, the I-76 Allegheny River Bridge (Fig. 9.7), near Pittsburgh, was built from the top down to keep traffic flowing. The long, sweeping spans deliver pleasing aesthetics and also protect the sensitive river environment. This was Pennsylvania's first concrete segmental balanced cantilever bridge.

The basic overall dimensions of the superstructure are based on structural requirements. For the 17th Street Bridge (Fig. 9.8) in Fort Lauderdale, Florida, the precast concrete segmental approaches to the movable main span consist of a closed box shape with sloping vertical webs. The shaping of a closed box girder was

Fig. 9.7 The I-76 Allegheny River Bridge is Pennsylvania's longest concrete span at 532′ and provides an environmentally friendly river crossing. The texture and color on the piers reflect the rock layers in the embankments leading up to the bridge and the stonework in the adjacent Oakmont Country Club. (Courtesy of FIGG)

Fig. 9.8 The 17th Street Bridge in the busy Las Olas area of Fort Lauderdale, Florida, rises above the riverbanks and creates park spaces in the urban area below. Space is opened up below the bridge with long variable depth spans and slender sculptural piers. (Courtesy of FIGG)

selected for its inherent visual appeal, derived from the smooth surfaces of continuous flat planes, while the cantilever wings at the top of the box section provide openness underneath and pleasing shadow effects. There are long spans over land on both sides of the river connecting to the main span crossing the Intracoastal Waterway. Additional landscaping, aesthetic lighting, and hardscape elements beneath these land spans have created new public gathering places. The sculptural shapes of the bridge and the smooth underside enhance these new parks and green spaces.

Pier shapes and their transition into the superstructure provide designers with an opportunity to further distinguish aesthetics. Tapered, slender piers with a relatively high height-to-width ratio make a graceful connection to the superstructure. Additionally, the piers' cross-sectional shape must be considered in light of the bridge's overall form. Simple, elliptical shapes provide a pleasingly classic look and also reduce the drag coefficient of high-wind conditions. Main piers, which must provide structural stability during construction and structural capacity under final service loads, can be sizable. Functional and visually pleasing main piers also can be developed using twin walls.

On the Wabasha Freedom Bridge (Fig. 9.9) in St. Paul, Minnesota, twin-wall, cast-in-place piers are visually appealing and also helped counter the unbalanced loads on the foundation during balanced cantilever construction. Additionally, these piers were cast with contemporary Art Deco-style reliefs that integrated them into the superstructure and hid the bearings at the interface. The shaping of both function and custom form creates a sculptural entrance to a city that features many historically beautiful Art Deco buildings.

Aesthetic design inspiration for the I-90 Dresbach Bridge (Fig. 9.10) between Minnesota and Wisconsin comes from the picturesque natural landscape of the surrounding area, which includes the Mississippi River and heavily forested bluffs and islands. The twin-wall piers are shaped to honor the old-growth trees that emerge from the water with great size and strength and to extend the forest environment across the river. The community selected the theme of celebrating the natural features at a series of design charettes as the design began. This bridge in function,

Fig. 9.9 The Wabasha Freedom Bridge crosses the Mississippi River with no piers in the main channel and is the centerpiece of the downtown in St. Paul, Minnesota. (Courtesy of FIGG)

Fig. 9.10 The I-90 Dresbach Bridge features a 508′-long main span over the Mississippi River that was built from above while maintaining commercial and recreational river traffic during construction. Piers are sculpturally designed to pay tribute to the old-growth trees. (Courtesy of FIGG)

form, color, and style respects its landscape connections and honors the community's spirit with their choices from a series of optional features.

When designing a structure, all visual vantage points must be considered, including those of passengers on and below the deck. On the Broadway Bridge (Fig. 9.11), which crosses the Intracoastal Waterway in Daytona Beach, Florida, particular attention was paid to the vantage point of the many boaters who pass beneath the bridge. At water level, the elliptical piers are wrapped with a 10-foot-tall mosaic tile mural of manatees and dolphins, which are part of the bridge's larger, ecologically inspired mosaic art program from the community-selected theme of "Timeless Ecology" celebrating marine life and wildlife in this region of Florida. Glass mosaic tiles in a style that complemented the color schemes and patterns of the downtown area's historic Art Deco district created 18 colorful mosaics of local marine life and wildlife on the walkways. Visiting school children experience these vivid images and learn about the natural habitat, coming away with a greater appreciation for

Fig. 9.11 The Broadway Bridge in Daytona Beach, Florida, features colorful glass tile mosaics that showcase wildlife and marine life indigenous to the area. School children are brought to walk across the bridge to learn about the ecology of the area with 18 different designs. All features fit the project budget. (Courtesy of FIGG)

local ecology. Created for a fraction of the construction cost and within the project budget, this program has yielded a valuable educational tool and lasting community landmark.

A unique feature of the I-91 Brattleboro Bridge (Fig. 9.12) in Vermont is the vaulted bottom soffit that runs the full length of the underside of the superstructure. The vaulted soffit is stained with a blue color to appear in landscape with the sky to travelers passing under the bridge. The four sweeping stonelike pier walls coincide with the true arching shape of the span across the river. The bridge foundations were brought above the ground to create an intentional outdoor observation platform at the river's edge—a place to enjoy an outdoor gallery of Vermont's picturesque natural landscape.

The shadows created by a bridge's shapes and contours provide depth, visual interest, and varying expressions as sunlight moves across the structure. Natural lighting must be considered in tandem with aesthetic lighting and roadway safety lighting, as the overall shape of the superstructure and substructure interacts, creating ever-changing shadows. On elevated roadways, tall, slender piers minimize shadowing and create more open space beneath and alongside the bridge.

When the primary structural form has been determined for a bridge, specific details can be considered to further enhance the aesthetics and develop the overall

Fig. 9.12 The design of the I-91 Brattleboro Bridge was centered around a theme of "A Bridge to Nature" inspired by the natural beauty of Vermont. (Courtesy of FIGG)

theme. The use of color, texture, native materials, and other details can add greatly to the beauty of the structure, make it unique to its community, and provide continuity between bridge elements.

Exploring new materials and technologies, while drawing on experiences from other industries, can create new bridges that reflect the time in which the bridge was built. An innate curiosity drives design with constant questioning, asking is there a better way? By asking that question, new thinking and possibilities are opened to better design. If the owner is willing to take that step with the design team, then meaningful breakthroughs can be achieved in design, materials, technologies, or financing, and something special for its time can be built. Many bridges can be one-of-a-kind solutions that enhance progress for the future.

In this regard, concrete—endlessly fluid and malleable—is a formidable communication tool. It can be shaped, stained, and placed to meet the spectrum of technical and aesthetic criteria. Today, enhancing the appearance of concrete with custom colors and textures is more economical and feasible than ever before. With foresight and attention, color and texture can be a lasting, low-maintenance addition to a bridge's aesthetics. Bridge coatings provide a uniform color and protectively seal the structure. Contrasting colors, achieved with a bridge coating, can be used to add visual interest to portions of the bridge and emphasize its three-dimensionality. Another option, concrete stain, permits subtle, variegated color treatments.

Concrete was the ideal material for the US 191 Colorado River Bridge (Fig. 9.13), located in the pristine natural setting of Moab, Utah, near Arches National Park. Utilizing long spans and staining the concrete to blend seamlessly with the region's famed red rock landscape yielded a bridge that appears to be born from the earth itself. Concrete's natural gray color is also attractive, especially if consistency is maintained by using single-source suppliers for aggregates, sand, and cement, as was done in AirTrain JFK (Fig. 9.14) in New York City.

The use of native materials can be explored as an opportunity to blend a structure with its natural environment, convey an environmental or earthen theme, or develop community pride. An example of this is the Smart Road Bridge (Fig. 9.15) near Blacksburg, Virginia, which contains Hokie Stone in the pier recesses. Hokie Stone is acquired from a quarry owned by nearby Virginia Tech and is prominently used throughout the Virginia Tech campus. Use of the stone on the bridge provides a visual link among the Virginia Tech Transportation Institute, the operators of Smart Road, and the Virginia Tech campus while adding visual interest by utilizing materials in the bridge that are consistent with the mountainous environment.

Bridges frequently offer unique vantage points for the user, opening new vistas when a new crossing is higher than any existing site or offers a unique view. In situations such as these, where drivers and pedestrians might appreciate a better view from the bridge, open railings that meet the required safety standards have been utilized.

Views from the Winona Bridge (Fig. 9.16) in Minnesota were considered a high priority by the community. Sidewalk overlooks at each of the main-span piers

Fig. 9.13 The US 191 bridges over the Colorado River blend seamlessly with the spectacular landscape of Utah's Canyonlands region fulfilling the community's vision of "A Bridge in Harmony with the Environment." (Courtesy of FIGG)

Fig. 9.14 AirTrain JFK consists of nine miles of elevated rail transit that carries passengers into JFK International Airport in New York. Consistent single sourcing of materials creating the concrete mix design results in a natural uniform color. (Courtesy of FIGG)

provide the community with gathering places to enjoy unmatched views of the Mississippi River, nearby marina, and historic downtown. An open pedestrian railing inspired by organic flow of the river was chosen along the west edge for the full structure length so pedestrians can enjoy open vistas of the surrounding landscape while crossing over the bridge. As the bridge connects to the land, the abutment walls capture the sculptural style of natural grasses.

Special nighttime lighting contributes to the creation of a signature design, enhances safety, and sets a bridge apart from other structures in a city skyline. Blue uplighting on the angular pylons of the I-93 Leonard P. Zakim Bunker Hill Bridge (Fig. 9.17) in Boston highlights the multiple planes of cable stays. The Cascades Connector Pedestrian Bridge (Fig. 9.18) in Tallahassee, Florida, creates shade across the bridge with solar tensile fabrics, which capture energy to support the LED night lighting. The white vertical bridge supports are also used to connect the canopies, which become the canvas for nighttime color. Programmable colors

Fig. 9.15 Smart Road Bridge. Preservation of the rural beauty of the Ellett Valley made aesthetics a major focus of the Smart Road Bridge in Blacksburg, Virginia. Long, sweeping, curved spans enhance openness in the natural setting. Use of Hokie Stone on the piers connects the bridge to the native material of the hillsides. (Courtesy of FIGG)

Fig. 9.16 Winona Bridge. Context-sensitive aesthetic features of the Winona Bridge in Minnesota celebrate the organic beauty of the Mississippi River and preserve the historical importance of the existing bridge. (Courtesy of FIGG)

celebrate the events and recognitions of the day. Programming nighttime lighting effects in the glass pylon of the I-280 Veterans' Glass City Skyway in Ohio has enabled the city of Toledo to create a limitless number of colored lighting schemes

Fig. 9.17 The I-93 Leonard P. Zakim Bunker Hill Bridge was the widest cable-stayed bridge in the world at 183′ with ten lanes when it was completed. The lighting of the pylon legs brings to life the bridge shape and gateway design at night. (Courtesy of FIGG)

Fig. 9.18 The Cascades Connector Pedestrian Bridge in Tallahassee, Florida, features a gateway of color to Florida's capital at night—reflected off of solar canopies and programmable for the color celebration of the day. (Courtesy of FIGG)

to celebrate holidays and other special occasions. Toledo is where the glass industry was born in the United States, so the theme of glass was selected by the community to celebrate the city's heritage. For the first time, four sides of glass were used on the upper 190 feet of the bridge pylon, and LED lighting behind the glass creates many patterns of artful celebrations at night from many vantage points. At the Lesner Bridge (Fig. 9.19) in Virginia Beach, Virginia, aesthetic lighting located

Fig. 9.19 The Lesner Bridge. Designed with the community in a theme of "Reflections of the Bay," the Lesner Bridge is a new signature bridge for the City of Virginia Beach. The sweeping arches of light on the sculptured piers are like dancing waves on the waterway. (Courtesy of FIGG)

within the inset curves of each pier column and along the barrier rail exterior uses color-changing LED technology to allow for fully programmable custom color palettes for holidays, seasons, and special events. The arching pattern was made to be like dancing waves across the water. At the main channel crossing, they are also directional, angling toward the direction boats traverse.

Art in the public realm enlivens and humanizes places. It provokes new understanding, stimulates greater creativity, strengthens local economies, and reminds us there is always something more to discover. While incorporating a public art program into a bridge is unusual, doing so contributes meaningfully to a bridge's capacity to endure. In every instance, the art has increased the public's enjoyment and its awareness of area history, culture, and distinctive environment.

In conjunction with local artists, an art program for the Four Bears Bridge (Fig. 9.20) at the Fort Berthold Reservation in North Dakota captured the spirit of the Three Affiliated Tribes—the Mandan, the Hidatsa, and the Arikara—who live there. The bridge's pedestrian walkway and railing, in effect a linear art gallery, incorporate the tribes' history, sacred symbols, and colors. For the Native American Indian community, these symbols created a spiritual corridor that expressed their sacred, ancestral history and cultural sense of place.

9.1.3 Connection

Sustainability is an indispensable aspect of good design. A successful bridge maintains a responsive and responsible connection to its environment and the community it serves. Sustainable bridge solutions are becoming more urgent in the face of challenges posed by climate change, diminishing energy resources, and aging and congested urban transportation networks. Taking bridge design to new levels of environmental responsibility requires exploring the many efficiencies inherent in concrete segmental bridges. Segmental design encourages ecologically aware land use and preservation, supports quality fabrication and local assembly, and enhances

Fig. 9.20 The Four Bears Bridge in North Dakota incorporates the history, sacred symbols, and colors of the Three Affiliated Tribes, embracing the culture of the Native American community. (Courtesy of FIGG)

a community's quality of life. Capturing the power of imagination, function, and technology, segmental bridges yield measurable social, economic, and environmental benefits—a "triple bottom line" for sustainable success.

Reaching higher levels of sustainability demands bold use of innovative technologies. In this regard, concrete offers tremendous versatility, allowing modular fabrication, top-down construction, and multiple concurrent operations. The nine miles of precast concrete segmental bridges constructed for AirTrain JFK, a mass transit link that has revolutionized commuting for millions of New Yorkers, were built in twenty months—adjacent to lanes carrying 160,000 vehicles per day—and utilized the same equipment design to build all spans.

In Minneapolis, the Minnesota Department of Transportation required that the New I-35W Bridge (Fig. 9.21) has a minimum design service life of 100 years, one-third longer than is typical. For the superstructure, a high-performance 6500 psi concrete mix containing silica fume and fly ash was used to ensure low permeability. As tested, the rapid chloride permeability was very low, with results averaging 250 coulombs, well below the 2000 coulomb maximum allowed. The use of fly ash (a by-product of coal) replaced cement and reduced the carbon dioxide (CO_2) by 3.5 tons per truckload, making the bridge construction better for the environment while ensuring a longer life structure. Other sophisticated new concrete mixes used on the New I-35W Bridge significantly reduce carbon emissions and utilize nanotechnologies that scrub pollutants from the air. With the use of similar mixes on other projects, concrete's carbon footprint is continuing to get smaller, while high-performance concrete reduces steel reinforcing corrosion and increases long-term service life.

Green design means using local materials wisely and with respect. Concrete lends itself to local fabrication and assembly. By using local labor and resources, less energy is needed to move materials and workers. Incorporating local aggregate and sand also ensures that the concrete will blend with its natural context. Segments for the Selmon Expressway, New I-35W Bridge, and Four Bears Bridge, to name a handful of bridges, were cast and stored at nearby sites. Concrete's inherently lower maintenance costs also offer longer life cycle cost benefits. Boosting a regional economy increases the quality of life, inspiring people to want to invest in a bridge that will be a part of their community for a long time.

Placing enormous trust in the power of human creativity can bring amazing design results. By adopting an interdisciplinary approach and engaging in constant dialogue between designers and the communities being served, a vast reservoir of

Fig. 9.21 The New I-35W Bridge in Minnesota was designed and built in 11 months with a 504′ precast concrete segmental main span across the Mississippi River. Nanotechnology concrete, which is a material that cleans pollution from the air, was used for gateway sculptures. (Courtesy of FIGG)

creativity is tapped into that catalyzes innovative ideas. Engineering is a humanistic, community-driven pursuit, and bridges have the power to improve people's lives in ways that go far beyond their functional presence.

Many years ago, a charette process began inviting area residents to have a say in the design of bridges that would be built in their communities, realizing that this was the best way to ensure that the bridge would be welcomed and fully integrated into the region reflecting the proper sense of place. Using the FIGG Bridge Design Charette™, a unique series of interactive listening and learning sessions, gives residents, local leaders, and business owners the opportunity to express their thoughts, needs, and hopes for a proposed bridge. This was a radical premise, since many bridges are built without any input from those who will use them on a daily basis.

One such charette, held to discuss the Penobscot Narrows Bridge and Observatory (Fig. 9.22) in Maine, yielded a theme of "Granite—Simple and Elegant," a nod to the region's granite, some of which was used to construct the Washington Monument.

Fig. 9.22 The 2120'-long Penobscot Narrows Bridge and Observatory in Maine features a 1161' main span and the world's tallest public bridge observatory at 420'. Horizontal articulations were recessed into the pylon as it was built giving the impression of large stone while at the same time hiding the construction joints. (Courtesy of FIGG)

That theme continued to evolve, and eventually a bridge was built with an obelisk-shaped pylon topped with the world's tallest public bridge observatory!

Aesthetic details of the Sarah Mildred Long Bridge (Fig. 9.23) between Maine and New Hampshire were selected during community design workshops based on the community's chosen theme of "Local Simplicity of the Working Waterway." The unique open-sheave lift-tower design selected by the community reflects the working waterway with simplicity and elegance. The dark-gray-stained precast concrete towers represent the sails of large ships and symbolically point to the navigational channel.

The Brooklyn Bridge, the George Washington Bridge, and the Golden Gate Bridge—to name some historic landmarks—each tell the story of its time and technology. During the past century, the erroneous idea arose that to create beauty costs more. However, looking outside the typical industry parameters and exploring the creative use of design and materials can create remarkable bridges that are also

Fig. 9.23 The Sarah Mildred Long Bridge carries US Route 1 Bypass and a heavy rail line that serves the Portsmouth Naval Shipyard over the Piscataqua River. Sculptural shapes with long open spans open new vistas. (Courtesy of FIGG)

cost-efficient. Inventive, new technologies can deliver real value and yield memorable aesthetics.

The I-280 Veterans' Glass City Skyway (Fig. 9.24) in Toledo, Ohio, is another example. The community selected a theme of glass because America's glass industry was born in Toledo, and they wanted the bridge to reflect this heritage. In exploring possibilities for the design with glass, the community was asked, "If you could put glass in a bridge, where would you put it?" The community's voices were recorded on an easel-sized writing tablet as designers listened and wrote down unique ideas. Some wild ideas emerged, but in that moment, wonder was inspired toward how glass could be used in a bridge for the first time. Ultimately, answering that question yielded a landmark bridge as well as an invention that has revolutionized cable-stayed design.

Incorporating glass was a challenge, and there were additional costs to invent a new type of glass for use in a bridge. To maximize the glass's visual interest while capturing the vision that the community wanted to celebrate, it was important to have a tall single pylon to emphasize the glass. Multiple ideas emerged, producing about 300 pages of sketches of ways to use glass in a pylon. Gradually, the design evolved into its final form: four sides of new composite glass on the upper 190 feet of the pylon reflecting the sky during the day and allowing dramatic nighttime lighting from behind the glass.

To do this, designing a slim, sculptural pylon with a distinctive, faceted form was required that would highlight glass's unique characteristics. To marry the slender pylon and the size of the cables needed to carry the bridge in a single plane, a new cable-stayed system called the "cradle" was invented. The cradle system allowed the pylon to be streamlined while maximizing the use of glass. The money saved by using the cradle system was invested in the glass. Ultimately, context and design worked together to create aesthetics and produce the cradle system, a transformative new bridge technology that increases design flexibility, reduces operations time and construction costs, and enhances future cable maintenance and monitoring.

Taking a team approach and engaging the creative spirit of many people to tap into the expression of the whole create a much higher realm of design. The community, when it comes together, has a lot of creative ability. Inviting public input

Fig. 9.24 The I-280 Veterans' Glass City Skyway in Toledo, Ohio, pays tribute to the area's rich heritage in the glass industry. (Courtesy of FIGG)

with the aim of creating a shared vision leads to stronger project support and com-mitment and also yields more intangible yet significant benefits, including the development of trust as well as a common language among stakeholders. That said, it takes courage for engineers to interface with the community in a hands-on interactive way designing a bridge together. It requires flexibility and bedrock confidence in one's creative design abilities to engage in that level of collaboration.

The charette's most important function may be the invitation to participate itself. Residents are empowered to think of themselves as decision makers in determining what will be built in their communities. Charettes create opportunities for individuals to voice their ideas and discuss various aspects of the bridge design. In effect, FIGG design charettes recreate the village green where citizens once gathered to discuss and debate with others who shared their concerns, business and home investments, and familiarity with the history, culture, and soul of a particular place (Fig. 9.25).

Fig. 9.25 The FIGG Bridge Design Charette™ pioneered a unique series of interactive listening and learning sessions that give residents, local leaders, and business owners the opportunity to express their thoughts, needs, and hopes for a proposed bridge in their community. The bridge is designed together, within a budget, with options determined by vote. The new bridge becomes a one-of-a-kind structure that reflects the spirit of the community and not a singular style by a designer. (Courtesy of FIGG)

9.2   Into the Future

By taking advantage of new technologies and materials, we can rebuild the world's bridges in ways that make sense for the planet. More durable bridges with smaller footprints and increased safety are emerging from the use of smart materials, including higher-strength concrete, fiber-reinforced polymers, nanotechnologies, and corrosion-resistant surfaces. Advanced technologies, such as the cradle system, enhance safety, permit better methods of analysis and forecasting, and allow new materials to be incorporated as they are developed.

Another challenge tethered to the aging infrastructure is the environmental impact of replacing old structures and recycling them. Increasingly, there are effective recovery systems for metals, concrete, and other mass components of bridges and buildings.

Designing in ways that will improve mobility, preserve the environment, and enhance the quality of people's lives are goals worth striving for. Long life and low maintenance, combined with enduring aesthetics, yield landmark bridges. A bridge that stands the test of time, measured in terms of its physical structure and resonance with the community, spurs economic and social development. Existing businesses are able to grow and expand. New industries and residential communities spring up around the new bridge and prosper. Parks and bicycle paths are built. A bridge enhances life in so many different directions; there really are no boundaries.

The story of each new bridge is an interactive journey with the community bringing people together both literally and emotionally to capture the culture and values of those who live there. The vision of the design becomes a celebrated functional bridge sculpture that is recognized for its special place. Creating a bridge as art is the art of design.

Linda Figg   received her BS in civil engineering from Auburn University in 1981 as one of three females in her class, was elected vice president of the engineering school by the student body, and was one of the "engineering girls" who chose to paint their concrete canoe pink. Today Linda is the president, CEO, and director of Bridge Art for FIGG Group, a family of companies that specializes in bridges and engages with communities to create functional and aesthetically pleasing bridges. She enjoys encouraging other female engineering students to explore the same joy she found to use math and science in uniquely creative ways through supporting Auburn's 100+ Women Strong program with a mission to "attract, support, and retain female students in Auburn Engineering." The young women are now over 24% (2017) of engineers and growing. Linda believes that these opportunities allow for diversely talented teams to design the best answers for the world's future infrastructure needs.

Linda spent her childhood ignoring boundaries as she secretly climbed to the top of tall trees to sit and daydream for hours about the beauty of building structures surrounded by nature. She loved creating new ideas while designing and sewing her own clothes starting at 9 years old. These patterns of thinking became the framework for Linda's passion to connect people and places with bridges that are in harmony with their landscapes.

Linda's father, Gene Figg (BS civil engineering, The Citadel, 1958), was a great inspiration to her, always repeatedly telling her that she could do anything she put her mind to. Her father founded the bridge company in 1987 while she was attending Auburn, and upon graduating, she joined this exciting bridge adventure, starting in construction inspection on new bridges being built in the Florida Keys. In 2002 Linda took over the reins upon her father's unexpected passing, and while it was devastating to lose her best friend and mentor, she knew that the company's team of outstanding engineers would continue to grow while achieving extraordinary new bridges important to helping others. The magic collaboration of creative engineers at FIGG Group has resulted in over 430 awards for communities and owners recognizing economy, innovation, sustainability, and aesthetics, including three Presidential Awards through the National Endowment for the Arts.

Dedicated to expanding public dialogue about the nation's infrastructure, Linda pioneered the FIGG Bridge Design Charette™, a custom series of interactive community listening, learning, and aesthetic development sessions with a cross section of local participants to create landmark bridges that reflect a sense of place with respect for the natural and built environment. To stimulate interest in bridge design and promote engineering among young people, Linda produced "Big Cable Bridges – How Did They Do That?" an award-winning educational video and teacher's guide. In 2017, Linda was featured in a nine-page interview in *"Bridges: A History of the World's Most Spectacular Spans"* authored by Judith Dupre.

Linda served as chair of the Construction Industry Round Table (2011), an advocacy group composed of one hundred CEOs of America's leading engineering, architecture, and construction companies. The American Road & Transportation Builders Association awarded Linda the Ethel S. Birchland Lifetime Achievement Award (2014). Linda served as the president of the American Segmental Bridge Institute (2012–2014) and was named one of Engineering News-Record's Top 22 Newsmakers in 1998. Concrete Construction magazine named Linda as one of the 13 most influential people in the concrete industry in 2007. In 2010, Linda was named to the Alabama Engineering Hall of Fame and a year later was inducted into the National Academy of Construction, which noted her "vision behind new technologies in bridges that are important to the long-term viability of our nation's infrastructure."

An article about Linda and the companies she leads says a lot about having passion: "Even though the accolades are many, with more sure to come, at its heart FIGG is run by a humble woman who doesn't like to speak of herself. She feels that she is simply doing what she loves best. Like other great works of art, perhaps it's because the bridges themselves say more about her and the firm than words ever could. Strong, resilient, and patiently spanning across the divide...." A bridge created in an artful way joins more than two banks of a waterway or great distances across land; it connects people to each other and their dreams, and sometimes you can see them far in the distance while sitting on the top of a tall tree.

Chapter 10
Inland Waterway Transportation

Sandra Knight, Erika Witzke, and Kate White

Abstract This chapter covers the history of the inland waterway system, its role in the growth of the nation over the past 200 years, and its continued importance to economic health. While the inland waterway system is often unknown to those who live away from the major waterways, the system is critical to the movement of goods and services upon which we all rely. Key topics presented include the need to measure system performance and to help guide investments for optimum returns (e.g., standardization of lock components, Waterways Action Plans). These in turn drive enhanced risk-based decision-making that is flexible enough to account for dynamic conditions, thus preparing the system for future challenges.

Keywords Inland waterway · River · Canal · Great Lakes · Lock and dam · Barge · Navigation infrastructure

10.1 Building the Nation and Its Economy

In many ways, the story of the evolution of the United States' navigation system is the story of the economic development of the nation. The inland waterway transportation system organically grew out of necessity, enabling people to move to new opportunities, retreat from threats, explore and discover new lands, find and trade resources, and connect with other people. These natural systems – rivers, streams, lakes, estuaries, and inlets – served a critical role in connecting cultures, developing economies, and expanding our nation.

S. Knight (✉)
WaterWonks LLC, Washington, DC, USA
e-mail: sandra@water-wonks.com

E. Witzke
CPCS, Chicago, IL, USA

K. White
U.S. Army Corps of Engineers, Norwich, VT, USA

© The Author(s), under exclusive license to Springer Nature Switzerland AG 2022
P. Layne, J. S. Tietjen (eds.), *Women in Infrastructure*, Women in Engineering and Science, https://doi.org/10.1007/978-3-030-92821-6_10

Over time, the backbone of inland waterways morphed in its role – and importance – with the intervention of man-made modifications to expand upstream transportation capabilities, accommodate new vessels, and facilitate increased commerce and economic development. These physical system investments were complemented by the introduction of policies and regulations designed to ensure that the system is maintained for continued efficient and effective use.

Today's US waterway system includes a large physical network of rivers, channels, canals, and supporting land and waterside infrastructure. Each major river system can operate independently, but the true benefit of the US waterway system is achieved as a broader part of the nation's multimodal freight and transport system that connects the waterway system with deep water ports, pipelines, roads, and railroads to convey raw goods to manufacturers and finished goods to customers.

The characteristics and nature of each section of the inland waterways are highly dependent upon the vessels it accommodates, the communities it serves and that service it, the commodities it carries, the mariners who operate the vessels, and the many organizations and people that operate and maintain the channels and infrastructure components. The survival and success of the inland waterways are dependent upon their resilience to the externalities of nature, the shifting supplies and demands of the economy, and the institutional policies and programs that support and finance them.

Today, inland waterways are a vital part of the multimodal freight system and the supply chains that drive the US economy, including those for agriculture, manufacturing, energy production, and other industries. The waterways provide a cost-effective means of transporting high volumes of bulky goods with a comparatively lower impact on the environment than other transport modes.

10.1.1   The Native American Story

The evolution of the current US waterway system began thousands of years ago. Before the colonization of America by European countries, Native Americans migrated to and settled along major lakes and rivers, sustained by nearby agriculture, hunting, and fishing. Crafting canoes and other conveyance vessels, they used the waterways for transportation and trade. In the Great Lakes region alone, more than 100,000 people had settled along major rivers such as the Saginaw and St. Joseph, and along the shores of the Great Lakes before the first white explorer in 1620 [1].

European and American explorers relied upon the knowledge and experience of the Native Americans for both land and water routes. Native Americans guided them on expeditions to establish new territories and chart new courses in what eventually became the United States.

The story of Sacagawea is well known and serves as an exemplar of the talents and strengths of the many unnamed Native Americans. An interpreter and guide, Sacajawea provided her navigational expertise of waterways to the Lewis and Clark

Expedition to explore the northwest following the Louisiana Purchase of 1803. Beginning with the expedition in 1805 traveling along the Missouri, Yellowstone, and Columbia rivers to the Pacific Ocean, her leadership, calm demeanor, and decision-making skills were invaluable in overcoming numerous obstacles and in making the vital decisions necessary to assure the success of the mission. She did all this as the only woman in the crew, at the remarkable age of 17 years old, while carrying her newborn son on the journey [2].

10.1.2   Water Transportation in the Nineteenth Century

Water transportation continued to grow in support of an expanding nation during the 1800s. With the spread of colonial settlers and requirements for transporting raw and finished products, the water routes became more and more important. This growth drove the need for larger vessels and more reliable waterways. Below we highlight the development of major US waterway systems and the technologies that helped shape them to become today's water transportation system and the laws and policies that govern them.

10.1.2.1   Eastern United States

The eastern United States was a focus of early colonial expansion. An early waterway was the Erie Canal (see Fig. 10.1) built between 1817 and 1825 based on the vision of DeWitt Clinton [3]. An engineering marvel at the time, this 363 mile-long man-made canal connected the Hudson River at Albany, NY, to Lake Erie in Buffalo, NY, with the aid of 34 locks having a total of 565 feet in lift. At the time of its completion in 1825, the Erie Canal was the second-longest canal in the world and provided transport of raw materials and finished goods at a time when roads were of variable quality and railways had not yet been constructed. The Erie Canal helped create opportunities for settlement to the west and industry and commerce along the route, and established the preeminence of the state of New York.

New York City became the most populous city and its port the busiest in the nation, in part because of the canal's connection between the Great Lakes and the Hudson River, increasing the flow of goods and materials. The canal was also the impetus for increased civil engineering in the United States, to complement the already existing military engineering field. An example is the Flight of Five Locks in Waterford, a major engineering feat constructed between 1905 and 1915. This project requiring blasting through rock to build five locks for a total lift of 169 feet in less than 7000 feet said to be the biggest lift over a short distance today [4].

After the grand opening of the Erie Canal in 1825, feeder canals were added to the system and the main canal route. As transport grew, the dimensions and infrastructure (locks, bridges, dams) were modified three times. The canal dimensions were ultimately upgraded from the original dimensions of 40 feet wide by 4 feet

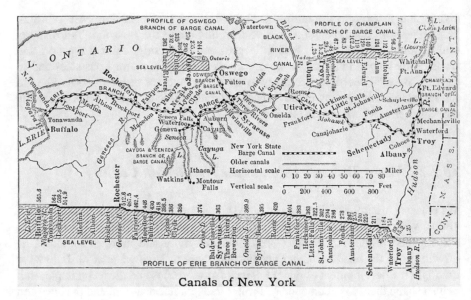

Canals of New York

Fig. 10.1 New York State Barge Canals including the Erie Canal. (Courtesy of New York State Archives)

deep to a 7 feet depth with twin locks each 110 feet long by 18 feet wide. The Erie Canal hit its peak tonnage in the 1880s. The completion of the New York State Barge Canal system in 1918, which accommodated self-propelled vessels, and the opening of the St. Lawrence Seaway in 1959, together, led to the obsolescence of the Erie Canal.

Further down the Atlantic coast, a waterway connecting the Potomac River and the Ohio River was envisioned by George Washington, who in 1785 formed the first publicly traded company in the United States, the Patowmack Canal Trust. This waterborne transportation route was completed in 1802, mostly through the hard labor of slaves and indentured servants. Fraught with construction and financing challenges, partially due to Washington assuming the new job of president of the United States, the company went bankrupt 26 years after it was formed.

As the Patowmack Canal Trust folded and the Erie Canal was being completed in New York, another effort to grow commerce along the Potomac River began. The Chesapeake and Ohio Canal (C&O) (see Fig. 10.2) was constructed between 1825 and 1850 [5]. Many engineering obstacles were overcome to build the C&O Canal. The first was to circumvent the huge rock outcrop at Great Falls on the Potomac. The finished canal encompassed 185 miles of canal and included 74 locks and numerous other structures such as culverts and aqueducts. Like the Erie Canal, C&O Canal boats were towed by mules on a path adjacent to the canal. While the canal construction continually competed with, and lagged, the expansion of the Baltimore and Ohio Railroad, the canal continued to be used until a devastating flood in 1924. The National Park Service designated 524 miles of the Erie Canal as a National Heritage Corridor in 2000 and the 184 length of the C&O Canal as a National Park in 1971

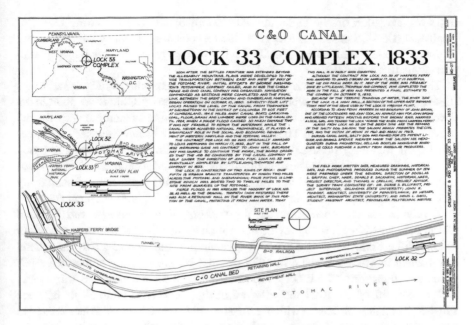

Fig. 10.2   Detail of C&O Canal dating to 1833. (Courtesy of Library of Congress)

Interestingly, and different from the Erie Canal, C&O Canal reports and local news indicated that women played an important role in operating the C&O and the canal barges that navigated the system [6]. Married men were favored to work on the canal and pilot the boats, as they were deemed more dependable than single men. But their wives and children provided valuable and unpaid labor. In their husband's absence or after they died, wives and widows took on positions as lock tenders. While women mostly prepared meals on canal boats, on rare occasions, they became captains by replacing a deceased husband or inheriting the position as a child of the captain.

The prevalence of women lock tenders diminished after the C&O board dismissed women from these positions in 1835, under the auspices that they were more inefficient than men. Nevertheless, several women persisted and defended their positions. An article in the Hagerstown Magazine also speaks about Eliza Reid, who bought her own boat in 1872 and successfully carried coal up and down the C&O Canal for the Central Coal Company [7].

10.1.2.2   The Ohio River System

As the Erie Canal was nearing completion, the US Supreme Court ruled that interstate commerce – to include navigation – was a federal authority. In response, Congress in 1824 passed the General Survey Act encompassing roads and canals of national importance from a military or commerce perspective and a related bill to improve navigation on the Ohio and Mississippi rivers. The US Army Corps of Engineers (USACE or Corps) was given the responsibility to carry out these laws, leading to the initiation of the USACE Civil Works program that continues to support navigation today [8].

The navigation work on the Ohio River included removing sandbars and snags (trees and other debris). The work to remove sandbars on the Ohio River spurred the development of a wing dam extending from the shoreline to increase velocity and move the sand naturally [9]. This 1825 experiment by Major Stephen Long near Henderson, Kentucky, presaged work more than 175 years later called nature-based or process-based engineering design, in which engineered structures enhanced natural processes to attain the desired effect, in this case the movement of sand away from a sandbar. Long's wing dam operated until 1872 with little maintenance.

A private corporation constructed the two-mile-long Louisville and Portland Canal, completed in 1830, to provide navigation around the Falls of the Ohio River at Louisville, Kentucky [10]. This project benefitted from assistance by the federal government in 1826 and again in 1829 through congressionally authorized stock purchases to prevent default [11]. The completion of the canal and its three locks with a total lift of about 24 feet allowed navigation on the Ohio River between Pittsburgh and the confluence with the Mississippi River. Continual issues related to high tolls and the operation of the locks eventually led Congress to authorize USACE to take over control of the canal in 1874 [11].

The USACE then began construction of a lock and dam at a second location on the Ohio River near Pittsburgh. The lock at Davis Island opened in 1885 and supported the movement of coal on the Ohio River after its movement along the Allegheny and/or Monongahela River. This was the first USACE-designed and USACE-constructed lock and dam on the Ohio River, and it used a wicket gate design in which the wickets were lowered at high water to allow open river navigation (Fig. 10.3).

10.1.2.3 The Mississippi River System

Man-made canals in the nineteenth century played an important role in transportation systems for a growing nation as described above, but it was the natural river systems in the United States that formed the backbone of the network. Even today's highly engineered navigation systems largely overlay these natural rivers. The mighty Mississippi and its tributaries reach deep into the heartland of the nation from the headwaters of the Missouri River to the Gulf. It is the longest river in the world, and its watershed encompasses more than 40% of the continental United States. Having long been a focal point for waterborne transportation and commerce, the Mississippi River stimulated the growth of many communities along its banks.

Engineers of the nineteenth century were emboldened by their self-confidence to attempt to control this highly volatile force of nature. In so doing, the Mississippi River and its tributaries became an engineering "test bed" for a myriad of experimental projects. The key objectives to engineering the river were to control its

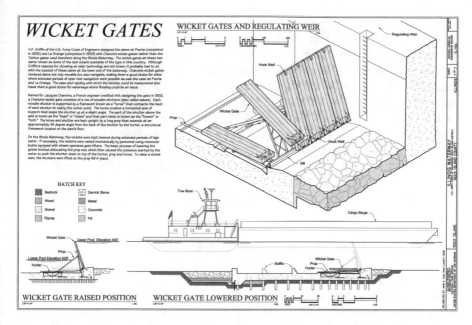

Fig. 10.3 Example of wicket gates used in navigation. (Courtesy of Library of Congress)

massive transport of sediment, minimize flooding, stabilize its banks for settlement, and allow for safe, navigable water transport. Two engineers with competing philosophical and scientific understanding of river mechanics – General Andrew Humphreys and James Eads – spent much of their lives competing to prove their solutions were the best [12].

As an officer for the USACE in 1850, Humphreys was directed to conduct a survey of the Mississippi River that led to one of the most influential scientific reports ever written, *Report upon the Physics and Hydraulics of the Mississippi River,* published in 1861 [13]. Co-authored with Henry Abbot, Humphreys' report supported various solutions such as levees, cutoffs, and outlets, as well as a prescient "no reservoirs" policy to control alluvial flooding. Upon being appointed the USACE Chief of Engineers in 1866, Humphreys used his position to promote a levees-only approach to reducing flooding and a canal for opening the outlet of the Mississippi River to the Gulf to promote commerce [12].

As Humphreys was increasing in celebrity, James Eads was also rising as a world-renowned engineer of the nineteenth century. Eads was responsible for the first steel bridge t in the United States, and it was located over the Mississippi River at St. Louis. He was successful, even though ordered to stop by Humphreys. Their feud continued in a battle for effective methods and approaches to reduce shoaling and open the outlet to the Gulf. Through creative financing (essentially betting on his own design), Eads won the battle to deepen the outlet of the river, constructing jetties in the south pass of the Mississippi River below New Orleans. The jetties were a remarkable success, providing a 30 foot navigable depth earlier than projected [12].

It was during this turmoil between Eads and Humphreys in 1879 that Congress established the Mississippi River Commission and created the US Geological Survey. Humphreys resigned as a result. The Mississippi River Commission was an oversight commission with members appointed by the president of the United States and vetted by the senate to recommend policies and project improvements to the Mississippi River and its tributaries. The membership was to include three officers of the USACE, one from the Geodetic Survey (now the National Oceanic and Atmospheric Administration or NOAA) and three civilians. It would not be until the twenty-first century that a woman would serve on the Commission. In 2012, Norma Jean Mattei was appointed by the president as a civil engineer civilian member, and in 2020, Maj. Gen. Diana M. Holland, commander of the USACE Mississippi Valley Division (MVD), served as president of the Commission [14, 15].

10.1.2.4   Connecting the Great Lakes to the Inland System

In 1836, upon completion of the Chicago Canal, Chicago became an important inland port connecting the Great Lakes to the vast expanse of the Mississippi River. Later, in 1900, prompted by an effort to divert the storm and sewer waters that were plaguing the drinking supply of Chicago in Lake Michigan, the 28 mile Chicago Sanitary and Ship Canal (CSSC) opened. This canal linked the Chicago River to the Des Plaines River following the path of a Native American portage and the previous smaller Illinois and Michigan Canal built in 1848. The CSSC essentially forced the Chicago River to flow away from Lake Michigan, reducing water quality problems that hampered Chicago's growth and prosperity.

Between 1911 and 1922, the development of Calumet Harbor and River and the Calumet-Saganashkee (Cal-Sag) Channel between the Little Calumet River and CSSC increased the capacity of a commercially viable link between Lake Michigan and the Mississippi River, via the 336 mile-long Illinois Waterway (See Fig. 10.4). The regulation of flow into the Illinois Waterway system is governed by an international treaty between the United States and Canada. By the twentieth century, upon completion of the St. Lawrence Seaway, the inland system at Chicago encompassed transit from the Atlantic Ocean to the Gulf of Mexico [16].

10.1.2.5   Laws and Technology

The Rivers and Harbors Appropriation Act, passed by Congress in 1899, was arguably the most important law governing inland navigation [17]. The original act prohibited the obstruction of navigable waters using dams, dikes, canals, etc. and prohibited the erection of bridges over these waters under state legislation without the approval of the secretary of the Army. The act and subsequent modifications importantly called for limiting construction, dredging, dumping, and discharging activities into navigable waters making it the first environmental act before the Water Pollution Control Act of 1948 and, subsequently, the Clean Water Act of 1976

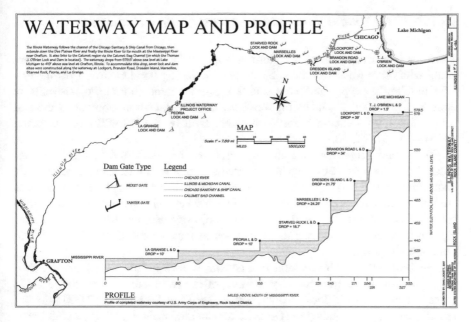

Fig. 10.4 Map of the Illinois Waterway. (Courtesy of Library of Congress)

[18]. This act provided the original regulatory authorities to the USACE and began the various arguments over the definition of navigable waters [19].

The development of the steamboat further transformed the uses and dimensions of inland waterways. The first steamboat, built in the United States by John Fitch, proved too expensive but was soon followed by a more affordable version designed by James Fulton and Robert Livingston. Steam engines continue to power several types of boats in the nineteenth century, but most steam engines supplied power to paddle wheels. As they evolved, steamboats assumed various uses including carrying passengers, removing snags in the river, providing entertainment venues (showboats), and pushing barges. As railways expanded in the nineteenth century, the popularity of river transport and the steamboat waned (USACE-SAM), but regained importance with major investments in water infrastructure and a shift to diesel-propelled push tow and barge configurations in the twentieth century [20].

The inspiration for the fictitious "Tugboat Annie" made popular in stories and on TV was a woman named Thea Foss [21]. A true pioneer for women in the maritime industry, she began her career renting rowboats to local fishermen. A businesswoman, she was able to ultimately build a large fleet providing tugboat services for transporting lumber at the turn of the twentieth century. Her company, Foss Launch and Tugboat, in Tacoma, Washington, grew and was passed on to her family. Still operating as Foss Maritime, it is now a part of a bigger network, Saltchuk Resources, Inc. [21].

10.1.3   The Twentieth-Century Infrastructure Expansion

The twentieth century brought major development of the inland waterway system along with other water resource objectives – including flood control, hydropower, irrigation, water supply, and recreation. Large-scale flood control projects built by the USACE and others, including the Bureau of Reclamation, dominated the era. More than 400 flood control reservoirs were built by the USACE alone between 1936 and 1970 [22].

While many projects were multipurpose, some structures or their missions conflicted with the needs of navigable waterways, such as the restrictions caused by dams on the upper Missouri River and the need to balance hydropower and fisheries with navigation needs on the Columbia River. Yet several major infrastructure projects and systems were built with the primary objective of navigation and only minor support to other non-flood mission objectives such as recreation, wildlife preservation, or hydropower.

The Tennessee-Tombigbee and Intracoastal waterways are examples of waterways built for water transport. By the end of the twentieth century, the USACE owned or operated 275 lock chambers at 230 sites and managed 12,000 miles of inland waterways (excluding the Great Lakes) on the Mississippi River and its tributaries, the Ohio River, the Columbia River, and the Intracoastal waterways.

10.1.3.1   Ohio River

While it ultimately drains into the Mississippi River, the Ohio River has often been considered its own system. This is largely due to separate USACE division offices that manage the two rivers, the advocacy of the stakeholders it serves, and the authorization and appropriation of funds by Congress. In 1910, Congress updated the Rivers and Harbors Act authorizing a series of fifty-one 600 foot by 110 foot locks that were completed in 1929 [23].

Following that, a major modernization was undertaken in the 1950s to accommodate even larger barge tow arrangements navigating up the Mississippi River and into the Ohio River system. The "stair-step" locks that enable navigation between Pittsburgh (710 feet above sea level) and the Mississippi River (250 feet above sea level) are shown in Fig. 10.5.

The modernized system includes non-navigable dams and a 1200 by 110 foot lock chamber along with a 600 by 100 foot chamber at each site [24]. Improvements to the Ohio River system continued through the 2010s, with the opening of Olmsted Dam in 2018 as a replacement for the wicket gate structures and Locks 52 and 53.

10.1.3.2   Mississippi River

Maintenance and improvements of the lower Mississippi River continued into the twentieth century guided by the concepts and engineering set by Humphreys and Eads. A navigable river could be maintained with bank revetments to stabilize the

Fig. 10.5 General plan and profile of the Ohio River. (Courtesy of US Army Corps of Engineers, Louisville District)

shoreline, dikes to keep the sediments moving, cutoffs to straighten the bends, levees to keep it within its banks, and dredging to maintain depth where shoaling persisted. While it was relatively doable to maintain at least a nine-foot navigable channel on the Mississippi River below St. Louis, MO, and a much deeper one below Baton Rouge, LA, the Upper Mississippi was a choke point. A 6 foot channel was authorized in 1907 by Congress above St. Louis. It was expected that this could be achieved by the USACE with a combination of channel outlet closures, wing dams, and dredging [25].

But the ultimate event that would forever change the infrastructure and ecology of the Upper Mississippi and Illinois rivers was the Great Depression of the 1930s. With the promise of creating jobs overriding the major environmental concerns, the 9 foot authorized depth approved by Congress in the 1930 Rivers and Harbors Act meant the addition of 37 locks and dam sites with 42 locks, to the Upper Mississippi and Illinois River System [26].

Collectively, the Ohio River locks and dams, the flood control dams on the Missouri River and the Upper Mississippi River, and Illinois River 9 ft navigation project not only formed the backbone of the inland waterway system but also irreparably altered the sociology, ecology, geomorphology, and hydrology of the US heartland to the Gulf of Mexico.

10.1.3.3 Great Lakes

Known as the longest inland deep-draft navigation system in the world, the Great Lakes-St. Lawrence Seaway (see Fig. 10.6) connecting US and Canadian ports, was opened in 1959 at a ceremony attended by US President Dwight Eisenhower and Her Majesty Queen Elizabeth II of the United Kingdom [27, 28]. With a new route that could handle larger deep-draft vessels, this waterway opened the Great Lakes from Duluth, Minnesota, on Lake Superior to the Atlantic Ocean. The system is 2342 miles long with 16 locks and a minimum depth of 27 feet. A specialized laker fleet dominates the vessels on the waterway with dimensions of 740 ft long by 78 ft beam and a 26.5 ft draft [29].

10.1.3.4 Gulf and Atlantic Intracoastal Waterway

The Intracoastal Waterway system of the United States extends for approximately 3000 miles in two distinct pieces – along the Atlantic Ocean and along the Gulf of Mexico. Originally envisioned as a single waterway from New York City, New York, to Brownsville, Texas, the link in Florida was not completed [30]. The inland coastal system, like others that evolved in the United States, began with a need for protected water transport from the harsh and unpredictable seas. Dating back to Thomas Jefferson, proposals for a national system of improved inland transportation that ultimately involved surveys by the USACE, the waterway was not completed in its current form until the twentieth century. The Gulf Intracoastal Waterway (GIWW) was authorized in 1925 in the Rivers and Harbors Act and later expanded under authorization in 1942 from Florida to near the Mexican Border [31]. The GIWW continues to be one of the busiest commercial waterways in the United States. The Atlantic Intracoastal Waterway is largely devoted to recreational use.

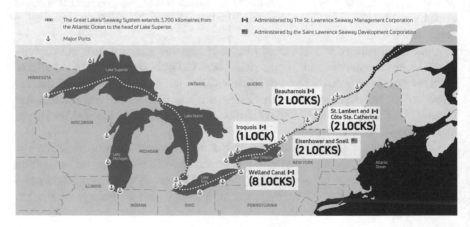

Fig. 10.6 Great Lakes-St. Lawrence Seaway System. (Courtesy of the St. Lawrence Seaway Management Corporation)

10.1.3.5   Other Waterways

The development of US waterways on other natural river systems such as the Arkansas, Columbia-Snake, Tennessee, Black-Warrior, and Ouachita-Black rivers followed a similar evolution as the Mississippi and Ohio rivers. Originally waterways used by Native Americans, they evolved into major regional routes for commerce as the nation grew. In the twentieth century, these systems were canalized with multiple locks and dams to regulate a navigable channel. Most locks and dams were of similar dimensions except on the Columbia-Snake River which needed to be significantly different to accommodate vessels in the Pacific Northwest [32]. The lock lifts along the Columbia-Snake averaged 100 feet which was also different from the low- or medium-lift locks on other systems.

In the late twentieth century, the Tennessee-Tombigbee Waterway was completed that connected the Tennessee River with the Old Tombigbee River and Black-Warrior River via a land-cut canal with locks and dams [33]. The project was intended to provide an alternative route to the Mississippi River from the Tennessee River to the Gulf outlet, but the project was politically charged, as local stakeholders did not think the economic benefits would be realized based on projected traffic volumes. It was not until a major drought on the Mississippi in 1988 that required traffic to reroute that usage of the waterway increased.

10.1.3.6   Legislation

Authorizations and appropriations for specific waterway projects through the twentieth century were primarily funded by the federal government and executed by the USACE. A notable exception was the establishment of the Tennessee Valley Authority (TVA) as part of President Franklin D. Roosevelt's "New Deal" in 1933. The Tennessee Valley Authority Act established TVA as a public corporation to provide navigation, energy, recreation, and flood control along the Tennessee River [34].

The establishment of the Inland Waterways Trust Fund (IWTF) has had a major impact in shaping the inland system in the twenty-first century. In 1978, the fund was established, and in 1986, the IWTF was authorized to expend funds through the Water Resources Development Act (WRDA). This legislation allowed the US Treasury to collect fuel fees from commercial barge users for improvements on the inland waterway system. However, Congress still had to appropriate the funds from the general fund to be expended on USACE projects on federal inland waterways. Until changes were made in later Water Resources Development Acts, it amounted to only 5–15% of USACE funding on construction and major rehabilitation of projects on the inland system.

Throughout the twentieth century, women played important roles in advocating and supporting legislation for the inland waterway system. Leading the Inland Rivers, Ports, and Terminals, Inc. as its executive director, Deirdre McGowan has dedicated her career to the inland system and has been influential in advocating for better funding at a national level. More recently (2021), though no newcomer to the industry, Mary Ann Bucci, executive director of the Port of Pittsburgh, pressed congressional committees to pass an updated WRDA and include more federal cost share on critical projects [35].

10.2   Today's Inland Waterway System

The US inland waterway system includes 12,000 miles of commercially navigable inland and intracoastal channels that directly connect 38 states to each other and to domestic and international markets. The system provides direct connections to deep water on the Atlantic, Pacific, and Gulf coasts (see Fig. 10.7) and to the Great Lakes-St. Lawrence Seaway System (see Fig. 10.6).

10.2.1   The Water

The inland waterway system and the 191 active lock sites with 237 operable lock chambers are maintained by the USACE. The locks are a distinguishing feature of the system and can be placed in three groups based on their dimensions: (1) 15% of the lock chambers are 1000–1200 feet long, (2) 60% are 600–999 feet long, and (3) 25% are less than 600 feet long. Some locks have more than one chamber, and in some cases, these chambers are different dimensions. The shorter locks (600 feet and less) are generally older; 50% of the locks and dams operated by the USACE are more than 50 years old [36].

The 1200 ft locks can accommodate a typical 15-tow barge, while the shorter locks accommodate less than half that number. In cases when a large tow approaches a 600 feet lock, the barge is cut into two sections, and each section passes through on its own. The multiple sections require extra time to pass through the locks. At some locks, the requirement to cut large tows can cause significant delays and queuing when multiple cuts are required for successive barges.

10.2.2   The Vessels

A variety of vessels use the inland waterway system. On the river system, the vessels are typically composed of barges and towboats – together referred to as "tows." Towboats push barges through the system and range in size and horsepower depending on where they operate. Tows on the Upper Mississippi River are restricted to tows with 15 barges lashed together due to the size of the locks (see Fig. 10.8). The Lower Mississippi River is much wider and deeper than the upper reaches of the system, and there are no locks or dams south of St. Louis; therefore, larger tow

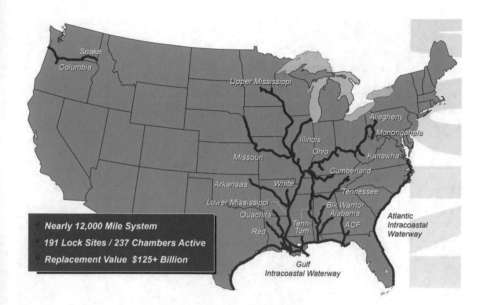

Fig. 10.7   The US inland and intracoastal waterway system [36]

configurations are seen in the southern reaches of the system – including as many as 30 or more barges.

Cargo on the inland river system typically moves on a barge which is 35 feet wide and 195 feet long, known as a standard hopper barge. There are several common variations of these barges including covered hoppers and jumbo hoppers. Liquid and tank barges have similar dimensions and allow for dry and liquid cargo to be carried in the same tow. Often liquids are transported in four-barge tows (see Fig. 10.9).

On the Great Lakes-St. Lawrence Seaway System, in addition to the variety of tugs, tows, barges, and icebreakers required to ensure close to year-round service, the primary vessels are boats generally nicknamed "lakers" and "salties."

A fleet of lakers (see Fig. 10.10) forms the backbone of domestic and transborder trade within the Great Lakes. US-flagged lakers were built specifically for Great Lakes trades and designed to fit through the Soo Locks. These "1000 footers" remain within the Upper Lakes as they are too large to transit the St. Lawrence Seaway locks to the east. Salties (see Fig. 10.11) provide a conduit for trade via the St. Lawrence Seaway from US states and Canadian Provinces that abut the Great Lakes. Many salties have been designed to serve this trade and to maximize seaway lock dimensions ("Seawaymax" ships). These ships can carry bulk and/or break-bulk (i.e., cargo in boxes, crates, bags, or barrels) cargoes and can unload one type of cargo at a Great Lakes port and reload another type of cargo for the outbound move.

Fig. 10.8 Typical 15-barge tow at Lock and Dam 12, Mixed Cargo, Mississippi River. (Courtesy of US Army Corps of Engineers, Rock Island District)

10.2.3 System Use

Marine transportation is generally slower than truck, rail, and pipeline transportation. Over long distances, however, marine transportation provides a lower ton-mile cost than rail and truck transportation. This is in large part due to the high economies of scale of marine transportation. A typical 1500 ton barge, for example, has the equivalent cargo-carrying capacity of 16 rail hopper cars and close to 70 truck trailers (see Fig. 10.12). Only pipelines offer a lower ton-mile cost of transportation over longer distances.

Marine transportation is particularly well suited to low value to weight cargo, such as coal, limestone, grain, iron ore, or other bulk and break-bulk cargo that are not particularly time-sensitive, since transportation costs represent a greater share of the total landed cost of these cargoes. Conversely, the marine mode is not particularly well suited to transporting time-sensitive, high-value goods, such as electronics, high-end fashion apparel, or parts destined for just-in-time automotive manufacturing processes. In 2019, nearly 515 million tons of cargo used the US inland waterway system, a volume equal to roughly 14% of all intercity freight and valued at $134.1 billion [37].

Fig. 10.9 Typical four-barge tow at Olmsted Lock and Dam, Liquid Bulk, Ohio River. (Courtesy of US Army Corps of Engineers, Louisville District)

While the US inland waterway system is vast, each corridor has unique characteristics and handling capabilities linked to the roles the corridors serve and the regions where they are located. Major commodity corridors for key inland waterways include the following [38]:

- *Coal corridor*: Ohio River system, including the Allegheny and Monongahela rivers
- *Food and farm corridor*: Upper Mississippi and Illinois rivers to New Orleans, Louisiana
- *Petrochemical corridor*: Mississippi River from St. Louis, Missouri, to New Orleans
- *Manufactured goods corridor*: Mississippi River from Saint Louis to New Orleans
- *Crude materials corridor*: Ohio and Upper Mississippi rivers (from Saint Louis) to New Orleans
- *Food and farm corridor*: Columbia River system, including Columbia, Snake, and Willamette rivers
- *Chemical and petroleum goods corridor*: Gulf Intracoastal Waterway

In addition to these major commodities, every corridor handles a variety of other commodities and has the potential to provide supply chain resiliency to select commodity groups [39].

Fig. 10.10 Typical Great Lakes "laker" vessel. (Courtesy of the St. Lawrence Seaway Management Corporation)

10.2.4 System Stakeholders

Inland waterway system stakeholders are numerous, and consequently planning, design, operating and maintaining, investment, and governance of the system are complex. These stakeholders, and highlights of their roles, are provided below.

10.2.4.1 Federal Agencies

- *US Army Corps of Engineers* (Department of Defense): As described throughout this chapter, the USACE is responsible for constructing and maintaining navigation channels and harbors and regulating water levels on inland waterways.
- *US Coast Guard* (USCG) (Department of Homeland Security): The USCG is responsible for maritime safety, security, and environmental stewardship in US ports and waterways, as well as for providing ice breaking and navigational aids.
- *US St. Lawrence Seaway Development and Canadian St. Lawrence Seaway Management Corporations*: These two corporations work closely together to manage and maintain the binational waterway system. Each corporation is responsible for the operation and maintenance of its portion of the waterway, including locks and dams, and maintaining channels and navigation aids. These corporations also actively market the seaway to develop trade and increase system use.

Fig. 10.11 Typical Great Lakes "saltie". (Courtesy of the St. Lawrence Seaway Management Corporation)

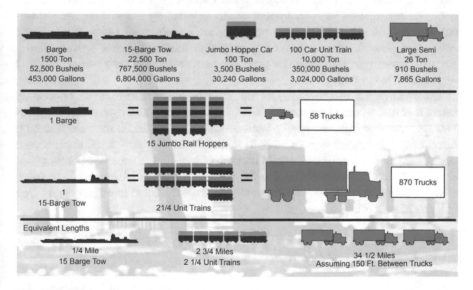

Fig. 10.12 Alternate transport mode comparison [36]

- *US Maritime Administration* (MARAD). This administration within the US Department of Transportation is focused on policy and promotion of all US ports and waterways for trade and broader integration in the multimodal transportation system.

10.2.4.2   Terminals, Ports, and Port Authorities

There are approximately 3700 marine terminals in the United States on all types of waterways including the Atlantic and Pacific oceans, the Gulf of Mexico, inland waterways, the Great Lakes, and surrounding island territories [40]. These terminals are largely privately owned, are operated on a commercial basis, and are often vertically integrated with transportation operations (e.g., barge operations) or shipper operations (e.g., steel mill).

Clusters of terminals form "ports," and often port authorities manage port and/or other transportation facilities in the area, as well as promote and finance local economic development initiatives. In some cases, a port authority's focus may not be on water port development but economic development more generally. These authorities are largely self-funded through operations revenues, and many have financing powers.

10.2.4.3   Vessel Operators

Although there are a variety of vessel operators on the inland waterway and Great Lakes systems, the industry is dominated by a small number of companies. Ownership and governance of the barge industry range from family-owned businesses, to subsidiaries of large conglomerates, to public corporations. Much of the goods on the inland waterway system are transported by the two largest national carriers: Ingram Barge Company and American Commercial Lines. While barge companies generally do not own or operate river terminals, a few operate single-purpose private river terminals that serve a product related to their parent company (e.g., ADM and grain).

On the Great Lakes, there are less than a dozen major marine carriers operating lakers that trade on the spot market. This means that operators generally only provide service when volumes justify, with operators aiming to fill their ships in both directions to generate revenues on both inbound and outbound trips. International carriers are all private companies and include Fednav, Polsteam, and Spliethoff that provide service to lake ports via the St. Lawrence Seaway. As with lakers, these ships typically do not operate "scheduled" service and tend to match inbound cargo movements with outbound cargo.

10.2.4.4   Shippers and Receivers

The shippers/receivers are those businesses that generate traffic and ultimately drive demand on the inland waterway system. The shippers/receivers are companies that use and produce the heavy, bulk commodities that the system is so adept at serving, including grain, construction materials, coal, and iron ore.

10.2.4.5   Multimodal Connections

A port's success and resiliency are in part determined by its connections. Railroads, trucks, and pipelines all provide services to/from inland waterway facilities and are typically private and commercially oriented. There are points on the inland system (e.g., New Orleans) where inland barge commodities are also transloaded to other marine vessels, such as deep-draft cargo ships for international import/export.

Other stakeholders also have an interest in ensuring multimodal connections are present and in a suitable condition for active use. State and local Departments of Transportation (DOT) currently have a limited role in maritime systems but do have the ability to influence broad transportation policy and engage in planning efforts that encourage the increased use of the maritime system. Also, some DOTs have provided funding for maritime-related investments including intermodal roadway/railway connectors and on-dock equipment.

10.2.4.6   Research and Coordinating Bodies

Several additional organizations conduct research, conduct advocacy, and guide all aspects of the maritime system. These organizations include (but are not limited to) the following:

- The World Association for Waterborne Transport Infrastructure (PIANC)
- Inland Rivers, Ports & Terminals, Inc. (IRTP)
- Inland Waterways Users Board (IWUB)
- US Committee on the Marine Transportation System (CMTS)
- Mississippi River Commission (MRC)
- National Waterways Conference (NWC)
- Transportation Research Board (TRB)

10.2.4.7   Waterways Action Plans

The inland navigation system is unique in that a coalition made up of USACE, USCG, and the towing industry has developed a cooperative system for safe and efficient navigation during periods of low water, high water, and ice conditions. The Waterways Action Plans (WAPs) are developed for each river section, with the first WAP for western rivers completed in 2007. The WAPs are live documents, frequently updated by the responsible parties, often including state and local governments and agencies such as the National Weather Service. The plans are tailored to specific conditions observed in each system (e.g., the Mississippi River and Ohio River and Tributaries Waterways Action Plan of 2020) [41].

The WAPs lay out roles and responsibilities for the parties to the plans, to include communications and agreed-upon actions under different conditions, termed watch phase, action phase, emergency phase, and recovery phase [42]. Safety zone

Table 10.1 Safety zone navigation restrictions for the Tennessee River Waterway [42]

Minimum horsepower requirements per barge
Maximum draft limits
Maximum tow sizes
Specific tow configurations
Length and breadth limits
Safe-speed zones, no-passing zones, or no-meeting zones
Helper or towboat requirements
Traffic separation schemes
Reporting requirements
Tank-barge prohibitions
Exclusion of all vessels from the safety zone

operations are implemented as necessary by the local USCG Captain of the Port (COPT) after consultation with the other parties. Safety zone operations include passive restrictions (Table 10.1) or active control of the river traffic. More severe conditions may warrant the implementation of a security zone with additional restrictions.

10.3 Engineering the Waterways

10.3.1 Engineering Complexity

In the development, upkeep, and evolution of the inland waterway system, engineering principles and practice have played a critical role. From the competing views of our nation's early waterways engineers – Humphreys and Eads – it was shown that while engineering infrastructure may rely on the technical mastery of the laws of physics and the mechanics of materials, the taming of a river for safe and reliable navigation relies on the art of understanding nature and communities.

Just as nature provides a ready-made opportunity for open river navigation, nature can also provide extremes that are difficult to control. Managing the geomorphology of the river, such as the deposit of sediments in the navigation lane, the sinuosity of the navigation channel, or the stability of its location within the channel, may require the use of river training structures, bank stabilization measures, dredging, and other man-made or engineered aids. Hydrologic and hydraulic conditions ranging from floods, to droughts, to tidal currents add another dimension of uncertainty. In canalized streams, such as the Ohio River or the Arkansas River, the available navigable depth and safe currents are managed by the construction and operation of a series of pools formed by locks and dams along the way. But this approach also has its limits. It is still subject to nature's extremes and can be costly to build, operate, and maintain. Likewise, building canals landside can improve reliability and expand the supply chain network, but the acquisition of land, a dependable water source, and upkeep can also be costly. And, as we have seen historically, many of these man-made systems became obsolete due to competing modes or replacement by bigger and better systems.

Even though challenging, engineers have had some success in managing the variability of geomorphic and hydrologic conditions, in part because they understood their stochastic nature. This allowed them to plan, design, and construct infrastructure systems that would be viable over many decades. However, the new epistemic uncertainties – imperfect knowledge – that have been introduced with climate change are impeding the ability to bound the uncertainties and are reshaping the way we must consider river management. More extreme events are challenging what is known and making it difficult to forecast or plan for major infrastructure investments that will be viable even in the next decade. Operating rules that were designed to accommodate multipurpose water infrastructure objectives can no longer deliver or meet their desired functions.

Advances in technology and supply chain demands also contribute to the complexities of engineering our waterways. Planning and designing inland waterways to accommodate specific cargo and/or vessels in many ways have been like the chicken and egg paradox. Do you build the channel for the vessel or the vessel for the channel? What has been observed is the navigable waterway system will be tested with the largest possible vessel it can pass – or cannot. Ignoring a "too big to fail" possibility, the grounding of a large deep-draft container ship, the *Ever Given*, in the Suez Canal in 2021 demonstrated that working on the margins of ship-to-channel dimensions is risky business. This paradigm also applies to the extraordinary sizes of the tow and barge configurations on our US inland system. With lock clearances of less than 5 feet of width – that is 2.5 feet on either side – a three-wide barge train will squeeze into a lock chamber built with a 110 feet width. The original locks on the Ohio River were built at 110 feet wide by 600 feet long. These seemingly huge chambers were designed to lock through a towboat and six barges in one lift. But operators realized they could push more barges (15 barges) up and down the Ohio River and thus began the process of double-locking. Essentially, the tow operator unleashes the first nine barges and pushes them into the chamber unpowered to be lifted or lowered to the next level. Then the remaining barges and towboat are locked through and reassembled up- or downstream along the approach walls. Needless to say, this adversely impacts others waiting in the queue going up or down river. This operational practice, larger horsepower vessels, and the growth in commerce led to the next-generation locks on the Ohio River having two 1200 feet chambers to accommodate both up and down river traffic.

Importantly, regardless of the type of waterway or the vessel, we have learned in the twenty-first century that the role of engineering must include evaluating not only the forces of nature on navigability and waterway design but also the impacts this expansive inland infrastructure and its vessels have on nature. The transformation of our inland systems during the twentieth century disrupted the natural ecology of river systems in ways no other infrastructure could have. Locks and dams, vessel traffic, development along the rivers, dredging, channel stabilization, increased agriculture, and deforestation all disrupted fish spawning and migration, endangered more species, and disturbed natural habitats. Large vessels with enormous propellers and wakes generate vessel effects that resuspend sediments, strand larvae near shore, and/or erode shorelines. In retrospect and going forward, engineers must respect nature and learn from it in planning, designing, constructing, and operating inland waterway systems and all of its water and landside components. More and more, water resource engineers are embracing an engineering-with-nature approach.

Finally, a lesson learned over the past 200 years from this highly engineered waterway system is that infrastructure has both positive and negative effects on the communities it serves and the workforce and human capital that are needed to support that system. Economies and communities were established along the waterways because of the need for water, sustenance, and transport of commodities and people. Coal, oil and gas extraction, agriculture, power, and other industries depend upon a reliable waterway transport system. But these economies/sectors often have disproportionately exposed the communities near them to water and air quality problems and flooding. While engineers have largely been devoted to the technical details of sustaining a navigable waterway, they can no longer ignore the social impacts of waterway design and management. Going forward, multidisciplinary teams must engage in the planning, design, and operations of the inland waterway system.

10.3.2   Engineering Requirements

Engineers have long played a role in the planning, design, construction, and operations of the inland waterways. As discussed above, engineers must consider navigability, infrastructure design, channel dimensions, vessel type and effects, and sediment management. Navigable inland waterways in the United States take on a range of shapes, sizes, and configurations. They typically include one or a combination of different types – open river, canalized streams with locks and dams, and land-cut canals (ASCE Manual of Practice No. 94) [43]. One or more of these types can often occur on the same waterway. The Tennessee-Tombigbee Waterway includes all types. Each type challenges the engineer with its own issues and solutions.

10.3.2.1   Considerations for Planning and Design

10.3.2.1.1   Natural Rivers

It may seem a natural river would require little intervention by engineers. Yet these systems often require major interventions such as channel improvements and navigational aids. The number of barges pushed by a tow (the predominant vessel in US inland waterways) is not only a function of the horsepower needed to overcome the currents and maintain steerage for its loading but also dictated by the width, depth, and curvature of the river. In straight reaches, where little shoaling occurs and velocities are less variable, the length and load of the vessel may only be limited by the horsepower and vessel draft. However, most rivers are a series of curves and straights, in which the radius of the natural river bends requires a wider channel for vessels to navigate the bend. Tow pilots approach open river bends with caution and skill as they crab through the helical effects of the crosscurrents, avoiding the shoaling on the inside and the banks on the outside of the bend. Single vessels can take over much of the navigable area as they move through a bend, and for particularly sinuous bends, two-way traffic may not be feasible. As might be expected, these bends must be navigated cautiously and can slow transit times. To improve navigation on natural rivers such as the Mississippi, the channel is often realigned by making a cut between the straight segments of the rivers on the inside of the bend. While improving maneuvering and vessel transit, the cutoffs introduce higher velocities and carry more sediment downstream to deposit in other areas that may ultimately adversely impede navigation. In fact, managing sediments and currents while maintaining navigation channel dimensions is the predominant challenge to engineers who design and maintain natural inland waterways. In a constant battle with the forces of nature, dredging, channel cutoffs, and river training and stabilization structures are the features most employed to keep the waterway viable. In the end, nature will find its way – whether through droughts, floods, shoaling, or rerouting itself, to challenge even the best-engineered systems.

Building land-cut canals and fairways (as they are known in Europe) reduces the impact of natural variables on the reliability of the navigable waterway, but these canals most often are connectors to natural rivers and/or a canalized system with locks and dams. Canals and man-made fairways are by design more restrictive in their overall dimensions than natural systems (rectangular or trapezoidal). This means that vessel maneuverability and speed are limited by canal effects driven by the size of the vessel relative to the size of the canal. As previously discussed, marginal channel designs can lead to groundings such as the *Ever Given* in March of 2021 in the Suez Canal (NYT July 17, 2021) [44]. Further, the design of inland waterway systems has not changed much since the USACE Publication of Layout and Design of Inland Waterways in 1980 (USACE EM 1110-2-1611). A more recent update is found in PIANC Design Guidelines for Inland Waterways (PIANC 2019, WG 141). In this document, the following fairway design parameters are considered:

- Fairway conditions (curvature, depth, navigable width, flow velocities and their direction, turbulence, water-level slope, bank course, training structures, etc.)
- Hydrologic conditions and weather (visibility, wind, raising or falling water level, low or high water)
- Vessel type, steering, and instrumentation (with or without bow thrusters, single or twin rudders, single or twin propellers, powering, radar, Global Positioning Systems (GPS), Electronic Chart Display and Information Systems (ECDIS), automatic identification system (AIS), autopiloting)
- Actual or aimed load and speed (deep draught, empty/ballasted, cargo type, fast or moderate ship speed)
- Driving situation and traffic (single lane or two way, meeting, overtaking, weak or strong traffic)

While technical methodologies may be updated in the newer PIANC version, neither manual considers the complexities discussed above to engineer with nature, consider the impact to communities, and/or adapt designs for climate change.

10.3.2.1.2 Canalized and Natural Rivers with Locks and Dams

Locks and dams within formerly natural streams and along connecting channels (landside canal) such as the Gulf Intracoastal Waterway are major infrastructure assets within the inland waterway system. To manage flow variations, meet multi-purpose objectives, and maintain a minimum navigable depth, dams often are built in tandem with a lock or series of locks. After determining the lift and number of locks required along a river reach, selecting the location(s) of each lock and dam site is critical. The location relative to a bend or crossing and whether it is located on the inside or outside of the bend can impact the tow operator's ability to align for safe entry and exit. The management of currents and sediments is related to these decisions. To aid in alignment, most locks have a guard and/or guide wall. A guide wall is a long wall usually landside and aligned with the inner lock wall. Operators can rest the tow along the wall while waiting to pass through the lock. The guard wall is usually shorter and used to keep the tow from moving into the dam. More complicated components of lock design involve the fill and empty system, the lock gates, and valves. These systems are less complicated for lower head lifts, but for high head lifts, fill and empty systems have to operate to minimize turbulence within the lock chamber that could increase hawser forces on the lashed barges. Engineers have to strike a balance between a fill system that is fast enough to minimize lock time and one that is slow enough to be safe. The designs have evolved over decades, and much of the planning and design guidance was completed in the twentieth century.

As system modernizations were considered on the Ohio and Mississippi rivers, new paradigms about lock design and, particularly, construction were introduced. Adding double lanes and bigger locks was important to a more efficient system, but closing down the system during construction was an unacceptable alternative for commerce. New studies and more research were needed to understand how in-the-wet construction or in-the-wet placement of components could be achieved. In addition, modernization allowed engineers to rethink safe approaches, fish passage, water savings, and sediment management.

10.3.2.2   Operations

In the United States, the inland system is operated and maintained by the USACE. The labor-intensive nature of the operations, the age and condition of the 270 plus locks and dams, and the repeated need for dredging and upkeep of river training structure (dikes, berms, wing dams, and other man-made structures in the waterway that redirect currents and maintain channel depth) are a labor-intensive and expensive proposition. In fact, the operating budget of the USACE is on the order of 60% of the total USACE Civil Works budget, year in and year out. Lock and dam operations require a round-the-clock workforce and a steady maintenance routine. But changing climatic conditions (droughts, floods, ice), major breakdowns in components, and accidents add more system downtime and result in a backlog of maintenance as discussed below.

To better accommodate the uncertainties and manage the system, a life cycle approach to the USACE massive infrastructure portfolio was needed. In 2005, driven by an executive order for government agencies to better manage their assets, the USACE stood up an asset management program. Beginning with an inventory followed by assessments of the various structures and their components, the USACE was able to better schedule, budget, and maintain its portfolio. The program harmonized the individual programs that were being used by each division and district on their own regional assets. It provided tools to identify priority needs and opportunities to share resources, standardize components, and generally improve operations.

10.3.2.3   Construction

With never-before-used concepts for in-the-wet construction came a need for different types of equipment and new construction techniques that challenge the certainty for contractors bidding on projects. This led to some innovations in contracting but also caused delays and more costs. Much has been learned from this work. Braddock Dam was a success story and Olmsted was a challenge to the new ideas.

10.3.3   Guidance

Much of the guidance for planning, designing, operating, and navigating the inland waterways were established during the growth of the system in the twentieth century and were often influenced by European waterway designs. As new technologies and methods and design objectives (such as water savings, accommodating larger vessels, mitigating environmental impacts, or in-the-wet construction) have been introduced, the guidelines are reviewed and updated. The following synopsis of guidelines from the ASCE, USACE, and PIANC reflect the engineering practices for various technical aspects of planning and design of channels, hydraulic structures, construction of navigation projects, and dredging operations and management of the inland waterway system. More current manuals of practice focus on adapting to changes in climate that are especially impactful to the hydrologic cycle and to greenhouse gas emissions reductions to mitigate the effects of climate change.

10.3.3.1 ASCE

Based on the design criteria from over 100 years of experience by the USACE, Manual of Practice (MOP) No. 94 pulls the various USACE engineering manuals preceding its release in 1998 together to provide a soup to nuts collection of state of the practice for hydraulic engineers. It characterizes the parameters of each type of waterway and the tow and barge systems of the waterway traffic. From river training works to detailed lock fill and empty system, the manual provides the basic tools for planning and designing a navigation project. It considers project costs, environmental impacts, lock and dam operation, and maintenance dredging. Since the release of MOP 94, updated information is also provided in MOPs 116 (2013), 124 (2013), and 140 (2018).

10.3.3.2 USACE

The USACE has a variety of ways to issue guidance for regulations, policies, standards, and users' guides. Policy and regulation publications fall into numerous categories ranking in importance from engineer regulations (ER), engineer manuals (EM), engineer technical letters (ETL), engineer circulars (EC), and engineer pamphlets (EP). There are also mandatory and nonmandatory standards and operating procedures. Additionally, various functional groups, organizations, or teams may establish more granular criteria or processes for management and use. The EMs are a primary source of guidance for planning and designing navigation infrastructure. Though not an exhaustive list, the engineering manuals and their titles in Table 10.2 reflect the basis for their use.

10.3.3.3 PIANC

Formerly known as the Permanent International Navigation Congress and now simply by its acronym, PIANC is the World Association for Waterborne Transport Infrastructure. Since 1885, PIANC has served as the major international technical association for both inland and maritime navigation and coordinated with other

Table 10.2 Examples of navigation guidance

Guidance issuer	Guidance document
American Society of Civil Engineers	Manuals and Reports on Engineering Practice (MOP) No. 94, Inland Navigation: Locks, Dams, and Channels
	MOP 116, Navigation Engineering Practice and Ethical Standards
	MOP 124, Inland Navigation: Channel Training Works
	MOP 140, Climate-Resilient Infrastructure: Adaptive Design and Risk Management
US Army Corps of Engineers	Engineer Manual (EM) 1110-2-1604, Hydraulic Design of Navigation Locks
	EM 1110-2-1605, Hydraulic Design of Navigation Dams
	EM 1110-2-1605, Hydraulic Design of Navigation Dams
	EM 1110-2-1606, Hydraulic Design of Surges in Canals
	EM 1110-2-1607, Tidal Hydraulics
	EM 1110-2-1610, Hydraulic Design of Lock Culvert Valves
	EM 1110-2-1611, Layout and Design of Shallow-Draft Waterways
	EM 1110-2-1613, Hydraulic Design of Deep-Draft Navigation Projects
	EM 1110-2-2602, Planning and Design of Navigation Locks
	EM 1110-2-2607, Planning and Design of Navigation Dams
	EM 1110-2-2610, Mechanical and Electrical Design for Lock and Dam Operating Equipment
	EM 1110-2-2611, Engineering for Prefabricated Construction of Navigation Projects
	EM 110-2-5025, Engineering and Design, Dredging, and Dredged Material Management
	EM 1110-2-6055, Inland Electronic Navigational Charts
PIANC	Inland Navigation Commission (InCom) Working Group (WG) 141, Design Guidelines for Inland Waterway Dimensions
	InCom WG 179, Standardization of Inland Waterways, Revision 2020
	InCom Task Group (TG) 204, Awareness Paper on Cybersecurity in Inland Navigation
	InCom WG 192, Report on the Developments in the Automation and Remote Operation of Locks and Bridges
	Environmental Commission (ENVICOM) WG 193, Resilience of the Maritime and Inland Waterborne Transport System
	EnviCom WG 178, Climate Change Adaptation Planning for Ports and Inland Waterways

international navigation-related associations including the International Association of Ports and Harbors, the International Council of Marine Industry Associations, the International Association of Dredging Companies, and Inland Waterways International.

PIANC has constantly adapted to changing conditions over time along with changes in navigation methods, vessels, regulations, laws, and stakeholder needs pertaining to waterborne transport infrastructure. Recent topics include application of ecosystem services in project planning, automation of container terminals, composites for hydraulic structures, sustainable development, and climate change.

10.3.4 Research, Modeling, and Testing

Research, modeling, and field testing have played a key role in planning, design, and operations of inland waterways. Scaled physical models, numerical and analytical models, simulators, and field studies are used to improve the hydraulic and engineering efficiency, assess the navigability and safety, enhance constructability, and mitigate environmental impacts. The various research and study methods track with the growth of the inland system.

10.3.4.1 Waterways Experiment Station

In 1929, soon after the great flood of 1927 on the Mississippi River, the US Waterways Experiment Station (WES) was established in Vicksburg, Mississippi, to be conveniently co-located near the newly relocated Mississippi River Commission. The hydraulic engineering studies performed at the WES hydraulics laboratory provided critical support to the USACE missions as established by the Rivers and Harbors Act of 1899 and the Flood Control Act of 1928 [45].

When the WES began its modeling efforts, it adopted the then-controversial methods used by laboratories in France, Germany (Karlsruhe), the Netherlands (the Delft), and others of small-scale physical models using Newton's principles of similitude. Likewise, WES was not the first USACE organization to apply computer modeling to the solution of hydraulic problems. That was done at Corps' Ohio and Missouri River Division offices in the early 1950s. However, the WES hydraulics laboratory began to develop its own numerical solutions for computer applications around the mid-1950s and then grew those capabilities in the late 60s and early 70s as talented new engineers and more capable computers were acquired [46].

As decades passed, the WES mission expanded to other technical areas including geotechnical, structures, and the environment leading to the establishment of other labs within WES at Vicksburg. In 1998, all the Corps' laboratories and missions were united under one organizational umbrella, the USACE Engineer Research and Development Center (ERDC) [45].

Small-scale physical modeling remains a viable tool used at the USACE Research and Development Center (ERDC) to support the water resources mission of the Corps. The physical models have been used to inform planning, design, and construction of inland waterways. Large-scale models at 1:1 and 1:25 scales have been used to design innovative fill and empty systems, minimize turbulence in the chamber, and improve gate operations and impacts to fisheries. Smaller scales (1:100–1:125) have often been used to model the layout and design of a whole river reach with locks and dams in place. These models, often used with remote-operated scaled vessels, helped engineers, biologists, and tow operators understand and improve the layout, design, construction, and operations.

The USACE operates the Army's only ship-tow simulator. Most simulators are designed to train pilots and boat operators, but this simulator is used to improve design. Professional pilots familiar with the inland waterway or channel are brought into the simulator to evaluate potential improvements and recommend alternatives for safer navigation through lock approaches, bridges, complicated river reaches, and turning basins.

Numerical modeling advanced with improvements in computing power and better physics-based algorithms. These models were particularly useful to address the hydraulics of riverine systems and, as the models advanced, the complicated flow field near hydraulic structures. These models can be stand-alone tools to improve waterway designs and operations but often are used in conjunction with physical models and field studies.

Discovering nature by observation was a primary tool of Eads when he designed the training structures in the Lower Mississippi. Field observations have always been critical to understanding and monitoring a system. These observations have been improved with new technologies that help us see underwater, measure flow and sediments, track vessels, and monitor the structural health, to name a few, of our inland waterway assets.

Even when there are vetted guidelines, the unique characteristics of each inland waterway and its supporting infrastructure may require using the applied research tools of physical and numerical modeling with field observations to validate our decisions. The USACE has not been the only contributor to the research and development of tools and guidelines for inland waterways. Other laboratories, such as the David Taylor Basin and the Northwest Hydraulic Consultants; many universities; engineering practitioners; and international laboratories have all contributed to advances in inland waterway systems.

10.3.4.2 Women in Inland Research

Women engineers and technicians have contributed to the many innovations and unique solutions presented while growing the nations' waterway network. At WES in the 1980s, women became more visible in their contributions to inland waterways studies. Often overlooked, these women were the backbone and continuity to many engineering advances. Research technicians like Dinah McComas, Debby George, and Karen Anderson-Smith spent their federal careers gathering data on physical models, analyzing it, and providing technical improvements to the art of the practice while also presenting the results needed by decision makers to modify and build new infrastructure. The ship simulator was supported by a cadre of women including Moria Fong, Donna Derrick, and Peggy Van Norman who contributed code, developed graphics, managed testing, and analyzed the findings for major studies on waterways around the United States. Female engineers, (including the authors Kate

White and Sandra Knight) though few and far between, worked their way up from conducting studies, to managing programs, to becoming leaders both within and external to the government.

10.3.4.3 Collaborations in R&D

In the mid to late 1990s, an informal group of interagency researchers from the USACE, NOAA, USCG, Navy, and MARAD regularly convened to share the latest methods and technologies in navigation research. The Federal Waterways Research and Development Coordination Committee, as it was known, also alternated hosting, with support from the Transportation Research Board, a biannual R&D conference. In 2005, the Committee on the Marine Transportation System (CMTS) convened for the first time. It was established as a cabinet-level interagency coordinating body to promote the CMTS. In 2010, the first and still serving executive director, Helen Brohl, promoted and convened the first CMTS biennial R&D conference. Supporting both inland and deep-draft research and development, this interagency conference built upon the previous informal group and has now held six biennial conferences.

10.4 Challenges and Drivers of System Use

The inland waterway system, used by Native Americans and explorers 400 years ago, has changed dramatically since that time in terms of the physical system itself, as well as how it is used, governed, and funded. Climate change also presents risks to the system that have not yet been fully captured in how the various stakeholders interface with the system. These changes are presented in the form of challenges in brief, below, and are ultimately drivers of how the inland waterway system is used.

10.4.1 Physical System

Physical system challenges relate to the navigation channels and waterways themselves but also the landside infrastructure that connects goods that arrive at terminals via road, rail, or pipeline. As supply chains across the United States continue to get tighter and focus on timely delivery, the maritime system will also need to provide more resilient infrastructure for the transportation users it serves. There are four key challenges highlighted here:

Deteriorating locks and dams and increased maintenance requirements/backlog The locks and dams on the inland waterway system are aged; however the most critical issue with the locks is that of deterioration where maintenance backlogs

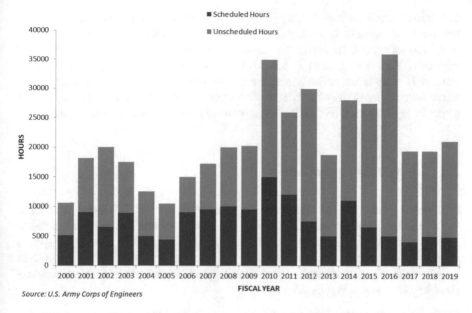

Fig. 10.13 National lock portfolio services trends, main chamber mechanical unavailable hours (events longer than 1 day) [47]

have not been addressed. When locks are not well maintained, the result is unplanned outages and risks to system resiliency. Figure 10.13 highlights the increasing unscheduled hours of delay at locks over the past two decades.

Maintaining channel draft Inland waterway system channels and terminals are also restricted in some areas in terms of draft. A combination of silting, a dredging backlog, and varying water levels due to a range of environmental factors may limit the depth and allowable draft of ships and barges that use the facilities. In order to continue moving goods in these draft-challenged areas, vessels must lighten their loads to pass, limiting the overall efficiency of the system and increasing costs.

Insufficient on-dock equipment Depending on the equipment that is present on terminal docks, cargo handled may be limited. On-dock equipment can hinder a terminal's ability to offer flexible service if equipment is not multipurpose or rated to handle higher weight limits. One inland waterway system market that has had difficulty initiating service is that of containerized goods. Special loading and unloading equipment is required at each terminal that will handle containers, and trade lanes with sufficient goods to transport between these points must be established with a volume of containers to make the service profitable for investment. At present, the commercial viability of this service has been limited.

Some individual terminals may have landside access issues Moving goods via the waterway system is an attractive proposition to many businesses due to the lower cost of a shipment and the lower emissions produced compared to truck or rail transport. However, getting goods to the water for shipment may be difficult. Terminals established at the turn of the twentieth century are often now located in dense urban areas with congestion on the surrounding transportation system, as well as not in my backyard (NIMBY) issues by abutting residential communities.

10.4.2 System Operations

Challenges related to system operations and maintenance are felt largely in overall use of the system – or, rather, declining use of the system. The waterways themselves have abundant capacity to handle more goods, but key aspects of system operations (including lock performance) will need to be honed before the system will be viewed as an attractive, year-round option for shipping goods. There are three key challenges highlighted here:

Overall declining use of the system The inland waterway system is a key component of the multimodal transportation system in the United States; however, the use of the system has been in gradual decline as a result of few key changes, including major investments in competing road and rail infrastructure that has pulled traffic from the water to those modes, the lack of waterway system resiliency (and alternative routing) that has done the same, and a structural change to a few key industries that have historically heavily relied on the maritime system.

Coal and iron ore are two of the top commodities carried on the inland waterway system, and these two commodities have been undergoing a continued structural decline as a result of a major shift in the fuel sources used to generate electricity. This shift is reflected in overall lower tonnage statistics, as well as lower funding receipts, both of which have major implications for the future of the system.

Status quo vessel fleet The vessels using the waterway system have not changed much since they were introduced in large part due to the lack of diversity or change in the commodities handled on the system. While inland vessels have taken steps to increase power; reduce emissions; and, in some cases, handle more cargo, these changes have not been transformative nor improved the efficiency of the maritime system. Longer vessels with increased cargo holding capacity (which may be viewed as a positive industry change) are in fact becoming more difficult to maneuver on parts of the system where shorter locks and low drafts are present, and they use more fuel than smaller and lighter vessels.

Seasonality, ice, flood, and drought While road, rail, and pipeline systems can transport freight essentially 24/7/365, the maritime system has seasonal restrictions placed on it that regularly limits portions of the Mississippi River System (northern reaches) and the Great Lakes-St. Lawrence Seaway System during winter months when the water freezes or when annual system maintenance is planned. Additionally, portions of the system are also restricted at times of high and low water that pose safety concerns and/or make the lakes and rivers unusable for loaded vessels.

10.4.3 Governance and Funding

The myriad stakeholder groups with various roles in the system contribute to the governance and funding challenges the inland waterway system faces. Three key challenges are highlighted here:

Insufficient funds and processes to address deteriorating infrastructure The inland waterway system has generally suffered from underinvestment, but this is not necessarily due to not having the right building blocks in place. Two key federal programs support the Great Lakes and inland river systems. First, the Harbor Maintenance Trust Fund (HMTF) is funded by the Harbor Maintenance Tax (HMT), a tax of 0.125% of the value of commercial cargo (except exports) loaded onto or unloaded from a commercial vessel at a harbor or port, and, second, the Inland Waterways Trust Fund (IWTF) (previously described).

Annual spending from the HMTF, which is controlled via the appropriations process, has not kept up with tax receipts and has rarely equaled receipts and interest in any year. Because of the lack of relationship between HMT receipts and HMTF spending, in 2018, about $1.7 billion in revenue was received from the HMT, but only $1.54 billion was appropriated. As a result of many years of underspending relative to revenue, HMTF has an estimated $9.3 billion surplus [48].

Conversely, the IWTF has faced revenue shortages in recent years that have prevented it from maintaining historical levels of expenditures. Revenues for the IWTF come from a tax, currently $0.29 per gallon, on commercial barge fuel. The funds are matched by federal appropriations from general federal government revenues. For FY 2020, Congress appropriated $1.29 billion for construction and operations and maintenance work, with $1.16 billion coming from the general fund of the Treasury and $131 million from the Inland Waterways Trust Fund.

The balance of the IWTF reached its highest level in FY 2002 at $413 million. It then declined from FY 2005 to FY 2010 due to a combination of increased appropriations, cost overruns at individual projects (especially the Olmsted Locks and Dam Project), and a decline in fuel tax revenues. The balance, however, has since stabilized. At the end of FY 2020, collections were $114 million, and the balance was $55 million [49].

Regulatory inconsistency and uncertainty In any system that transcends states and nations, there will inevitably be inconsistencies in regulatory standards or operating protocols. For the Great Lakes and inland river system, inconsistent ballast water regulations and the designation and regulation of national marine sanctuaries are among the regulations that create uncertainty (and potentially additional cost) for the marine industry. Where interstate commerce is involved, there need to be consistent standards concerning those regulations affecting the barge industry.

Barriers to improved modal connectivity From a planning standpoint, the inland waterway system has been poorly reflected in state and regional transportation plans and economic development plans. As a result, road connectivity issues (e.g., over-sized/overweight corridors to/from maritime facilities) have largely been unaddressed or given consideration on an ad hoc basis, rather than a systematic approach to yield multimodal connectivity and improved system resiliency. However, from a business perspective, the competitive tensions between road, rail, and maritime within the multimodal transportation system should be recognized.

10.5 Transforming the Inland Waterway System

The inland waterway system has a long history and has a significant role in shaping the US economy and competitive position in several industrial markets. But, strategic direction for the system, collaboration on priorities among stakeholders, and overall investments have been challenged and not kept pace with users' needs. This has resulted in a few critical outcomes on the system including the following:

- Looks very similar today to what it looked like 50 years ago
- Has changing usage patterns in different systems
- Has declining reliability and limited resiliency
- Does not fully consider the surrounding conditions and communities
- Is deeply affected by the changing climate

However, based on the role the waterways serve in critical industrial supply chains, domestic and international trade, their untapped capacity, and the potential redundancy they could provide for other transportation modes, we know that the waterway system will remain a critical component of the multimodal freight system for decades to come.

Key question: If the US economy and global supply chains depend on inland waterway systems, how can we ensure waterways blend within the transportation system of the future?

The continued use of the inland waterway system and even a growing role in domestic and international commerce will depend on our ability to transform the current siloed approach to system planning, design, operations and maintenance, and investment; we can, in turn, transform the inland waterway system itself, as *Waterways Journal Weekly* reports:

There are a few opportunities to plan for and provide resiliency to the inland waterway and other modal systems by considering the location of waterways, the available parallel transportation systems (rail and pipeline have similarities in terms of cost of use), and commodities handled. The unplanned closure of the Colonial Pipeline, the US's largest pipeline system for refined oil products, highlighted that communities with access to inland waterways/ports that handled petroleum products, were impacted less [39].

In 2018, the *National Strategy for the Marine Transportation System: Channeling the Maritime Advantage* was developed by the US Committee on the Marine Transportation System members through interagency engagement. The National Strategy provides strategic guidance to enhance the federal understanding and support of the marine transportation system under five priority areas [50]:

- Optimize system performance
- Enhance maritime safety
- Support maritime security
- Advance energy innovation and development
- Facilitate infrastructure investment

While this strategy is a true step in the right direction for the maritime system, it considers only the federal role and stakeholders and not necessarily the stakeholders with boots on the ground that use the system daily and have a vested interest in making it "work." A complement to this National Strategy should be developed (*National Strategy for All Maritime Stakeholders*) that embraces all system users' needs, as well as builds on several key principles, to truly transform the system including (1) modernizing the system, (2) exploring innovation, (3) acting on climate change, (4) dedicating funding and making strategic investments, and (5) embracing diversity, equity, and inclusion.

References

1. Davis, Charles Moler, *Readings in the geography of Michigan*. Ann Arbor Publishers: Ann Arbor, Michigan. 1964.
2. Potter, Teresa and Brandman, Mariana, "Sacagawea c. 1788 – c. 1812/1884?", *National Women's History Museum*. https://www.womenshistory.org/education-resources/biographies/sacagawea. Accessed September 18, 2021.
3. "A National Treasure", *Erie Canalway National Heritage Corridor*. https://eriecanalway.org/learn/history-culture. Accessed September 18, 2021.
4. Williams, Stephen (2011-09-06), "Waterford Flight to receive honors as engineering marvel", *The Daily Gazette*. Schenectady, New York. https://www.asce.org/project/flight-of-five-locks/. Accessed July 23, 2021.
5. "History of the Canal", *C&O Canal Trust*. https://www.canaltrust.org/about-us/about-the-co-canal/history/. Accessed September 18, 2021.
6. "Women on the C&O Canal", *C&O Canal Trust*. https://www.canaltrust.org/2020/04/women-on-the-co-canal/. Accessed September 18, 2021.
7. Biggins, Carolyn, "Women of the C&O Canal: Exploring the vital role women played in the construction and operation of the Chesapeake & Ohio Canal", Hagerstown Magazine. https://hagerstownmagazine.com/news/women-co-canal. Accessed September 18, 2021.
8. Lowry, Pam, "How the Army Corps of Engineers Helped to Build the American Nation", *American System Now*, July 16, 2017. https://americansystemnow.com/how-the-army-corps-of-engineers-helped-to-build-the-american-nation/. Accessed July 23, 2021.
9. "Improving Transportation", *U.S. Army Corps of Engineers*, https://www.usace.army.mil/About/History/Brief-History-of-the-Corps/Improving-Transportation/. Accessed July 23, 2021.

10. Parrish, Charles, "Ohio River", *The Encyclopedia of Louisville*. University Press of Kentucky, Ed. John H. Kleber, pp. 667–668.
11. Trescott, PB (1958) "The Louisville and Portland Canal Company, 1825-1874". *The Mississippi Valley Historical Review*. 44(4) pp. 686-708. https://www.jstor.org/stable/pdf/1886603.pdf. Accessed September 18, 2021.
12. Barry, John, *Rising Tide: The Great Mississippi Flood of 1927 and How it Changed America*. Simon & Schuster: New York City. 1997.
13. Humphreys, Captain A.A. and Abbot, Lieut. H.L., *Report upon the Physics and Hydraulics of the Mississippi River*. Accessed via http://name.umdl.umich.edu/AHE3908.0013.001. Accessed July 23, 2021.
14. "Dr. Norma Jean Mattei", *U.S. Army Corps of Engineers*, Published April 17, 2014. https://www.mvd.usace.army.mil/About/Leadership/Bio-Article-View/Article/473882/dr-norma-jean-mattei/. Accessed September 18, 2021.
15. "Major General Diana M. Holland", *U.S. Army Corps of Engineers*, Published June 29, 2020. https://www.mvd.usace.army.mil/About/Leadership/Bio-Article-View/Article/2240852/major-general-diana-m-holland/. Accessed September 18, 2021.
16. Eagan, Dan, "A Battle Between a Great City and a Great Lake", *The New York Times*, July 7, 2021. https://www.nytimes.com/interactive/2021/07/07/climate/chicago-river-lake-michigan.html?referringSource=articleShare. Accessed September 18, 2021.
17. "Title 33 – Navigation and Navigable Waters", *US Code of Federal Regulations*. https://www.govinfo.gov/content/pkg/USCODE-2010-title33/pdf/USCODE-2010-title33-chap9-subchapI.pdf. Accessed September 18, 2021.
18. "History of the Clean Water Act", *U.S. Environmental Protection Agency*. https://www.epa.gov/laws-regulations/history-clean-water-act. Accessed September 18, 2021.
19. "Regulatory Program Overview". https://www.usace.army.mil/Portals/2/Regulatory_Program_Overview.pdf. Accessed September 18, 2021.
20. "A History of Steamboats". https://www.sam.usace.army.mil/Portals/46/docs/recreation/OP-CO/montgomery/pdfs/10thand11th/ahistoryofsteamboats.pdf. Accessed September 18, 2021.
21. "History", *Foss Maritime, a Saltchuk Company*. https://www.foss.com/about-us/history/. Accessed September 18, 2021.
22. Lonnquest, John, et al., *Two Centuries of Experience in Water Resources Management* (2014). https://www.iwr.usace.army.mil/Portals/70/docs/iwrreports/Two_Centuries.pdf. Accessed September 18, 2021.
23. "History of navigation development on the Ohio River", U.S. Army Corps of Engineers. https://www.lrl.usace.army.mil/Missions/Civil-Works/Navigation/History/. Accessed September 18, 2021.
24. "List of locks and dams of the Ohio River", https://www.wikiwand.com/en/List_of_locks_and_dams_of_the_Ohio_River. Accessed July 5, 2021.
25. O'Brien, William Patrick, Rathbun, Mary Yeater, and O'Bannon, Patrick, *Gateways to Commerce: The U.S. Army Corps of Engineers' 9-Foot Channel Project on the Upper Mississippi River*, (1992). http://npshistory.com/series/archeology/rmr/2/chap1.htm. Accessed September 18, 2021.
26. *Illinois Waterway, Lock & Dams, Rock Island District* (2018). https://usace.contentdm.oclc.org/utils/getfile/collection/p16021coll11/id/2968. Accessed September 18, 2021.
27. "Great Lakes-St. Lawrence River Shipping", *Chamber of Marine Commerce*. https://www.marinedelivers.com/great-lakes-st-lawrence-shipping. Accessed July 12, 2021.
28. "St. Lawrence Seaway officially opened in 1959", *Great Lakes Seaway Partnership*, Published July 17, 2019. https://greatlakesseaway.org/st-lawrence-seaway-officially-opened-in-1959/. Accessed July 12, 2021.
29. "Great Lakes/St. Lawrence Seaway (Highway H2O) Facts", *Great Lakes St. Lawrence Seaway System*. https://greatlakes-seaway.com/en/the-seaway/facts-figures/. Accessed September 18, 2021.

30. "Intracoastal Waterway", Britannica. https://www.britannica.com/topic/Intracoastal-Waterway. Accessed September 18, 2021.
31. Alperin, Lynn M., *History of the Gulf Intracoastal Waterway* (1983). http://libraryarchives.metro.net/DPGTL/us-army-corps/1983-history-of-the-gulf-intracoastal-waterway%20(1).pdf. Accessed September 18, 2021.
32. "Navigation", *Pacific Northwest Waterways Association*. https://www.pnwa.net/navigation/. Accessed September 18, 2021.
33. "Tennessee–Tombigbee Waterway". https://en.wikipedia.org/wiki/Tennessee%E2%80%93Tombigbee_Waterway. Accessed September 18, 2021.
34. "Our History", *Tennessee Valley Authority*. https://www.tva.com/about-tva/our-history. Accessed September 18, 2021.
35. Statement of Mary Ann Bucci, Executive Director, Port of Pittsburgh Commission before the Subcommittee on Water Resources and Environment Water Resources & Development Act of 2020: Status of Essential Provisions, U.S. Congress, March 23, 2021. https://transportation.house.gov/imo/media/doc/Bucci%20Testimony.pdf. Accessed September 18, 2021.
36. *Inland Waterway Navigation – Value to the Nation*, U.S. Army Corps of Engineers (2010). https://www.mvp.usace.army.mil/Portals/57/docs/Navigation/InlandWaterways-Value.pdf. Accessed September 18, 2021.
37. "Waterways System", *Waterways Council, Inc*. https://waterwayscouncil.org/waterways-system. Accessed September 18, 2021.
38. *Failure to Act: Ports and Inland Waterways–Anchoring the U.S. Economy*, American Society of Civil Engineers (January 2021). https://infrastructurereportcard.org/wp-content/uploads/2020/12/failure-to-act-2021-ports-inland-waterways.pdf. Accessed September 18, 2021.
39. Byrne, Shelley, "Pipeline Shutdown Research Shows Importance Of Barging", *Waterways Journal Weekly* (June 25, 2021). https://www.waterwaysjournal.net/2021/06/25/pipeline-shutdown-research-shows-importance-of-barging/. Accessed September 18, 2021.
40. "TSA provides support to Coast Guard to secure U.S. ports", *Transportation Security Administration* (August 22, 2016). https://www.tsa.gov/news/press/top-stories/2016/08/22/tsa-provides-support-coast-guard-secure-us-ports. Accessed September 18, 2021.
41. *Waterways Action Plan: Mississippi River & Ohio River & Tributaries* (2020). https://www.irpt.net/wp-content/uploads/2020/10/2020-UMR-WAP.pdf. Accessed July 26, 2021.
42. *Tennessee River Waterway Management Plan* (2020), https://tva-azr-eastus-cdn-ep-tvawcm-prd.azureedge.net/cdn-tvawcma/docs/default-source/environment/20203428_tn_river_waterway_management_p1.pdf?sfvrsn=76257868_2. Accessed July 26, 2021.
43. *ASCE Manual of Practice No. 94 – Inland Navigation: Locks, Dams, and Channels (MOP 94)*, American Society of Civil Engineers (1998).
44. Yee, Vivian and Glanz, James, "How One of the World's Biggest Ships Jammed the Suez Canal", *The New York Times* (July 19, 2021). https://www.nytimes.com/2021/07/17/world/middleeast/suez-canal-stuck-ship-ever-given.html. Accessed September 18, 2021.
45. Fatherree, Ben, *The First 75 Years: History of Hydraulics Engineering at the Waterways Experiment Station*, US Army Corps of Engineers, Engineer Research and Development Center (2004).
46. Cotton, Gordan A., *A history of the Waterways Experiment Station, 1929-1979* (1979). https://hdl.handle.net/11681/15231. Accessed September 18, 2021.
47. "Inland Waterways", *2021 Infrastructure Report Card*. ASCE and ASCE Foundation https://infrastructurereportcard.org/wp-content/uploads/2020/12/Inland-Waterways-2021.pdf. Accessed September 19, 2021.
48. "U.S Harbor Maintenance and Tax Fund", *Blue Accounting*. https://www.blueaccounting.org/investment/us-harbor-maintenance-tax-and-fund. Accessed September 18, 2021.
49. "Inland and Intracoastal Waterways: Primer and Issues for Congress", *Congressional Research Services* (July 2020). https://crsreports.congress.gov/product/pdf/IF/IF11593. Accessed September 18, 2021.

50. "National Strategy for the Marine Transportation System: Channeling the Maritime Advantage 2017-2022", U.S. Committee on the MTS (October 2017), Washington, DC. 42 pp. https://www.cmts.gov/downloads/National_Strategy_for_the_MTS_October_2017.pdf. Accessed September 18, 2021.

Sandra Knight, PhD, PE, D.WRE, D.NE, is the president of WaterWonks LLC, District of Columbia, and has an appointment as senior research engineer, University of Maryland. She started her college education in the summer of 1975 at Memphis State University where she grew up. She had dreamed of attending college out of state, but even with scholarships, tuition was out of reach. Making the best of it, she attempted to sign up for a freshman English class in the summer prior to her full-time enrollment. When classes were not available to accommodate her schedule as a waitress, she signed up for Engineering Computer Programming 1001. That decision landed her in the engineering department where, surprising herself and others, she demonstrated her aptitude for engineering and was inducted into Tau Beta Pi. In 1980, while working two jobs throughout her education-waiting tables and in the sanitary engineering lab on campus, she graduated cum laude with a BSCE.

Joining the US Army Corps of Engineers upon graduation, Knight began her career as a hydraulic engineer. Seeking out every opportunity available to her, she steadily gathered her credentials and promotions. Along the way, she earned her professional engineering license, a master of science and a PhD in civil engineering, while working full time and raising her son. For her significant contributions to her professional fields, Knight was certified as a Diplomate Water Resources Engineer and Diplomate Navigation Engineer. An honor of pride for her was being elected as the first woman chair of the Inland Navigation Commission (InCom) in the 125-year history of an international navigation association – PIANC.

Knight continued to accept and excel in positions of increasing responsibility. While at the Corps, Knight had leadership roles as Technical Director for Navigation Research, Engineer Research and Development Center; Acting Chief of Engineers, Mississippi Valley Division; and Acting Deputy for Research and Lead Asset Manager, USACE Headquarters. Bumping the glass ceiling, in 2006, she accepted a Senior Executive Service (SES) position – the highest career level for federal employment – with the National Oceanic and Atmospheric Administration. At NOAA, she was responsible for corporate policies and planning strategies for research. Her roles at USACE and NOAA prepared her for the culmination of her federal career as the Deputy Associate Administrator for Mitigation at FEMA, leading the nation's floodplain mapping, management, and mitigation grants and supporting the National Flood Insurance Program, environmental compliance for FEMA, and the National Dam Safety Program.

Upon leaving federal service in 2012 after 32 years, Knight simultaneously established her own consulting firm and the Center for Disaster Resilience, University of Maryland. She considers her most important career contributions to the field of science and engineering in the support and mentoring she has provided to others, particularly women. She is a sought-after science communicator and speaker, having appeared on or contributed to various news media including TV, radio, magazines, and newspapers. She speaks authoritatively at conferences and forums around the world on disaster resilience, hydraulic engineering, flood risk management, and marine transportation. She has authored or contributed to dozens of technical reports, papers, and policy documents. Knight serves as a member of the Marine Board, National Academies of Sciences, Engineering, and Medicine (NASEM) and the District of Columbia Climate Change and Resiliency Commission, as US Commissioner for PIANC, and on the Advisory Board for the Women's Aquatic Network. She is a member of the American Society of Civil Engineers, the American Meteorological Society, the Women's Aquatic Network, and Sigma Xi and a Fellow for PIANC.

Erika Witzke, P.E., earned a bachelor of science in civil engineering degree, focused on transportation planning, from Marquette University in 1995. The importance of transportation and how it can positively impact a person's life was instilled in her at a young age. While growing up on the outskirts of Chicago during a building boom in the early 1970s, her father always recalled that he had the option to work on constructing the Sears Tower or the Standard Oil Building, ultimately selecting to work on the latter due to the parking availability.

While Erika initially selected a path in her university coursework to design transportation systems, including highways, interchanges, and even parking structures, her interests ultimately led her to the strategic planning, not engineering, of these systems.

Today Erika works for a global management consulting firm that specializes in the planning and financing of transportation and power infrastructure. She has worked in several countries in Southern Africa and has had the opportunity to see firsthand the benefits that basic infrastructure can provide, including connecting people to jobs and raising them out of poverty when they have access to bike or bus transportation. In Southern Africa, she has led multimodal transportation system strategic planning mandates and has provided guidance so that these plans for developing countries can be realized, for example, through supporting the establishment of legislation in Botswana to generate new transportation system revenue.

Currently based in Chicago, the majority of Erika's work is in the US Midwest and is focused on improving infrastructure systems and related policies and regulations to facilitate goods movement. As part of this, she has developed strategic freight, rail, and maritime plans for the states of Minnesota, Illinois, Ohio, Kansas, and others. A common thread shared by each of the states is how their economies are closely tied to the inland river or Great Lakes systems that carry the goods of each states' key industries. Her work that focused on applying best practices from project to project has led Erika to become more deeply involved in the research community so she can better understand what comes next and the new innovations her work should lean into. Erika currently serves as chair of the Transportation Research Board's Standing Committee on Inland Waterway Transportation and is former vice chair of the Board's Standing Committee on Agriculture and Food Transportation.

Early in Erika's career, while she was working for a large engineering firm in Minnesota, she was encouraged to attend meetings of Women's Transportation Seminar (WTS), a group that welcomed both women and men to support the advancement of women in all career pursuits related to transportation. She attended meetings to get better acquainted with women in her field and has used the organization as a springboard when she moved to Columbus, Ohio, and then to Chicago. While in Columbus, she served as chapter president and spearheaded a reinvigoration of the local chapter and supported establishing the Cleveland chapter.

Kate White, a registered professional engineer, leads the US Army Corps of Engineers (USACE) Climate Preparedness and Resilience Community of Practice and has over 28 years of experience in the USACE.

Her work includes development of policy, technical guidance, methods, and tools to support climate preparedness and resilience. Her work reflects extensive interagency and expert collaboration in the areas of climate change that are related to the USACE's mission and operations, particularly the Civil Works program and water resources management. Topics include sea-level change impacts and adaptation, hydrologic nonstationarity, changing reservoir sedimentation, and climate vulnerability analyses affecting coastal projects, watersheds, reservoirs, and supply chains. She is also responsible for implementation of climate preparedness and resilience in USACE projects, including agency technical review and oversight of policy and guidance.

White received a 2013 GreenGov Presidential Award: Climate Champion for her role in the interagency team that developed the Sea Level Rise Tool for Sandy Recovery. She was selected as the USACE 2014 Elvin R. "Vald" Heiberg III "Engineer of the Year" and as a 2015 Top Ten Federal Engineer of the Year by the National Society of Professional Engineers. She is a member of the National Academy of Engineering.

Dr. White holds BS and MS degrees in civil engineering and a PhD in civil and environmental engineering.

Chapter 11
Seaports

Geraldine Knatz and Katherine Chambers

Abstract Seaports are part of the global transportation system that moves goods and people around the world. A port is the node that connects the maritime link in the global logistic chain to inland transportation systems. Ports can be described by the types of cargo handled and governance structure. Port development is driven by technological changes in maritime transportation, including the ever-increasing size of oceangoing vessels and the impact of those changes on the physical infrastructure at seaports. Responding to those development pressures, ports have undertaken activities to improve water quality and protect the natural resources within their jurisdictions. More recently, the focus of port environmental strategies centers on reducing emissions of primary pollutants and addressing the long-term impacts of global climate change. Advanced technology is being used to monitor the performance of our marine transportation system (MTS) to make better decisions about investment and operations and the variety of mechanisms that are available for funding port improvements. Four key factors will shape the progress of our seaports and their role in the marine transportation system in the future: cleaner technologies, resilience to disruption, energy supply, and information flow.

Keywords Seaports · Marine transportation system (MTS) · Port performance · Water resource infrastructure · Green ports · Climate change · Women in maritime · Seaport future trends

G. Knatz
University of Southern California, Los Angeles, CA, USA
e-mail: knatz@usc.edu

K. Chambers (✉)
U.S. Army Corps of Engineers, Washington, DC, USA
e-mail: Katherine.F.Chambers@usace.army.mil

© The Author(s), under exclusive license to Springer Nature   241
Switzerland AG 2022
P. Layne, J. S. Tietjen (eds.), *Women in Infrastructure*, Women in Engineering
and Science, https://doi.org/10.1007/978-3-030-92821-6_11

11.1 Introduction

For centuries seaports have played a vital role in connecting regions of the world together – supporting economies and the exchange of commodities, people, and ideas. Ports are a part of a larger marine transportation system (MTS). The MTS includes waterways, ports, and intermodal landside connections that facilitate these movements of goods and people. In more recent history, the globalization of the world economy has resulted in consumer access to products produced anywhere in the world. Globalization has also brought challenges to ports. For example, the advent of containerized shipping and the ever-growing size of oceangoing vessels has resulted in increased demand for deeper channels, longer berths, and faster cranes. This rapid growth in ship capacity has meant that ports must be adapted – deepening, expanding, and maintaining navigation channels and adjusting terminal operations and intermodal connections to manage large influxes of cargo.

Ports have not driven globalization; rather they have had to react and respond to the forces that shape world trading patterns. This chapter seeks to summarize the various types of ports and the roles they play in the MTS, the drivers of change that ports have adjusted to over time, financing challenges and opportunities to facilitate these adjustments, and new trends in performance management and environmental issues. Finally, several recommendations will be made for adapting ports and the MTS to future needs as the twenty-first century continues to present the global community with challenges to overcome and opportunities to improve.

11.2 Seaport Typology

Seaports are nodes in a logistic chain that moves goods or people. A port is a place where land and water transportation modes intersect. The ability to berth ships necessitates that the port be situated upon a navigable waterway. Access to navigable waterways was and continues to be the most important site consideration for a successful port [32], although today a port's position in the global supply chain along with landside access, both highway and rail, warehousing, and other factors contribute to a port's success.

Ports are typically characterized by the type of cargo they handle, such as a container, dry bulk, liquid bulk, or passenger cruise port. While ports may be characterized by their predominant cargo type, it is not uncommon to find ports that handle multiple types of cargo. For example, the Port of Los Angeles, known as the largest container port in the United States, also handles liquid and bulk cargoes, as well as passengers. There are also ports in the United States that serve other functions or industries such as commercial fishing and recreation.

In addition to characterizing a port by cargo handled, a port can also be characterized by its function. For example, a major urban port like Port of New York/New Jersey may be a designated hub for the international movement of cargo by the world's major ocean carriers (vessel operators). A port hub might also be characterized as a "gateway" port, meaning it has extensive rail, highway, or waterway access to serve a deep hinterland. For example, US ports on the east coast and west coast, along with Canadian ports, have rail connections reaching to Chicago. In a region with multiple ports, smaller ports may serve as "feeder" ports that collect and transport goods from their locality to a larger hub port for international distribution. This is common in a situation where not all the ports in a particular region have the market strength, facilities, or navigation channel depth to handle larger ships. Instead, they serve to ferry goods to a larger hub port or gateway that has sufficient facilities and capacity for larger ships that travel in the global trading lanes. US waterborne trade can be characterized as international or domestic. Domestic service is between two US ports, such as trade from Philadelphia to Jacksonville. Such trade is reserved for US-flagged vessels by federal law.

Third-way ports can be characterized by a governance model, such as private or government-owned. Public port authorities were first established at the start of the twentieth century [32]. Large public ports tend to be multipurpose ports and handle many different types of cargo. Many public ports recognize a role in contributing to the economy of the region where they are located, and often economic development and job creation are a part of their mission (Fig. 11.1). Ports have the lead responsibility of identifying waterside infrastructure needs including navigation channels,

Fig. 11.1 For more than 100 years, Port Houston has owned and operated the public wharves and terminals along the Houston Ship Channel. Port Houston is the advocate and a strategic leader for the Channel. The Houston Ship Channel complex and its more than 200 public and private terminals, collectively known as the Port of Houston, are the nation's largest port for waterborne tonnage. (Photo courtesy of the Port of Houston)

berthing, wharves, and terminal facilities. They may also be actively promoting or participating in the improvement of highway and rail infrastructure in their region to facilitate the movement of goods through their port and to regional warehousing and retailers.

Fully private ports tend to be single-purpose ports, often a single terminal that specializes in one particular cargo type such as liquid bulk (crude oil, petroleum products, or dry bulk cargoes like coal or aluminum). Similarly, some smaller government-owned ports, typically non-container ports, are often referred to as "niche" ports. The market segment of niche ports tends to rely on traditional, non-containerized cargoes, such as automobiles or seasonal fruit moved as break-bulk cargo. An example of a niche port would be the Port of Hueneme in California which specializes in agribusiness and vehicles.

Seaport operations typically involve multiple actors charged with the management, regulation, and/or operation of services that utilize a navigable waterway. Globally, there are various models of port governance ranging from privately owned, managed, and operated (the United Kingdom) to government-owned but managed by the private sector (Sydney, Australia), government-owned but managed by a corporatized entity (Canada), or regional or locally owned and managed (the United States). The way a port is governed can reflect the tension that sometimes arises between commercial operators and local and national governmental authorities or regulators.

With a significant role in a nation's economy, it is not surprising to find that in many other countries, ports are under the control or management of a national authority. In some ways, the United States is unique in having its major seaports under the control of local governments. In the majority of US ports, the port authority does not operate the cargo terminals but rather serves as a landlord. The port authority then leases out terminals to private companies, often stevedoring, terminal-operating companies or ocean carriers who carry out port operations on port-owned property through a contractual arrangement.

For every generality about seaport typology, exceptions exist due to the complexity of port ownership, management, and operational control and the global variations in market.

11.3 Drivers for Port Growth

Any conversation around progress and industry includes planning and investing for the future. The MTS will continue to provide supply-chain services across the globe, but the challenges and requirements that the system will encounter in the future will be drivers for transformation. Transformation is not new to the system – in the past, there have been many examples of sudden shifts in trade balance or trends; major crises like conflict and oil shortages; the conception of Panamax vessels; natural disasters like hurricanes, tsunamis, and earthquakes; and many others. The nature of growth for the MTS and the drivers that prompt it range broadly from global demographics and governance, technological advancements, changes in business models and financing, environmental regulations, and challenges to infrastructure systems – including competition for space and outside hazards and threats [31].

11.3.1 Global Demographics

As the backbone of the international supply chain, ports play a major role in the global economy, and as such, they are vulnerable to the cyclical and often volatile global market. Demand is the primary driving force for carrier companies who are constantly shifting to meet the logistical requirements of their customer base. Tracking trends in spending and economic health across the globe is a key part of this effort. Since the 1950s, trade has outpaced gross domestic product (GDP), and only in recent years has the ratio between trade and GDP begun to even out [26]. Future global demographics predict that while the world's population will continue to increase, the rate of growth will decline, resulting in less demand than historically experienced. As the rapid globalization of the 1900s and early 2000s begins to taper, maritime shipping is projected to experience a compound annual growth rate of 3.8% and continued volatility from 2018 to 2024 [45]. Additionally, as economic balances shift, the role of global politics and governance may play a greater role.

The consolidation of the ocean carriers has resulted in a majority of international cargo being routed to specific trading lanes. This has increased the collaboration of US container ports and terminal operators to enhance their regional market power. Such trends toward regionalism may bring with it governance changes like the 2014 merger of the cargo operations of the ports of Tacoma and Seattle [20, 21, 29, 44].

Those nations that profit from economic changes or drivers will continue to invest in maritime infrastructure and fleet development. Unfortunately, politics and governance may also have negative consequences on trade. These impacts could include political stability, unwillingness to follow international law, and increased protectionism (e.g., increased tariffs and sanctions). These effects may be particularly felt if the disruption happens at key choke points in the global MTS (e.g., the Suez Canal, Panama Canal, Strait of Hormuz; [12]).

11.3.2 Technological Advancements

Competition among players in the maritime sector is high, and companies are aggressively investing in solutions to increase capacity, lower costs, and provide them a competitive edge. From 2003 to 2013, the world's fleet of merchant maritime ships doubled in size and increased its capacity by 94% – an amazing statistic that is owed to capital investment and the increased buy-in of a small number of dominant ship-owning nations (e.g., Greece, Japan, China, and Germany; [36]). Not only have the numbers of vessels increased but also the size of ships has increased at a rapid pace. In the last 15 years, a major topic of conversation for the maritime industry is the trend of gigantism in container ships. In 2020, 17.5% of vessel calls for US West Coast ports involved in ships with capacities between 10,000 and 15,000 TEU (twenty-foot equivalent unit which refers to the size of a standard 20 foot shipping container) – an increase from 8.6% in 2014 [25]. In 2020, the 15,000+ TEU CMA CGM Brazil made a call at Savannah: the largest ship to arrive on the East Coast. By early 2022, it is estimated that 15% of the global fleet will be in the 12,000+ TEU size range and 16% of the fleet will be in the 15,000+ size range (Fig. 11.2; [34]).

These fleets have resulted in an increased demand for streamlined and efficient dockside operations. Insights and advanced decision-making are being made possible by maritime data analytics and models. These analytics are offering many benefits including increasing the efficiency of port and terminal operations, predicting future channel use, environmental protection, and ship safety. Another solution to meet competitive demands is the Internet of Things (IoT): one of the largest topics in any discussion about the port of the future. The IoT is a broad term that represents anything connected to the Internet that is enabled to transfer data within human-to-human or human-to-computer intervention. These technological

Fig. 11.2 The vehicle carrier vessel Luminous Ace awaits cargo at the Port of Baltimore. Economies of scale and shipping alliances have hastened the production of massive vessels. As of 2020, the Port of Baltimore is expanding terminals and deepening its berths to accommodate Neopanamax vessel sizes. (Photo courtesy of Katherine Chambers)

advancements range from virtual reality training, supply-chain analytics and optimization, IoT-enabled remote operations, remote sensing, robotics, supply-chain tracking through radio frequency identification, smart metering for fuel and energy consumption, autonomous transport, and predictive analytics, etc. The maritime industry has embraced these advancements more than any other sector [17]. They can manifest in any subsection of the MTS due to their availability, decreasing cost, and increased gains in efficiency and real-time communication, collaboration, and analytical capabilities.

11.3.3 Impacts on Infrastructure Systems

The pace of vessel upsizing, along with the new trends in mergers, acquisitions, and shipping alliances (three main alliances account for more than 84% of global container transport capacity; [33]), has resulted in strains on the maritime and port infrastructure systems that support the movement of goods – including navigable channel depths, terminals, chassis, warehouses, and intermodal infrastructure. Particularly for ports that are trade partners with Asia, Neopanamax vessels have resulted in the need for major investments in bridge raising, navigation channel deepening, rail capacity expansion, additional chassis and warehousing capacity, and upgrades to container yard, crane, and terminal operating systems to handle the massive cargo surges. It is expected that these trends and upgrades to vessel size and infrastructure will continue if the capacity utilization and freight rates remain high [15].

11.4 Financing Port Improvements

The American Society of Civil Engineers (ASCE) hosts a quadrennial national assessment of investment needs and conditions for various types of infrastructure including air, land, and water transportation systems. In 2017, the ASCE rated the nation's overall infrastructure condition as D+, with ports scoring slightly higher at [46]. In 2021, the nation's overall score improved incrementally to a C− with coastal and inland ports scoring a B− [47]. The ASCE attributes these increased port scores to increased multimodal competitive grant programs, like the US Department of Transportation's Maritime Administration's Port Infrastructure Development Grants. Through these discretionary grants, port authorities are encouraged to apply for funding for port and intermodal infrastructure projects. Despite these improvements, ASCE estimates a funding gap of over $12 billion for waterside infrastructure, especially for smaller and inland ports that have trouble competing for economic justification of federal investments. As trading partners across the globe continue to invest in upgrades to their infrastructure systems, the United States must continue to invest in its own deficient infrastructure to maintain competitiveness.

11.4.1 Water Resource Infrastructure Investments

In the United States, the US Army Corps of Engineers (USACE) is the agency responsible for operating and maintaining water resource projects. Navigation was USACE's earliest Civil Works mission, with federal laws signed in 1824 that authorized the improvement of safety of the Ohio and Mississippi rivers. Blossoming from these early requests is USACE's modern navigation mission: to provide a safe, reliable, efficient, and environmentally sustainable waterborne transportation system for the nation. This includes the maintenance of navigation channels and harbors to their federally mandated depths and the upkeep of navigation infrastructure like locks, dams, jetties, and other channel training structures. The majority of USACE's navigation budget is spent on the dredging of navigation channels and harbors in the coasts; the Great Lakes; and, to a lesser extent, inland rivers.

Requirements for dredging vary significantly between ports and harbors. Some ports are major commerce centers that require large annual investments in dredging (i.e., Port of New Orleans) or have long entrance channels with newly deepened authorizations to accommodate Neopanamax vessels (i.e., Port of Savannah). Other large ports generate massive import income but are naturally deep and require little dredging (i.e., Port of Los Angeles and Port of Long Beach). Still others are medium- or low-use ports that provide very limited services to the national supply chain but may be critical for a regional economy or a particular commodity (e.g., soybean exports at the Port of Kalama). Investments in dredging enable US ports to remain competitive globally. Some ports and local or regional governments contribute to the cost of navigation improvement beyond their required local share as a way to accelerate benefits to their region. The trade-off decisions between what to dredge and when have major consequences for regional economics. At the center of these trade-off decisions is the Harbor Maintenance Trust Fund (HMTF). The fund was created by Congress in 1986 and is supported by a tax on imports of waterborne cargo ($1.25 per $1000 worth of cargo). In 2019, the tax collected $1.6 billion, investment earnings totaled $214 million, and annual appropriations for maintenance projects to the USACE totaled $1.6 billion [10]. Recently, as imports have increased, the HMTF has not been consistently fully appropriated by Congress, and a surplus has built up. The 2020 CARES Act ensured that the surplus is spent fully to accommodate the backlog in maintenance and that the HMTF will continue to be fully appropriated for its intended purpose.

11.4.2 Port Infrastructure Investments

The HMTF is one in a sea of many sources of federal funding for port infrastructure projects. For example, there are over 25 agencies and organizations that have interest and engagement in marine transportation. As such, there is a wide array of mechanisms for assistance. The US Committee on the Marine Transportation System

(CMTS) is a federal interagency coordinating committee for these 25 agencies that is directed by a subcabinet-level coordinating board and endeavors to improve communication and awareness and make recommendations to improve federal policies relating to the MTS. In an effort to gather and promote these funding opportunities, the CMTS identified over 75 authorized federal multimodal transportation infrastructure funding, financing, and technical assistance programs [40]. These programs and funding sources are divided into seven broad categories: infrastructure, economic development, energy, resilience, safety and security, environment and sustainability, and research and development.

Beyond federal funding, US ports are constantly investing in new upgrades and looking for ways to provide services safely and reliably. There are many different investment approaches for port projects as they can range from security investments to terminal upgrades; industrial property developments; and road, rail, and waterside infrastructure connections. Funding can come from a wide variety of sources including formula funding from metropolitan planning organizations or state transportation programs, public-private partnerships for studies or capital development, and private sector investment. Some ports have taxing authority and can use general-obligation bonds to finance improvements. Ports that generate sufficient revenues can finance their own improvements by issuing revenue bonds.

The US Department of Transportation's Maritime Administration, in conjunction with the American Association of Port Authorities, developed the Port Planning and Investment Toolkit, a resource intended to provide ports with a common framework and best practices for planning, evaluating, and funding freight transportation, facility, and other port-related improvement projects [2]. The financing module of this toolkit identifies several different financing strategies for any planned project that range from public (e.g., pledged security, public tax exemptions, port facility tariffs) to private (e.g., taxable debt, special-purpose facility bonds, leases). This framework adds to many existing resources to help ports prepare and become more competitive for a variety of investment and financing options.

11.5 New Trends in Performance Management

Performance management is the practice of understanding how well a company or organization has achieved its goals and provides some guidance in how to course-correct and make improvements. A solid effort to measure performance will result in more accurate understanding of the health of the company, and performance metrics are key to this goal. Recent changes in logistics have motivated those interested in understanding port performance to look beyond year-over-year comparisons of a single port or even comparisons between ports in the same region. Instead, forward-looking logistical performance management must meet the global demands of a diverse set of customers in a maritime environment that is competitive, complicated, and easily disrupted by hazards and threats [43]. Considering the entire supply chain, planning, and adapting to disruptions is a major challenge that, if met, will ensure that management decisions incorporate a broader perspective that is ready for future changes that lie ahead.

11.5.1 Emerging Datasets and Analytics

In the past decade, the proliferation of GPS and Internet-enabled devices, updates to safety regulations, and increased use of remote sensing and monitoring tools have resulted in an explosion of new datasets. Maritime analytics has grown as an industry because of an emerging need to provide advanced solutions and customizable insights that increase efficiency across ports. In 2019, the global maritime analytics market was valued at $894.28 million and is expected to reach $1833.50 million by 2027 [18, 38]. These technologies are not limited to private or port authority applications. Federal agencies are adopting new technologies like USACE's structural health monitoring [39], the National Oceanic and Atmospheric Administration's Physical Oceanographic Real-Time System [28], and the Bureau of Transportation Statistics' annual Port Performance Freight Statistics Reports [8]. With the installation of mobile technology and sensors and the accurate and sensible selection of performance metrics, big data can transform the industry by increasing safety, improving operations, and enhancing environmental protection.

One of the emerging datasets that is of great use to supply chain and predictive models is automatic identification system (AIS) data. The AIS vessel data was originally intended for maritime safety and domain awareness and for ship-to-ship communication. It was used in this way for 20 years before satellites and receivers were installed to collect and store AIS data across the globe. Global AIS data can be purchased from many vendors (e.g., exactEarth, Spire, ORBCOMM, MarineTraffic), or US-based data can be downloaded in batches for free from several federal sources (e.g., Marine Cadastre, USCG National AIS Database). Once acquired, AIS data can give insights to vessel movements, type, and unique identifying numbers in near real time [9, 42]. This information quickly expands the awareness of connectivity across the globe – vessel tracks can be measured and their timing or delays quantified from their points of origin to destination. This detailed information can inform more accurate models of traffic predictions and collision risk in narrow channels [35] and dissect questions about big topics like oil trade [1] by understanding first the trends in vessel traffic.

Another way that AIS data can be useful is in understanding the response of the MTS to a broad range of changes and disruptions like oil spill risk [23], hurricanes (Fig. 11.3; [14, 37], [48]), new green shipping policies [3, 16], and the opening of new Arctic shipping routes [11, 41].

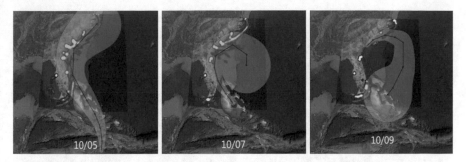

Fig. 11.3 Vessel density heatmaps derived from AIS data (high = white, low = blue). Vessel densities are overlaid with hurricane cone projections for Hurricane Matthew (2016). (Source: Touzinsky et al. (2018))

11.5.2   Selecting Port Performance Metrics

There is no doubt that ports have had to adapt their range of activities to adapt to a global, competitive, and supply-chain-driven market. They are driven to provide their customers with efficiencies and value greater than their competitors. Port performance indicators, therefore, must be selected to detail (or help improve) these competitive advantages and to account for these new realities. A study by Woo et al. [43] identified multiple aspects of evolutionary change that drive the selection of performance metrics. These fall under three major categories: (1) service (e.g., timeliness, reliability, cargo damages, responsiveness, flexibility, price), (2) operations (i.e., throughput, ship waiting time, cargo time, compliance, accidents prevented), and (3) logistics (i.e., cargo waiting time between modes, port cooperation and networking, cargo accruing from value-added service). These categories may not describe every application under which the owners, managers, or MTS stakeholders may want to understand performance, but they give a good start for where to apply these new and emerging datasets to directly affect decisions. It is clear that ports have much to consider when evaluating their performance. Service quality, responsiveness, and reliability through disruptive events are important – price and timing are no longer enough to fully justify their selection over competition.

11.6   The Greening of the Port Industry: A Continuously Evolving Challenge

The consciousness of port authorities to environmental concerns somewhat mirrors the development of the major environmental laws in the United States. The passage of the National Environmental Policy Act (NEPA) in 1969 brought a mandate to examine federal actions through the critical eye of an environmental lens, including the navigation projects of the USACE. Many states followed the lead of Congress and subsequently adopted their own environmental legislation requiring similar environmental review of state and local actions, including projects undertaken by port authorities.

Prior to NEPA, the major environmental focus of port authorities was water pollution, not only discharges from ships but also land-based sources of pollution that were often discharged into the harbor waters. The passage of significant environmental legislation brought major changes to the typical operations of a port authority that traditionally focused on constructing port facilities to serve the movement of goods. Soon ports across the county began to focus on environmental concerns, recognizing their role as "stewards" of the natural resources within their purview.

The top environmental concerns of ports have evolved over time since the early days of a focus on water quality and refuse. Throughout the 1980s and 1990s, port navigation channels were deepened, and landside areas were configured to handle containerized cargoes. These changes resulted in significant impacts on marine and coastal habitats. Ports began to create or restore coastal and wetland habitats within and outside their jurisdiction to compensate for the impacts of dredging and landfilling. The words "mitigate" and "compensate" became ingrained in the port vernacular. Federal resource agencies like the US Fish and Wildlife Service and the National Marine Fisheries Service developed models to evaluate in a more quantitative manner the impacts on habitats and the required amount of mitigation to compensate for habitat losses. Ports began to conduct comprehensive environmental baselines, which often revealed the presence of endangered or other species with special biological importance within their jurisdiction (Fig. 11.4). Ultimately, the process for addressing water quality and habitat impacts became well established.

The increased sophistication of the environmental regulatory process, coupled with the awakening of an environmental consciousness of residents who live in

Fig. 11.4 A great blue heron in the Port of Los Angeles. A 2018 survey identified eleven special status birds among the 87 species of birds that are found in the port. (Photo courtesy of the Port of Los Angeles)

close proximity to ports, began to challenge the ports' parochial view of their environment. Traffic and air emissions created by port activity knew no boundaries and impacted neighborhoods well beyond port jurisdictional borders. Located at the node of landside and waterside transportation systems, ports have a multitude of different air emission sources: ships, harbor craft like tug boats, terminal equipment, trucks, and railroads. Air quality regulators focused their attention on industrial sources and mobile sources like automobiles. Ports were off the radar screen of many local air quality agencies. Many of these air emission sources had escaped intense regulation, owing to their nature as "federal" sources that local air authorities had no ability to regulate. Even at the federal level, regulation of air emissions from oceangoing vessels was limited to US-flagged vessels. The publication of "Harboring Pollution, the Dirty Truth about U.S. Ports" [27] was a call to action that resonated with local communities, governmental officials, and port authority environmental staff.

Many ports viewed the regulation of emissions from oceangoing vessels as an international responsibility, through the actions of the International Maritime Organization (IMO), the specialized United Nations agency with the responsibility for safety and security of international shipping. The IMO's first International Convention for the Prevention of Pollution from Ships, also known as MARPOL, was adopted by the IMO in 1973. Initially the treaty dealt with oil, sewage, and refuse pollution from ships. In 1978, the treaty was revised as a result of a series of tanker accidents and oil spills. In May 2005, Annex VI, Prevention of Air Pollution from Ships, was added to set limits on sulfur oxide and nitrogen oxide emissions.

The challenge with waiting on international regulation to remedy concerns about health impacts on port communities was not a viable strategy for urban ports that wanted to expand. A major breakthrough in the "business as usual" came from the development of the Clean Air Action Plan (CAAP) by the Ports of Los Angeles and Long Beach. The CAAP included a combination of tactics that included incentive programs, tariff measures that mandated compliance and rewarded customers for implementing aggressive emission reduction strategies. The most controversial aspect of the plan was the Clean Truck Program (CTP), adopted in 2007, which banned class 8 heavy-duty trucks by model year to ensure turnover of the trucks to newer, cleaner models. The CTP was litigated all the way to the US Supreme Court but still stands today as a model of how a port authority can significantly reduce air emissions from all port sources and reduce community health risk exposure from diesel particulates. From a baseline year of 2005, diesel particulate emissions were reduced by 87%, and NOx was reduced 62% and SOx by 98% [30]. The program's success motivated other ports to also adopt locally relevant Clean Air Action Plans and Clean Truck Programs. The 2017 update of the San Pedro Bay Ports CAAP calls for all terminal equipment to be zero emission by the year 2030 and all heavy-duty trucks to be zero emission by 2035.

11.6.1 Climate Change

In the United States, environmental priorities have evolved over time, representing the progress made in environmental management and changes in the maritime business culture. The European Seaport Organization has surveyed the evolution of port authority top environmental concerns from 1996 to 2020 [13]. That survey indicated that in the 1990s, water quality, dredging, and disposal of dredged material were the three top primary concerns. By 2020, the top three concerns had shifted to air quality, climate change, and energy consumption.

Global temperature averages are projected to increase as a result of the upward trend in greenhouse gas emissions [24]. Across the globe, the awareness and concern about the impacts of climate change are spreading, and the demand for a reduction in the environmental footprint of maritime shipping services is also increasing. Societal emphasis and pressure on industry to adopt greener practices have resulted in regulations on sulfur emissions, environmental protection, greenhouse gasses, and marine recycling efforts.

Global shipping, however, was not included in the 2015 Paris Climate Agreement. Under pressure from the European Union, in 2018, the IMO adopted its initial greenhouse gas strategy with two primary objectives: (1) reduce the carbon intensity of ship emissions by 40% by 2030 and 70% by 2050 and (2) reduce total greenhouse gas emissions from global shipping by 50% by 2050. Meeting these goals will require significant changes in global shipping in the near term, particularly in the types of fuels used. Already many new vessels are being built that will use liquified natural gas (LNG) as a fuel. LNG is still a fossil fuel, and while it is being adopted by many shipping companies, it will not meet the longer-term objective of zero emissions due to it being a fossil fuel. Other potential fuels being considered for the future include hydrogen and ammonia.

On the landside, ports are pushing to eliminate the use of fossil fuels, converting equipment to electricity and focusing on multiple ways to move toward zero emissions. More ports are now providing electrical power at their berths to allow vessels to shift to shoreside power rather than having the vessels run auxiliary engines while at berth. In addition to using greener power sources, ports are also generating green power to support their operations through the installation of solar panels and various electrical and battery-operated equipment. The World Port Sustainability Program, a program of the International Association of Ports and Harbors, through its website at sustainableworldports.org, showcases how ports are devising ways to be leaders in meeting the Sustainable Development Goals of the United Nations.

11.7 Women in the Maritime and Seaport Sector

The maritime sector has traditionally been a male-dominated industry in all its disciplines from management to engineering and operations. An increase in the ranks of women in our nation's seaports coincided with the rise of environmental consciousness in the 1970s. Since the 1970s, when females were rarely found on port boards, seaports have made inroads in addressing the gender bias on board appointments. Early female appointments to port boards often had a legal or business background. But over time, communities surrounding the port became more vocal about port activities and environmental conditions. This led to board appointees with more diverse interests, including environmental or activist backgrounds.

In those early days, it became common for an appointing authority to put one woman on a port board. In 2020, 28.3% of the US port board members were female, with those boards ranging from 5 to 14 members. Eleven of the US ports have only one female board member. But seven ports have at least three female members. Canadian ports fare better than US ports, with 44.8% female board members. In Latin America, board gender data was sought for 21 major ports, but not all published the names of board members or may not have port boards, especially at nationalized ports. Of the 11 ports where board members were identified, females held 27.4% of the posts.

Recognizing the need to prepare women for leadership positions within the maritime industry, the IMO launched its Women in Maritime program in 1988 to help foster the training and recruitment of women in the maritime fields, including ports. Similarly, in 2013, the International Association of Ports and Harbors launched the Women's Forum, providing scholarships to assist women in advancement within the port sector. Individual ports have also developed their own initiatives to increase diversity among their employee ranks. Data for female seafarers indicated even less progress than at the ports. The IMO maintains statistics on the number of women seafarers. In 2020, only 2% of the world's 1.5 million seafarers were women, and 94% of them worked in the cruise industry.

11.8 The Future of US Ports: What Trends Will We See?

Beyond the "drivers for port growth" outlined in sections above, there are several factors that will influence the activities of ports in the future. These are (1) developing and implementing cleaner and more efficient technologies, such as zero-emission cargo handling; (2) improving resilience to disruptions to the supply chain whether from natural or man-made disasters, including cyberattacks and the long-term impacts of climate change; (3) the necessity to be part of the solution to the changing paradigm on energy supply and distribution; and (4) the need for ports to be proactive in addressing supply-chain issues, particularly when it comes to increasing visibility within the supply chain and improving the flow of information among supply-chain stakeholders through increased digitization and data sharing.

11.8.1 Developing Cleaner and More Efficient Technologies

Ports are moving beyond their traditional roles in several ways, such as assuming a greater role in green energy supply and production and advancing digitization of the supply chain. Over the past decade, ports and the maritime industry have made significant strides in improving efficiency and reducing emissions. US policy on climate change will impact the direction that ports pursue to achieve their greenhouse gas reductions. While earlier efforts focused on the federally defined criteria pollutants such as NOx and SOx, the focus is shifting to greenhouse gases. Although greenhouse gases from oceangoing vessels are regulated on an international level, ports are embracing strategies to move toward zero-emission operations. For example, the Ports of Los Angeles and Long Beach's 2017 update to their Clean Air Action Plan calls for all port terminal equipment to be zero emissions by 2030 and all heavy-duty trucks serving the port to be zero emissions by 2035. The Environmental Protection Agency's Ports Initiative provides technical resources for all ports to utilize in developing and implementing specific emission and greenhouse gas reduction goals.

11.8.2 Improving Resilience to Supply Chain Disruptions

As vital nodes in the movement of goods, ports are working to improve their resilience. Many ports have developed plans to address the impacts of global climate change, particularly sea level rise. Seaports can be disrupted by extreme weather, natural or man-made disasters, infrastructure failures, labor disputes, cyberattacks, and recently a global pandemic (COVID-19). During these events, ports may be unable to perform their primary function, or they may experience high levels of congestion. The global pandemic of 2020–2021 dramatically impacted our nation's seaports. The efficiency of the supply chain which has allowed retailers to limit their inventories by using a "just in time" supply chain model resulted in significant shortages of essential goods during the COVID-19 crisis. Vessels and seafarers were stranded in ports or at anchorage at the start of the pandemic, and a shutdown of transportation services stranded individuals far from home. Reduced demand for consumer goods was replaced by sharp increases in demand. Dock workers impacted by the virus coupled with a greater than normal demand for consumer goods led to major congestion at the nation's ports, with hundreds of vessels waiting at anchor to enter ports around the country in the first half of 2021. To improve their resilience, importers and retailers are likely to diversify their supply chains while reconsidering a switch in inventory strategy from "just in time" to "just in case."

11.8.3 Improving Energy Supply and Distribution

As part of the global supply chain, ports need to accommodate the needs of the next generation of vessels that will call at their port. Large container ships powered by LNG are already moving into the transpacific trading lanes. The development of new fuels to meet the 2050 greenhouse gas goals of the IMO will mean that ports need to consider their ability to provide new types of fueling services. The marine fuels of the future include cleaner fossil fuels and biofuels, hydrogen, and ammonia. What combination of fuels and zero-emission strategies will be used by vessels will depend on location. Yet it will be the port that often has to address the public perception of any risks associated with new types of fuels used by ships that call at their port. Ports have always provided support services for offshore industries, particularly offshore oil development. Going forward, ports will provide the landside services for the development of wave energy and offshore wind farms.

11.8.4 Increasing Visibility and Flow of Information

The maritime sector has lagged behind other transportation modes in data integration and digitization. The lack of visibility into the maritime supply chain is often cited as an obstacle to improving the efficiency of the flow of cargo. In April 2019, the Convention on Facilitation of International Maritime Traffic made it mandatory for ships and ports to exchange data electronically. The treaty also promoted the use of a "single window" for electronic transmission of data for participants in a port community's supply chain. Ports that have implemented single windows or other electronic data interchange systems among their supply-chain partners are referred to as "smart ports." The COVID-19 crisis demonstrated clearly how the maritime transport sector was hampered because of a continued reliance on systems that are not fully integrated. The harmonization of data standards that will optimize port calls and movement of cargo is necessary to improve efficiency and can be expected to be a primary focus of seaports and their customers in the coming decade [19].

The role of seaports and the expectations of the level of service they provide have changed in the past and will continue to change in the future. The balance of their ownership and operation between public and private entities, their vulnerable locations on exposed coastlines with often critical habitats, their exposure to global economic volatility, and their massive importance to the movement of goods and people have created quite a complex and competitive environment to operate. The traditional focus of ports was on the development of facilities to handle cargo or passengers while also serving the economic needs of the surrounding community. There is growing focus on the port-city interface as a place for the public to experience, recreate, and learn about their local ports. Today's gateway ports also recognize the need to organize around their role in the global supply chain. No matter the outcome, the role that seaports play in supporting the nation and future transportation needs will adapt and continue to evolve.

References

1. Adland R., Jia H, Strandenes SP (2017) Are AIS-based trade volume estimates reliable? The case of crude oil exports. Maritime Policy & Management 44(5):657–665.
2. American Association of Port Authorities (AAPA) (2019) Port Planning and Investment Toolkit. U.S. Department of Transportation and Maritime Administration and AAPA [DTMA-91-H-2013-0004]. https://aapa.cms-plus.com/files/PDFs/Toolkit/Final%20toolkit.pdf. Acceessed 19 Apr 2021
3. Andersson P, Ivehammar P (2017) Green approaches at sea – The benefits of adjusting speed instead of anchoring. Transportation Research Part D: Transport and Environment, 51:240–249.
4. Brooks, Mary R. (2011), Seaports, *Intermodal Transportation: Moving Freight in a Global Economy*, Chapter 9 in L. Hoel, G. Giuliano, and M. Meyer (eds), Washington: Eno Transportation Foundation, Inc, 270-301.
5. Brooks M., & Faust P (2018) 50 Years of Review of Maritime Transport, 1968-2018: Reflecting on the Past, Exploring the Future (No. UNCTAD/DTL/2018/1).
6. Brooks, M. and Knatz, G. (2021) Seaports. In: Vickerman, Roger (eds.) International Encyclopedia of Transportation. vol. 3, pp. 299-304. UK: Elsevier Ltd.
7. Brooks, M. and Pallis A. (2011) Port governance. In W. T. Talley (Ed.) *Maritime Economics- A Blackwell Companion* (pp. 491-516). Blackwell Publishing, Ltd., Oxford, GB.
8. Bureau of Transportation Statistics (BTS) (2021) Port Performance Freight Statistics in 2019: Annual Report to Congress 2020. U.S. Department of Transportation. Washington, DC. https://rosap.ntl.bts.gov/view/dot/54022. Accessed 18 Apr 2021.
9. Dong Yang, Lingxiao Wu, Shuaian Wang, Haiying Jia & Kevin X. Li (2019a): How big data enriches maritime research – a critical review of Automatic Identification System (AIS) data applications, Transport Reviews, DOI: https://doi.org/10.1080/01441647.2019.1649315
10. Congressional Research Service (2020) Distribution of Harbor Maintenance Trust Fund Expenditures. In Focus. Washington, DC. https://crsreports.congress.gov/product/pdf/IF/IF11645. Accessed 19 Apr 2021
11. Eguíluz, V. M., Fernández-Gracia J, Irigoien X et al. (2016) A quantitative assessment of Arctic shipping in 2010-2014. Scientific Reports, 6:30682.
12. Emmerson C, Stevens P (2012) Maritime choke points and the global energy system. Charting a Way Forward, Chatham House Briefing Paper, London, 4.
13. European Seaport Organization (2020) ESPO Environmental Report 2020, Accessed May 21, 2021. https://www.espo.be/publications/espo-environmental-report-2020Environmental.Report
14. Farhadi N, Parr S, Mitchell K et al. (2016) Use of nationwide automatic identification system data to quantify resiliency of marine transportation systems. Transportation Research Rec-ord: Journal of the Transportation Research Board 2549: 9–18.
15. Ge J, Zhu M, Sha M et al (2019) Towards 25,000 TEU vessels? A comparative economic analy-sis of ultra-large containership sizes under different market and operational conditions. Maritime Economics & Logistics, https://doi.org/10.1057/s41278-019-00136-4
16. Hensel T, Ugé C, Jahn C (2020) Green shipping: using AIS data to assess global emissions [Grüne Schifffahrt: Verwendung von AIS-Daten zur Bewertung globaler Emissionen]. In Nachhaltigkeits Management Foruml Sustainability Management Forum 28(1):39-47
17. Inmarsat. 2018. "Industrial IOT at Sea". https://www.inmarsat.com/en/insights/maritime/2018/industrial-iot-at-sea.html. Accessed 20 Apr 2021
18. The Insight Partners (2020a) Maritime Analytics Market Forecast to 2027 – COVID-19 Impact and Global Analysis By Application (Optimal Route Mapping, Predictive and Prescriptive Analytics, Pricing Insights, Vessel Safety and Security, and Others) and End User (Commer-cial and Military) and Geography. Research and Markets. https://www.researchandmarkets.com/reports/5178588/maritime-analytics-market-forecast-to-2027? Accessed 22 Apr 2021

19. International Association of Ports and Harbors, Policy Statement dated June 01, 2020. Retrieved from http://www.iaphworldports.org/news/7797 on March 13, 2021.
20. Knatz, G. (2017). How competition is driving change in port governance, strategic decision-making and government policy in the United States. *Research in Transportation Business and Management* Vol. 22 67-77.
21. Notteboom, T., Knatz G., Parola F., Notteboom, T, Knatz G., Parola F. (2018a) Port co-operation: types, drivers and impediments, *Research in Transportation Business and Management*, 26:1-4.
22. Lim S, Pettit S, Abouarghoub W et al. (2019) Port sustainability and performance: A systematic literature review. Transportation Research Part D: Transport and Environment, 72:47-64.
23. Longépé N, Mouche AA., Goacolou M, et al. (2015) Polluter identification with spaceborne ra-dar imagery, AIS and forward drift modeling. Marine Pollution Bulletin, 101(2): 826–833.
24. Marchal V, Dellink R, Van Vuuren D et al. (2011).OECD environmental outlook to 2050. Or-ganization for Economic Co-operation and Development, 8:397-413.
25. Mongelluzzo B (2020) Increasing vessel sizes a red flag for US ports. Journal of Commerce Maritime News. https://www.joc.com/maritime-news/container-lines/increasing-vessel-sizes-red-flag-us-ports_20201221.html. Accessed 20 Apr 2021
26. Murnane J (2017) Ports and shipping: the need for solutions that cross lines. McKinsey & Company.   https://www.mckinsey.com/industries/travel-logistics-and-infrastructure/our-insights/ports-and-shipping-the-need-for-solutions-that-cross-lines#. Accessed 20 Apr 2021
27. National   Resources   Defense   Council   (2014)   *Harboring   Pollution: The   Dirty Truth   about   U.S.   Ports*, Accessed   May   13,   2021   https://www.nrdc.org/resources/ harboring-pollution-dirty-truth-about-us-ports.
28. National Oceanic and Atmospheric Administration (NOAA). (2020) PORTS (Physical Oceanographic Real-Time System). National Ocean Service, Center for Operational Oceanographic Products and Services. https://tidesandcurrents.noaa.gov/ports.html. Accessed 22 April, 2021
29. Notteboom, T., Knatz, G., Parola, F. (Eds.) (2018b) Port co-operation: types, drivers and impediments, Vol 26, Special Issue *Research in Transportation Business and Management*.
30. Port of Los Angeles. (2020) *2019 Inventory of Air Emissions*, Technical Report, APP# 191122-551   A.   https://www.portoflosangeles.org/environment/air-quality/air-emissions-inventory. Accessed 15 April, 2020.
31. Rodrigue J-P. (2010) Maritime Transportation: Drivers for the Shipping and Port Industries" In-ternational   transport   Forum.   Paris,   France.   https://www.researchgate.net/publica-tion/238689612_Maritime_Transportation_Drivers_for_the_Shipping_and_Port_Industries. Accessed 20 Apr 2021
32. Rodrigue J-P. (2020) The Geography of Transport Systems, Hofstra University, Department of Global Studies & Geography, Section 6.3 Port Terminals, https://transportgeography.org.
33. Sanchez R, Perrotti D., Gomez Paz A (2020) Ongoing challenges to ports: the increasing size of container ships. Facilitation of Transport and Trade in Latin America and the Caribbean. Bulletin 379:3. 1564-4227. https://www.cepal.org/sites/default/files/publication/files/46457/ S2000485_en.pdf
34. Sánchez R, Perrotti D, Fort, A (2021). Looking into the future ten years later: big full container-ships and their arrival to South American ports. Journal of Shipping and Trade, 6(1), 1-20.
35. Schultz MT, Bourne SG (2019) Using Automatic Identification System (AIS) Data to Assess Col-lision and Grounding Risk in US Coastal Ports. ERDC Vicksburg United States.
36. Scott R, (2014) An exhilaration expansion: the world merchant ship fleet's heady growth over the past decade. Greenwich Maritime Institute, Maritime Publications. https://maritimeat-greenwich.wordpress.com/2014/11/03/an-exhilarating-expansion-the-world-merchant-ship-fleets-heady-growth-over-the-past-decade/ University of Greenwich. Accessed 20 Apr 2021
37. Scully BM, Chambers KF (2019) Measuring Port Disruptions with Automatic Identification Sys-tem Data. American Society of Civil Engineers Ports 2019: Port Planning and Development, 311-321.

38. The Insight Partners. (2020b). "Maritime Analytics Market Forecast to 2027 – COVID-19 Impact and Global Analysis By Application (Optimal Route Mapping, Predictive and Prescriptive Analytics, Pricing Insights, Vessel Safety and Security, and Others) and End User (Commercial and Military) and Geography. Research and Markets. https://www.researchand-markets.com/reports/5178588/maritime-analytics-market-forecast-to-2027?
39. Treece ZR, Smith MD, Wierschem NE et al. (2015) USACE SMART Gate: Structural Health Monitoring to Preserve America's Critical Infrastructure. Structural Health Monitoring.
40. U.S. Committee on the Marine Transportation System (CMTS) (2019a) Federal Funding Hand-book for Marine Transportation System Infrastructure. Washington, D.C., p. 60
41. U.S. Committee on the Marine Transportation System (CMTS) (2019b) A Ten-Year Projection of Maritime Activity in the U.S. Arctic Region, 2020-2030. Washington, D.C., 118 p. Accessed 18 Apr 2021
42. Yang D, Wu L, Wang S et al (2019b) How big data enriches maritime research – a critical review of Automatic Identification System (AIS) data applications, Transport Reviews. DOI: https://doi.org/10.1080/01441647.2019.1649315
43. Woo SH, Pettit S, Beresford A (2011) Port evolution and performance in changing logistics environments. Maritime Economics and Logistics 13:
44. Yoshitani, T. (2018) PNW Seaport Alliance: Stakeholder's benefits of port cooperation. *Research in Transportation Business and Management* 26: 14-17. 250–277 https://doi.org/10.1057/mel.2011.12
45. United Nations Conference on Trade and Development (UNCTAD). 2018 "Review of Maritime Transport". UNCTAD Secretariat, Geneva Switzerland. UNCTAD/RMT/2018
46. American Society of Civil Engineers (ASCE). (2017). 2017 Infrastructure Report Card: A Comprehensive Assessment of America's Infrastructure. www.infrastructurereportcard.org
47. American Society of Civil Engineers (ASCE). (2021). 2021 Report Card for America's Infrastructure: A Comprehensive Assessment of America's Infrastructure. www.infrastructurereportcard.org
48. Touzinsky, K. F., Scully, B. M., Mitchell, K. N., & Kress, M. M. (2018). Using empirical data to quantify port resilience: Hurricane Matthew and the southeastern seaboard. Journal of Waterway, Port, Coastal, and Ocean Engineering, 144(4), 05018003

Geraldine Knatz is professor of the practice of policy and engineering, a joint appointment between the University Of Southern California Schools of Public Policy and of Engineering. She was the first female director of the Port of Los Angeles, serving from 2006 to January 2014, and prior to that was managing director at the Port of Long Beach. She is past president of the American Association of Port Authorities and the International Association of Ports and Harbors and founding chairman of the World Ports Climate Initiative, the precursor to the World Ports Sustainability Program. Knatz was the driving force behind the development of the Port of Los Angeles Clean Air Action Plan and its Clean Truck Program. Today the plan serves as a model for programs in ports around the globe. In 2014, Knatz was elected to the National Academy of Engineering in recognition of her international leadership in the development of environmentally clean urban seaports. That same year, she was honored with Containerization & Intermodal Institute's Lifetime Achievement Award.

Knatz did not start out with the idea of a career in maritime. She was drawn to the ocean growing up in New Jersey and spending time at the Jersey shore. Captivated by the color and diversity of marine organisms, Knatz originally set out to be a marine biologist. She became one, earning her PhD in biological sciences from the University of Southern California (USC). But along the way, she got a master's in environmental engineering, and it was the engineering degree that helped her land her first job working in the environmental office at the Port of Los Angeles. When she wasn't transplanting kelp from Catalina Island to the breakwater of Los Angeles harbor in an effort to restore habitats, she was diving to the floor of the harbor to measure oil pools that resulted from the explosion of the Liberian-flagged oil tanker *Sansinena*. Her career at the Ports of Los Angeles and Long Beach spanned 37 years, despite being told many times that "shipping is a man's world." Knatz became known as a problem solver with the ability to produce results in a highly politicized environment while balancing the needs of diverse stakeholder groups. The opportunity to lead the largest container port in the western hemisphere gave Knatz a national and global platform to drive change in the maritime industry, consistent with her long-held reverence for the environment.

Knatz's remarkable career has been featured in several popular books, most notably in Pulitzer Prize-winning journalist Edward Humes's *Door to Door*, David Helvarg's *The Golden Shore*, and Bill Sharpsteen's *The Docks*.

She has also authored two prize-winning books on the Port of Los Angeles, *Port of Los Angeles, Conflict, Commerce and the Fight for Control* and *Terminal Island: Lost Communities of Los Angeles Harbor*. Knatz is on the board of Dewberry Engineering and vice-chair of the board of trustees of AltaSea at the Port of Los Angeles.

Katherine Chambers is a research scientist with the US Army Corps of Engineers Research and Development Center (ERDC) at the Coastal and Hydraulics Laboratory. For the past 6 years, she has focused on understanding the concepts of resilience as they pertain to the navigation mission of the US Army Corps of Engineers (USACE). She has leveraged her location in Washington D.C. to lead and participate in a variety of interagency teams and initiatives including the Resilience Integrated Action Team (RIAT) as a part of the US Committee on the Marine Transportation System, Transportation Research Board committees on ports and extreme weather, and several World Association for Waterborne Transport Infrastructure working groups focused on both resilience and ecosystem goods and services.

Katherine's initial interest in research and the environment stemmed from childhood summers spent exploring her neighborhood creek and participating in youth fishing tournaments in Central Ohio. She went on to pursue a BS in biology from Wittenberg University in Springfield, Ohio. Her early exposure to undergrad research and summer internships with the US Forest Service inspired her to pursue a career with the federal government that would balance applied research with solutions derived through policies, regulations, and management techniques. After college, Katherine pursued an MS from Purdue University's Ecological Science and Engineering Interdisciplinary Program. Her graduate thesis focused on Asian carp: an invasive species problem that requires an understanding of watersheds, fisheries ecology, state and federal policies, and a variety of expanding mitigation techniques. Exploring this challenging dynamic led her to apply to a Knauss Marine Policy Fellowship sponsored by National Oceanic and Atmospheric Administration in Washington, DC. There, she was matched with a position at the ERDC and realized that USACE was an excellent place to improve decisions through enhancing the resilience of our environment and infrastructure systems. Since then, she has dedicated her career to better understanding how to preserve the critical services that ports and waterways provide while accounting for a wide variety of future uncertainties.

Her work has been published in trade and academic journals, and she has been recognized by her agency through several Department of the Army Achievement Medals for leading the RIAT and for serving in Puerto Rico after Hurricane Maria.

Chapter 12
Tunnels

Sanja Zlatanic

Abstract Throughout human history, tunnel construction has straddled the practical and spiritual (mystical), attaining a status somewhere between science and art. Ancient civilizations independently developed methods for excavating underground structures, and that practice has continued and evolved throughout the ages, right up to the present day, fulfilling many different functions and adapting to meet civilizations' requirements as accelerating population growth demanded. These days, the ever-increasing use of underground space – be it for mining, utilities and communications tunnels, road and rail tunnels, underground mass transportation (metro) systems, underground storage, or (commercial, residential, recreational, and mixed use) subterranean structures – is placing a great strain on the tunneling industry's global resources and has created the need for a wider understanding of the practical uses of this complex construction method.

This chapter explores the practice related to tunnels as facilities, serving many different purposes; the evolving methods used to construct them; and the necessary planning and subsurface discovery works required to manage the design and risks associated with these unique structures, which never fail to captivate the attention of both engineers and constructors alike.

Keywords Tunnels · History of tunneling · Bored tunnels · Mined tunnels · Tunnel design · Tunnel construction · Urban tunnels · Risks to tunneling · Costs of tunneling

S. Zlatanic (✉)
HNTB Corporation, New York, NY, USA
e-mail: szlatanic@hntb.com

12.1 Notable Events in the History of Tunneling

Mankind's first attempts at tunneling were likely made in prehistoric times by people striving to enlarge existing natural caverns for shelter, storage, and other basic uses. This gradually evolved into excavating more complex irrigation and drainage tunnels to improve harvests, as well as underground cisterns, cellars, and passageways to provide storage, safety, and refuge. One of the earliest documented tunnels is the legendary Euphrates River tunnel, in Babylonia, which was a 3000 ft-long brick-lined tunnel that was purportedly built by Queen Semiramis more than 4000 years ago (between 2180 and 2160 BC) to connect the royal palace with the great temple of Belos (one of the great wonders of the world at the time). In an impressive early feat of engineering, construction of the "cut and cover"-style structure was made possible by damming and diverting the river into a series of lakes during the dry season, restoring the river's flow once construction was complete.

Elaborately mined temples, tombs, and underground networks were excavated by ancient civilizations, with notable examples including the temples of Abu Simbel, in Egypt (Fig. 12.1); Lalibela's monolithic underground churches, in Ethiopia; the intricately carved Kailasa Temple at the ancient Ellora Caves, in India (Fig. 12.2); the valley city of Petra, in Jordan; and the underground cities of Cappadocia, in Turkey (Fig. 12.3).

The ancient Greeks and Romans used tunnels and underground caverns extensively, particularly for drainage and water aqueducts but also for many other uses, including temples, catacombs, and even theatres and underground markets. One of the largest ancient tunnels in Naples, Italy – a city with a vast legacy of ancient underground structures beneath its streets thanks to the residence of both Greek and Roman societies and a geology highly suited to tunneling – is the 4800 ft-long, 25 ft-wide by 30 ft-high Crypta Neapolitana road tunnel, which was built over 2000 years ago (Fig. 12.4).

By that time, common surveying methods had been developed, using "string line and plum bobs," and tunnels were progressed by connecting regularly spaced shafts (provided for ventilation) mostly in strong self-supporting rock to avoid the need for excavation support. Such rock was usually broken off, or "spalled," using a

Fig. 12.1 The ancient temple of Ramesses II at Abu Simbel, Egypt

Fig. 12.2 The intricately carved Kailasa Temple guards the Ellora Caves, in Maharashtra state, India

Fig. 12.3 One of the underground cities of Cappadocia, Turkey

technique called "fire quenching," which included superheating the rock with fire and then suddenly cooling it, by dousing it with water. Such primitive excavation and ventilation methods (such as waiving a canvas at the top or mouth of a shaft) claimed the lives of thousands of slaves that were used to build these structures.

Following the fall of the Roman Empire, tunnel construction was primarily limited to mining and military fortifications during the Middle Ages. It wasn't until the start of the Industrial Revolution that the next set of advances in the field of tunneling were made. Europe's growing transportation needs during the seventeenth century, as industry and export grew, saw numerous canal tunnels being constructed to transport goods. The first of these was the 515 ft-long (22 ft by 27 ft cross section) Malpas Tunnel (on the Canal du Midi), in Languedoc, France, which was completed in 1681, to link the Atlantic and the Mediterranean. That project introduced the first use of explosives for public works tunneling (gunpowder was placed in holes drilled by handheld iron drills). In the eighteenth and early nineteenth centuries, many more challenging, often brick and masonry-lined, canal tunnels were constructed across Europe and North America.

The use of canals subsided and eventually fell into disuse following the introduction of railroads in about 1830. However, this new form of transportation caused a huge increase in tunneling globally and, in many ways, has continued to do so to

Fig. 12.4 Entrance to the
Crypta Neapolitana tunnel,
in Naples, Italy

this day. Early railroad tunneling was pioneered in England and the United States. The 3 mile-long Woodhead Tunnel, on the Manchester to Sheffield railroad, was the world's longest steam railway tunnel when it was completed in 1845 and was driven from five shafts up to 600 ft deep. It was the United Kingdom's first TransPennine tunnel, preceding the famous Standedge and Totley tunnels, and the human cost of the project was high: 30 people lost their lives, 200 workers were maimed, and 450 suffered some form of injury in the harsh working conditions.

The first railroad tunnel in the United States, the 900 ft-long Staple Bend Tunnel, was built in 1831 for the Allegheny Portage Railroad. Although it only remained open for 21 years, the 36 mile railroad provided an innovative solution in an era of roads and canals as the most common form of transportation. Using a combination of animal-powered towing and steam engine–powered windlasses, fully loaded wheeled river boats and barges were hauled over the steep grades of the Allegheny Mountain, rising almost 2300 ft above sea level at its summit, with both ends connected to the Pennsylvania Canal.

Subaqueous tunneling was largely considered impossible until the tunneling "shield" was developed by Marc Brunel in 1818. Brunel, and his son, Isambard Kingdom, used the shield to construct the Thames Tunnel, a 1300 ft-long carriage and pedestrian tunnel that was excavated 75 ft below the River Thames, in London

Clay. The horseshoe-shaped cast-iron shield consisted of 12 frames, divided into three stories (resulting in 36 cells for workers to excavate from), and was closed at the front with moveable boards. The shield was advanced by screws, which pushed against the finished brickwork lining and impelled the structure forward (a technique echoed by modern-day tunnel boring machines, albeit with hydraulic rams pushing against rings of concrete lining segments). Following several inundations and a 7-year shutdown, the Brunel's completed the tunnel in 1841 and in doing so changed the face of soft-ground tunnelling forever.

In 1869, a second Thames Tunnel was completed in just a year, with an 8 ft-diameter circular shield, designed by James Henry Greathead, and a lining of cast-iron segments. Greathead made subaqueous and soft-ground tunneling practical by mechanizing the shield and adding compressed air pressure inside the tunnel to hold back the water in the ground. In addition, he invented the concept of sprayed concrete grout to stabilize earthworks and a gritting pan that hydraulically injected reinforcing grout into the annulus between the lining and the tunnel wall (techniques that are still in use today). So effective was the "Greathead Shield" that it was used successfully for the next 70 years with very few developments.

Construction of the Hoosac Tunnel (Fig. 12.5) began in 1855 under the Berkshire Mountains, in Western Massachusetts. Although originally planned as a canal tunnel, railroads had rapidly established themselves as a superior mode of transport, and the double-track 4.75 mile-long tunnel, which was 24 ft wide and 22 ft high, was eventually built as part of a new railroad link from Boston to the Hudson River. The project frustrated authorities, requiring 21 years to finish (3 years were originally estimated), at five times the original cost. As the rock proved too hard for traditional hand drilling, the tunnel contributed significant advances in terms of hard rock tunneling methods, including the first large-scale commercial use of nitroglycerin and electric blasting caps. The project also pioneered the use of power drills, initially by steam and later compressed air, and even saw a prototype rock tunnel boring machine (TBM) trialed on the project (unsuccessfully, it drilled 10 ft before breaking down). Again, large numbers of workers lost their lives as a result of construction accidents, with 196 killed, largely due to explosions caused by gunpowder and the more powerful, but less stable, nitroglycerin.

At approximately the same time of Hoosac tunnel construction, works commenced on the 8.5 mile Mont Cenis Tunnel (Fig. 12.6), in the Alps, between France and Italy. Many pioneering techniques were introduced during its construction, which began in 1857 and required 14 years to complete, including the use of rail-mounted drill carriages, air drills, hydraulic ram air compressors, and a complete project camp (including housing for workers, medical facilities, and a mechanical repair shop), a practice still used today for isolated projects. Surveying techniques were also advanced, and ventilation techniques were improved through the use of forced air provided by water-powered fans and an improvised exhaust duct in the tunnel. By 1872, works for numerous notable and challenging Alpine base tunnels had begun, including the 9.5 mile-long Gotthard Tunnel. Many of these suffered from major water inflows and weak rock formations that saw many contractors fall into financial troubles.

Fig. 12.5 The Hoosac Tunnel was the second longest transportation tunnel in the world when it was completed. (Source: Library of Congress)

Fig. 12.6 Nineteenth-century illustration of the Mont Cenis Tunnel's south portal and site camp

The Gotthard Tunnel introduced the first large-scale use of dynamite (a recent innovation at the time), mechanized tunneling machines (a number of which had been developed during the previous decade), and compressed-air locomotives to transport materials in and out of the tunnel. The 12-mile Simplon Tunnel followed, in 1898, and the 9 mile Lötschberg Tunnel, in 1906. Simplon, built 7000 ft below the mountain peak and driven as two parallel tunnels with frequent cross-cut connections, was a site of major tunneling problems including rock bursts from highly stressed rock flying off the walls; very high stresses in weak schists and gypsum with swelling tendencies, mandating a 10 ft-thick masonry tunnel lining; and 130 °F water inflows continuously mitigated by cold spring sprays. The Lötschberg Tunnel witnessed a major tunneling disaster in 1908. Although it was predicted that the tunnel would be built through solid bedrock, while crossing under the Kander River Valley, a sudden inundation of water, soil, gravel, and broken rock filled the tunnel for a length of 4300 ft, burying the entire crew of 25 men. In reality, the bedrock was located at a depth of 940 ft and, at 590 ft, the tunnel had tapped the Kander River, causing large sinkhole at the surface. Later, the tunnel was rerouted about one mile upstream to cross the valley in solid rock, but the lesson prompted major improvements in the geological investigation of future tunnels.

Years before the Lötschberg disaster, Dewitt Clinton Haskin first attempted to build the Hudson Tunnel (Fig. 12.7) under the Hudson River, between New Jersey and Manhattan Island. Having founded the Hudson Tunnel Company, he began construction on the tunnel in 1874, commencing work on a shaft in Jersey City. Dewitt

Fig. 12.7 The Hudson and Manhattan railroad tunnel was the first major underwater railroad tunnel in the United States

Haskin's plan included sealing the tunnel and filling it with 35 pounds of air pressure to oust water and hold the tunnel's iron-plate liners in place. Workers would enter through a concrete wall equipped with an air lock. The compressed air, however, could not keep the tunnel walls sealed; a blowout occurred in 1880 killing 20 workers, then again, in 1882, flooding the work site. These unfortunate events, combined with a loss of financing, halted the project in 1887. After a few restarts, the project was finally completed in 1908 and incorporated into the current PATH (Port Authority Trans-Hudson) rapid transit system.

One of the most infamous projects to encounter major water inflows was the first Tanna Tunnel, on a rail line between Tokyo and Kobe, in Japan, in the 1920s. The 5-mile tunnel was driven through the Takiji Peak and suffered a number of heavy inundations and collapses that each buried dozens of workers (67 men were killed in total). The excavation of a parallel drainage tunnel for the entire length of the main tunnel, along with the use of compressed-air tunneling with shield and air lock techniques, unheard of for mountain tunneling at the time, finally mitigated the problems and the tunnel was completed in 1934.

It wasn't until the early 1950s that "drill and blast" tunneling ceased to be the only viable method for constructing tunnels in rock. There had been previous attempts to develop mechanized rock tunnelling machines, in particular the percussion drill locomotive, used on the Mont Cenis Tunnel, and the cast-iron machine that was fabricated for the Hoosac Tunnel. However, following repeated failures, interest in such machines had faded. This remained the case until 1952, when the Oahe Dam Project, on the Missouri River, in South Dakota, began. One of the Oahe Dam Project's contractors, FK Mittry, had witnessed a rotary pre-cutter being used on the nearby Fort Randall Reservoir Project and approached the designer of the unit, James S. Robbins, to ask if he would develop a similar machine for use on his diversion tunnel contract.

Instead of a rotary pre-cutter, Robbins ended up creating the world's first mechanized rock TBM, the 25.5 ft-diameter "Mittry Mole" (Fig. 12.8). Robbins, and his son, Richard, would go on to cultivate the modern-day hard rock TBMs and, in 1964, also fabricated a 33.8 ft-diameter compressed-air TBM for the Paris RER Metro, in France, which was designed to work below the water table in soft ground. Quite separately, Japanese engineers had also begun to develop mechanized TBMs, and by the mid-1980s, a number of European TBM manufacturers had also established themselves as major players in the TBM market. Nowadays, TBMs (ranging in diameters from 3.3 feet to 58 feet, and growing) dominate the global tunneling industry as a method of excavation and operate in a wide range of ground conditions (see Fig. 12.9 for Robbins' large rock TBM used on Niagara Project, Canada).

Fig. 12.8 The "Mittry Mole," developed for the Oahe Dam Project, in South Dakota, by James S. Robbins, was the world's first mechanized rock tunnel boring machine (TBM). (Source: Robbins)

Fig. 12.9 A 47.5 ft-diameter Robbins TBM "breaks through" on the Niagara Tunnel Project, in Canada, in 2011. At the time, it was the world's largest rock TBM. (Source: Robbins)

12.2 Tunnel Classification

Tunnels are generally classified by the service(s) they provide, their location, the methodology utilized for their construction, and the type of ground they pass through. Man-made tunnels and caverns are generally excavated or mined underground with the overlying material (overburden) left in place; they are "lined," where necessary, to provide temporary support during construction and for permanent support of the finished structure in respect to the surrounding ground. Tunnel excavations are generally initiated (or launched) from the bottom of a vertically excavated opening (shaft) or from the end of a horizontal access trench. So-called "cut and cover" tunnels are constructed by excavating down from the surface, temporary supporting the excavated box opening, constructing a permanent structure, and then covering with backfill. Subaqueous tunnels can be constructed by a number of methods, including the use of "immersed tube" tunnels, where prefabricated reinforced concrete box sections of "tube" are floated to the tunnel site, sunk into a prepared trench in the river or seabed, and covered with backfill. All tunneling and underground construction risks and difficulties tend to increase with the size (diameter or cross section) of the tunnel and are greatly influenced by depth and the natural features of the ground (geography, geology, and geotechnical properties), including groundwater pressure and potential water inflow.

12.2.1 Tunnel Service Classifications

- *Highway tunnels* usually accommodate all types of vehicles permitted on public roads, with the exception of bicycles, horse-drawn vehicles, and occasionally buses and/or large trucks, which may be limited or completely prohibited, as per state and/or local ordinances.
- *Rail tunnels* serve standard railroad vehicles (trains) and must accommodate vehicle clearances; these tunnels often require spatial provisions for electric traction power, which can be delivered through a third rail or overhead catenaries, or both, in the case of shared use by various railroads.
- *Rapid transit tunnels* serve urban and metropolitan rapid transit system trains and must accommodate the particular standards of each transit system.
- *Water and wastewater tunnels (aqueducts and sewers)* convey potable water and sanitary waste and/or storm water, respectively; local conditions and requirements usually result in widely varying sizes and construction methods of these tunnels.
- *Underground caverns* vary significantly in accordance with their end purpose/ service needs and local conditions; they can be used for underground hydroelectric power plants, pumping stations, water treatment plants, underground transit stations, storage, defense facilities, mine processing, manufacturing, vehicle

parking, housing, recreational and commercial use, and a multitude of other functions.

- *Shafts* comprise vertical or inclined excavated openings serving as access to mines or tunnels or for ventilation, emergency egress, utility corridors, drainage, or other specific uses as per the requirements of the facility they serve.
- *Utility and special tunnels* are used to convey electricity and communications cables, water, oil and gas pipes, and other utilities (they can also be mixed use, housing two or more different pipelines or services); tunnels are also often built for specific needs, such as pedestrian tunnels (particularly at airports), different types of access tunnels, conveyance of mining ore, nuclear storage, ship and submarine tunnels, and other specific needs.

12.2.2   Tunnel Location and Methodology Classifications

- *Underwater tunnels* can be constructed by various methods under rivers, harbors, straits, or other waterways to serve any purpose listed above; often, these tunnels are built when clearance requirements, topographic features, land use, environmental impacts, or strategic security objectives preclude use of bridges.
- *Mountain tunnels* are usually constructed for transportation or water transfer purposes.
- *Shallow urban tunnels* are primarily used for utilities, conveying water and wastewater, pedestrian use, and rapid urban transit systems. They are often connected in a network under city streets and surface utilities and are sometimes under or adjacent to public or private properties and facilities.
- *Mined tunnels* are constructed by "drill and blast" methods, via mechanical excavators and tools (sometimes implementing sequential excavation methodologies), or they can be excavated (bored) by tunnel boring machines (TBMs). They require a minimum overburden that is dependent upon the size (diameter or cross section) of the tunnel and the anticipated ground conditions.
- *Cut and cover tunnels* are open cut structures, usually shallow when most economically constructed, and are commonly used for urban road tunnels or to serve as station boxes for rapid transit lines. The construction impacts of this tunneling method within dense urban zones can be significant and, when deemed too disruptive, can drive rapid transit lines deeper in favor of the utilization of mined (bored) stations/tunnels that require a greater overburden.
- *Jacked box tunnels* are prefabricated reinforced concrete box structures that are gradually hydraulically jacked horizontally through the soil; they are very shallow and are generally only used where surface structures cannot be disturbed.

12.2.3   Tunnel Ground Classifications

- *Rock tunnels* are excavated through a firm medium, which can vary from relatively soft marl, shale, chalk, and friable sandstone to very hard igneous rocks such as granite. Rock layers, bedding and jointing, and the presence of groundwater, greatly influence the tunnel methodology, risks, and costs.
- *Soft-ground tunnels* include tunnels in soft, granular, and plastic soils or very weak rock. Again, interlayering, interbedding, and/or perched layers of different materials, and the presence of groundwater, will greatly influence the excavation method selected, as well as the risk and costs.
- *Mixed-face tunnels* can be partially in rock and partially in soft ground (or a mixture of each or both) and represent a particular challenge for tunnel construction, often requiring mixed or multiple excavation methods across different sections of a single project.

12.3   Important Considerations for Tunneling

Considering the broad tunnel classifications noted above, differing subsurface structures often require very different methods of excavation and ground support, based on the ground conditions they are likely to encounter. In contrast to mining, where tunnels are generally designed and built to satisfy a temporary purpose, civil tunneling works demand the assurance of permanent safety and stability, both for the end users of the facility and the protection of adjacent and overlying structures. For this reason, tunnels are usually designed conservatively with a view to long-term service life.

For all tunnels, subsurface conditions (including geology and hydrogeology) play a critical role in determining the construction method utilized. More often than not, the ground actively participates in ensuring (or not) stability of the excavated opening; therefore, the design of tunnels is very much dependent on the subsurface conditions and site situation, the ground characteristics, and the excavation and support methods used. Encountering unanticipated conditions, however, had often been the main contributing factor to failures during tunnel construction that claimed the lives of workers, and caused project schedule extensions along with unacceptable cost increases. In the 1960s, the 10 mile-long Awali Tunnel, in Lebanon, was the site of an enormous inundation of water and sand that filled over 2 miles of the tunnel; this was an event that more than doubled the project schedule.

History has provided important lessons for tunneling professionals, both for those in the tunnel planning and design realm and also for those involved in their construction. There are commonly adopted industry standards and general processes and procedures that must be followed to ensure successful outcomes. The following sections provide a brief summary of common processes followed by the tunneling industry, including planning, subsurface exploration, constructability considerations, design, risk management, and cost considerations.

12.3.1 Tunnel Planning

Tunnel planning and development include project feasibility studies, site and ground investigation, assessment and evaluation of project alternatives, and the identification of preferred tunnel options. Allowing sufficient time and budget to fully investigate alternatives (options) and demonstrate technical viability of the preferred alternative, prior to proceeding to tunnel design and construction, is of utmost importance. In addition, the execution of adequate design to advance selected tunnel alternative (appropriate to the chosen type of project procurement) is a prerequisite for the success of a project.

Throughout the process of selecting tunnel alternatives, environmental requirements need to be satisfied and potential construction impacts minimized, such as traffic congestion, impacts on pedestrian movement, air quality, noise pollution, and aesthetic or visual intrusion. Of particular concern are areas and objects of special cultural and/or historical value that need to be preserved, impacts on natural habitats, and surface or subsurface rights-of-way that need to be maintained. The alternatives should include tunnel route analysis, subsurface data, geological and hydrological conditions, constructability, long-term environmental impacts, seismicity, land use restrictions, tunnel design service life expectancy, economic benefits and life cycle cost, operation and maintenance, safety and security, and a sustainable approach to tunnel design, construction, and service. In dense urban areas with high property values, development air rights may offer significant source of future income to public agencies, and this can be used to offset the construction cost of tunnels, either partially or fully.

Therefore, planning is the key step that ensures the success of all other phases of tunnel project. As noted, one of the first steps in the planning process is environmental planning and decision-making. Environmental planning is the process of evaluating how social, political, economic, and governing factors affect the natural environment. The goal of environmental planning is to come up with a win-win situation for society and the environment. There are three main components of environmental planning, and they are as follows: first, the evaluation of the current status of the natural environment where the planned project is located; second, the vision of a solution that would best serve society and the environment (tunnel alternative analyses); and third, the implementation of the plan. Environmental decision-making can be defined as the process of evaluating the way we go about making choices that impact upon the natural environment; in the tunnel planning domain, this largely relates to creating criteria for the selection of a preferred alternative and its implementation. Environmental planning and decision-making work together to create sustainable outcomes.

In the United States, planning is regulated by laws. A national commitment to the environment was formalized through the National Environmental Policy Act (NEPA) of 1969. NEPA establishes a national environmental policy and provides a framework for environmental planning and decision-making by federal agencies. NEPA directs federal agencies, when planning projects or issuing permits, to conduct environmental reviews and consider the potential impact of their proposed actions on the environment. To meet NEPA requirements, federal agencies must prepare an environmental impact statement, or EIS, which is a document required for the construction of a tunnel or underground project that may significantly impact the environment. A completed EIS is submitted to the Environmental Protection Agency, or EPA, which is an independent federal agency that works to reduce pollution and protect the environment.

The EIS must discuss all aspects of a project that could potentially impact the environment. This includes direct and indirect, temporary and permanent project impacts, mitigation measures identified, public and third-party inputs, and costs. During the preparation of an EIS, the public and any interested parties may provide input. Once the final EIS is created, a public record is prepared by the federal agency. The lead federal agency works cooperatively with other federal and state agencies during the review process. This coordinated review process includes input from the public as well as other agencies to address important environmental and related social issues the project might generate.

12.3.2 Subsurface Exploration

Geology is the single largest factor in the planning, design, and construction of tunnel projects. There is a direct relationship between the subsurface conditions, including geology and hydrogeology, and construction costs and schedule. The purpose of subsurface explorations including geotechnical investigations relates primarily to risk reduction; as more subsurface information is obtained and properly interpreted, reductions in construction risks are materialized. The overall investigation objectives are to characterize the soil, rock, and groundwater conditions and use this understanding to predict ground behavior in relation to tunnel construction. It is well known that a proper subsurface investigation program limits construction claims and provides for increased certainty of the project scope, schedule, and costs.

The tunneling industry widely accepts that the owner "owns" the ground. It is therefore incumbent upon the owner to provide and properly represent the subsurface conditions on a project. In tunneling, this is accomplished using a comprehensive subsurface investigation program and geotechnical data report (GDR). Also, the owner has the ability to set limits on the material properties he "represents" as conditions the contractor should anticipate and price. This is done using a geotechnical baseline report (GBR). The owner should seriously consider providing the engineering rational that went into the tunnel design to potential tunneling contractors. The GBR is one of the key contract documents on a tunnel project to allocate construction risks and provide assumptions and baseline values for ground conditions; it is primarily based on subsurface investigation data obtained during the subsurface exploration process. The document provides the basis for any claim of differing site conditions and must be written by an experienced team, ideally with specialists in engineering geology, geotechnical engineering, tunnel engineering, construction, and contracting law. It sets the baseline values that reflect the owner's desired risk allocation strategy; usually, conservative baselines lead to higher bid prices but fewer claims; adversely, optimistic baselines lead to lower bid prices and more construction claims.

A properly planned geotechnical investigation program consists of (1) planning phase, where subsurface exploration needs, geotechnical data gap analysis, and specific data objectives are identified and presented as a geotechnical investigation plan; (2) data collection phase, where existing data, field explorations, and laboratory testing are identified, collected, and summarized in a GDR; and (3) data interpretation phase, where summarized data is interpreted and baselined in a GBR as interpretative geotechnical profiles, ground characterization, and geotechnical parameters for the design and construction.

There is a correlation between the project development phase and subsurface investigation phase (Fig. 12.10). For design-build projects, for example, it is important to have subsurface investigations and ground interpretation and characterization substantially complete and geotechnical conditions baselined, for proper risk allocation and management before the design-build team is procured (the team would undertake the final design and construct the project).

In order to establish a proper subsurface investigation program, the owner must ensure that its selected consultant follows subsurface investigation guidelines and establishes clear objectives that need to be accomplished, with all data collected at the start of each investigation phase. Consultants should ensure that no unnecessary or duplicate data are collected, that each boring is selected with an objective in mind, that sufficient data are collected to meet design and construction planning needs, that the data collected are of the appropriate type and of acceptable quality, that accepted professional standards of care are met, and that specific issues forming potentially adverse geological conditions are specifically addressed.

It is customary for a tunnel project to establish a subsurface exploration depth of two tunnel diameters below the horizon of the future tunnel. Specific geotechnical properties of interest include soil and rock types and their distribution, soil properties, intact rock properties, rock mass discontinuities, the depth and nature of soil/rock interfaces, zones of weakness and/or groundwater inflow, abrasiveness, ease of excavation, and soil/rock permeability as well as groundwater conditions. An appropriate ground classification system should be established considering geology, engineering properties, anticipated ground behavior, and tunnel support requirements.

The nature, scope, and extent of site and ground investigations should be based on the nature, scope, and extent of the project; its specific location, and the nature of both its geological and hydrogeological settings. The investigations should be planned and procured by qualified and experienced personnel according to established standards and codes of practice that are clearly identified; they should also be

DEVELOPMENT PHASE	INVESTIGATION PHASE
1. Planning / Project Development	**1. Regional Reconnaissance**
• Feasibility studies	• Map studies
• Conceptual design	• Field reconnaissance
• Evaluation of alignment alternatives	• Review of existing data (aerial photos, remote sensing, historical research)
• Environmental assessments	• Initial review of existing subsurface structures & foundations
• Draft Environmental Impact Statement	
• Selection of locally preferred alternative	• Surface-based geophysical surveys
	• Limited field exploration
	• Limited laboratory testing
2. Preliminary Engineering	**2. Localized Investigations**
• Alignment refinement	• Expanded field exploration program
• Construction methodology	• Expanded laboratory testing program
• Definition of requirements for licensing, permitting, real estate, financing schedule	• Initiate monitoring for baseline data (groundwater levels, vibrations)
• Range of potential adverse subsurface conditions	• More detailed field reconnaissance
• Final Environmental Impact Statement / Record of Decision	• Existing subsurface structures and foundations
3. Final Design & Construction Planning	**3. Detailed Investigations**
• Final engineering	• Detailed site- and structure-specific subsurface exploration
• Construction planning	• Laboratory testing to develop robust data set
	• Investigation of potential adverse conditions
	• Data collection to support design
4. Construction & Post-Construction	**4. Detailed Assessments**
• Project completion	• Geologic mapping during construction
• Project operation and maintenance	• Records and evaluation of advance rate, installed support, water inflow, other construction records
	• Comparison of predicted and actual ground behavior
	• Construction and post-construction monitoring

Fig. 12.10 General correlation between tunnel project development and subsurface investigation phases

phased appropriately and identify both man-made and natural hazards (such as methane or radon gases or hydrocarbons) to enable assessment of related construction risks (See Fig. 12.11 for natural gas occurrence near Buffalo in State of New York, United States).

The investigations should provide sufficient information on ground and groundwater conditions, as well as the previous history of the project site, including any subsurface constraints that may impact engineering decisions in relationship to the nature of the future works. Ground contamination or subsurface obstructions (underground facilities, tiebacks, foundations, buried passageways, and others) must be explored and defined for the purpose of identifying different tunnel methodologies and safety issues. The tunnel's temporary and permanent support requirements should also be identified, including their technical feasibility, as well as potential impacts to project cost, schedule, and third parties.

These investigations should enable the preliminary design, from which decisions relating to the project's technical and financial viability can be confirmed; in

Fig. 12.11 Natural gas seep, Eternal Flame Falls, Chestnut Ridge Park, near Buffalo, NY

addition, they should enable alignment options to be compared in terms of feasibility and constructability, as well as cost, schedule, and third-party impacts.

In the course of subsurface explorations, existing data, being readily available and at a low cost, should be obtained early on. This information can be instrumental in the course of forming a solid basis for concept designs and may comprise published geological and topographic maps that provide useful regional and site information including surficial (soil) geology; bedrock geology (formations and contacts); faults, folds, intrusions, orientation of faults, bedding, and foliation; outcrops and historic boring, and well locations. Other sources of existing data might include government agencies, academic institutions, aerial photographs, public utilities, professional societies, previous project investigations, and investigations for other projects. Additional data that might be instrumental include well records, historical records of flooding, and information on slope instability, sinkholes, settlement and ground behavior during previous construction projects, property ownership information, and foundations for existing structures.

It is very important to have a clear focus, not only on data collection but also on known or suspected potentially adverse geological conditions that form potential risks for tunneling. Those conditions are site specific and must be addressed early on; as an example, New York State risks to tunneling would include irregular bedrock surface, particularly for mixed-face zones; buried valleys or channel deposits; abrasive, sticky, or expansive soils; soft, compressible soils; boulders in glacial till; mixed materials; exceptionally hard or weak materials; zones of weakness; solutions in carbonate rock; high in situ stresses in rock; areas with potentially high groundwater inflows; aggressive groundwater or soil chemistry, including saline; hazardous contaminants in soil or groundwater; and methane and hydrogen sulfide gases.

Focusing on critical structures that form part of the larger tunneling project and addressing their specific conditions early on may also bring great value in the process of subsurface investigations and ground characterization. Tunnel portals require special attention as they are more prone to construction difficulties; these might be exposed to increased weathering, reduced rock/soil cover, and stress relief, and might experience past or future slope instability (See Fig. 12.12 for illustration of difficult portal conditions related to location and ground characterization of Devil's Slide Tunnel portal). Other critical project structures requiring increased attention are underground rail or transit stations (if any), cross-passages, shafts, adits, and retained excavations.

The results of the subsurface investigations should allow the definition of tunnel alignment, reached on the basis of the geological features of the subsurface within the tunnel horizon, proper ground classifications, definition of tunneling conditions (soft ground, rock, mixed face), geotechnical design parameters in terms of soil and rock properties, groundwater conditions as well as rock mass indices, as applicable.

Properly using this information can significantly aid in advancing the project, by providing the definition for modeling and predictions of ground behavior during and after construction, selection of excavation method(s), constructability assessments, determining the project configuration, scheduling and estimating, and risk assessment and management in terms of occurrence and mitigation of adverse geological

Fig. 12.12 Devil's Slide Tunnel portal on Route 1, in California. (Source: HNTB)

and general subsurface conditions. The laboratory and field testing should always be recorded accurately in accordance with recognized standards and codes of practice and the method of reporting identified and stated explicitly and unambiguously; any departure from standard reporting practices should be clearly annotated to avoid uncertainties in the factual data presentation and reporting.

12.4 Selection and Constructability Considerations

Selection of tunnel type depends primarily on the geometrical configurations required to fulfil the desired function of the structure (defining the tunnel cross section), the ground conditions the tunnel will go through, and the specific site conditions of the tunnel alignment; it also depends on environmental considerations, including public and community concerns, as well as the concerns of affected businesses, property owners, and third parties the project will impact. Considerations of programmatic requirements in terms of cost and schedule also play an important role. For example, a cut and cover tunnel might be a practical solution for an urban rapid transit system; however, the significant surface impact of this construction method might be unacceptable. Therefore, during the planning process, it is of outmost importance to perform a tunnel study as early as possible to select the most appropriate type of tunnel that would meet the specific project and community needs. A possible high-level preliminary tunnel selection process is shown in Fig. 12.13.

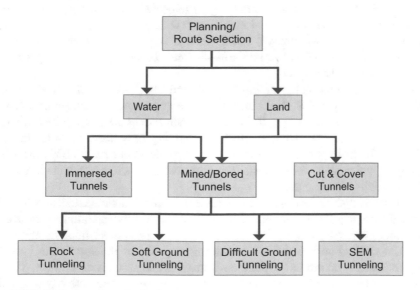

Fig. 12.13 Preliminary tunnel selection process

Fig. 12.14 NYC MTA's Second Avenue Subway cut and cover construction of 96th Street Station. (Source: cc. Metropolitan Transportation Authority/Patrick Cashin)

The main types and methods of tunnel construction in use include "cut and cover" tunnels (Fig. 12.14), which are built by excavating a supported underground box using temporary and/or permanent support, often including a waterproofed permanent concrete structure, and covering it with soil (utilities are temporarily supported while the construction is underway, or relocated in advance of construction). These tunnels can be constructed in place or with precast sections using a "top-down" or "bottom-up" method; the former is implemented when permanently "closing" the box structure with the top slab and reopening the overlying street is expected early in the project.

In self-supporting rock, traditional "drill and blast," excavator mined, or rock TBM-bored (using single shield or main beam/gripper machines), tunnels can be constructed without disturbing the surface. When linings are not required, the tunnels readily display the character of the ground they have been excavated through (Fig. 12.15). In fractured or unstable rock, double-shield rock TBMs are often deployed, which allow for the installation of precast concrete segments in parallel with tunnelling operations.

The sequential excavation method (SEM) is often implemented in situations where the use of a TBM is not practical, such as for shorter-length tunnels or irregular tunnel geometries that serve specific functions. The proper implementation of the SEM method utilizes the self-supporting capacity of the surrounding ground while effectively controlling ground deformations and groundwater inflow. SEM is very versatile and can be implemented for various openings in a variety of ground

Fig. 12.15 Unlined rock tunnel on the Epping to Chatswood rail link, in Sydney, Australia. The tunnels were excavated by two 23.6 ft-diameter Robbins Main Beam TBMs. (Source: Robbins)

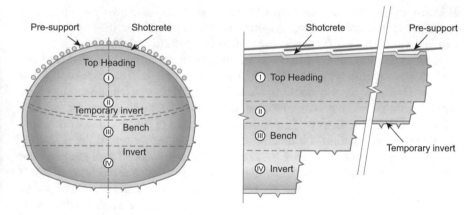

Fig. 12.16 Sequence of SEM excavation in soft ground

conditions ranging from hard rock to soft ground (and "mixed-face" conditions). Excavation is carried out in increments (headings or rounds) in a numerical sequence (Fig. 12.16); these are immediately supported with sprayed concrete (shotcrete), followed by the installation of additional steel reinforcement and shotcrete support elements until a safe and stable opening is provided.

Instrumentation is a key element of this observational method; the behavior of the ground when excavated, as well as surface settlement, is constantly monitored.

Fig. 12.17 SEM excavation on the Devil's Slide Tunnel, in California. (Source: Tunnelling Journal)

Effective on-site supervision of the excavation sequences and timely application of ground support are of the utmost importance (See Fig. 12.17 for SEM method application during Devil's Slide tunnel excavation in California, United States).

Soft-ground tunnels are often excavated through soil using a pressurized-face TBM (either an earth pressure balance (EPB), slurry, or hybrid machine) or by SEM mining. Over the past few decades, the ability of TBMs to excavate through challenging ground conditions has advanced tremendously. Closed, or pressurized-face, TBMs maintain the stability of the ground at the front of the machine while allowing safe working conditions within the shield to construct the tunnel lining (formed by installing rings of reinforced precast concrete segments). There are two main types of TBMs for soft-ground tunneling; earth pressure balance machines (EPBM), best suited to cohesive soils with high clay and silt contents and low water permeability, are in essence operated by pressurizing the excavated soil within the TBM's excavation chamber so that it equals (or balances) the pressure of the surrounding soil and groundwater. Slurry, or mixshield, TBMs are best suited to heterogeneous ground, such as sand and gravels with high water permeabilities, and use a compressed-air "cushion" within the excavation chamber to provide tunnel face support in concert with a bentonite clay fluid (slurry) that mixes with the excavated material and carries it in suspension away from the machine. There are also hybrid machines that can work in both modes with some modification.

As successful TBM tunneling records and achievements have built up and ever more sophisticated TBM control systems have become available, the ability of

Fig. 12.18 (Left) "Bertha," the 57.5 ft-diameter TBM that excavated the Alaskan Way SR-99 tunnel, in Seattle, WA. At the time, this was the world's largest EPBM. (Right) The tunnel lining in place, in January 2017, 2 years before the tunnel opened to traffic. (Source: HNTB)

Fig. 12.19 Virginia's Midtown Tunnel immersed tube construction. (Source: Tunneling Journal)

TBMs to excavate larger and larger diameter tunnels in more and more challenging ground has increased (See Fig. 12.18 for the world's largest TBM deployed in Seattle, Washington State, United States).

Immersed tube tunnels are made from large reinforced, precast, concrete, box-shaped elements (see Fig. 12.19 for Virginia's immersed tube construction, United States) that are fabricated in the dry and then floated to the tunnel site and sunk into a prepared trench below the water, where they are permanently connected to the elements already in place, and then covered with backfill.

Jacked box tunnels are prefabricated, reinforced, concrete, box structures that are hydraulically jacked horizontally through the soil using methods to reduce surface friction (Fig. 12.20); they are usually very shallow and are only used when the surface must not be disturbed (beneath runways, highways, or the embankments of operational railroads).

Constructability evaluations of a potential tunnel type should include the identification and evaluation of the associated hazards and consequent risks for each method. In addition, such assessments should take into account the project geology and the hydrogeology as characterized by subsurface investigations; implementation of various tunnelling methodologies and other methodologies associated with related underground works, such as caverns, shafts, and adits; temporary and

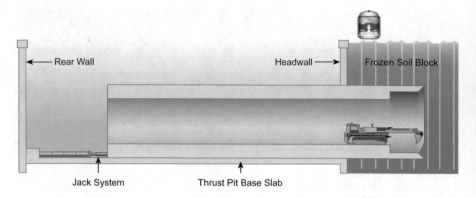

Fig. 12.20 Typical jacked box tunneling sequence under existing rail tracks

permanent ground support systems (e.g., sprayed concrete linings, rock bolts/dowels, precast concrete segmental linings, cast-iron segmental linings, cast in situ concrete linings); and ground and groundwater treatment and improvement measures (e.g., grouting, dewatering, ground freezing) along with their impacts on the public, communities, traffic, utilities, and third parties. Ground movement and settlement at the surface and their impacts should also be considered as well as any impacts on buried structures such as utilities, adjacent tunnels, underground structures, or foundations. Environmental considerations including dust, noise, vibration, and traffic impacts should be considered as well as associated costs, health (including occupational health considerations), safety, and project implications regarding the impact of hazardous materials including gasses, chemicals, other pollutants, or naturally occurring substances that could affect health or project structures and system durability.

Thorough constructability analyses should assess the relative risks of different locations as well as measures to reduce the uncertainties of ground and water conditions at the location chosen. Key factors, for instance, might include the behavior of the rock mass and zones of weaknesses including faults, shear zones, and areas deteriorated by weathering or thermal action; size of rock block between joints; weak beds and zones; groundwater, including flow patterns; and the potential for gas.

For mountain regions, thorough aerial and surface surveys assist constructability assessments as deep site investigation borings require a large upfront investment and prolonged schedule. In general, constructability issues are often approached assuming flexibility and decision-making at the tunnel face (especially for SEM methods) or by mining a pilot bore in advance of the main tunnel.

For large caverns in rock and very large tunnels, such as those constructed as single bores, the size of the opening tends to increase constructability issues exponentially, and adverse geology can make the project impractical or extremely costly. Constructability assessments in these cases should be carefully conducted and contingency measures adopted on a timely basis.

Shallow tunnels are often constructed in soft ground, and as such, site investigation borings become very practical. Hence, most subways involve borings at intervals of 100–500 ft on centers and at the location of every significant structure (stations, shafts, entrances, cross-passages, etc.). It is important to obtain undisturbed samples in order to test the engineering properties of the soil including strength and permeability; observing the long- and short-term groundwater levels is also important. For depths of 30–40 ft, cut and cover is generally believed to be more economical than mined or TBM tunneling; economy assessments usually do not take into consideration "soft costs" related to increased street, traffic, utility, and general community disturbance. Designed as a rigid frame box structure and often housing rapid transit and rail services in urban areas, the cut and cover box tunnel is usually constructed within a neat excavation line using rigid system of braced or tied-back excavation-supporting walls. The depths of these structures can go as deep as 100 ft when required. Recently, wherever possible, dense urban areas are giving way to mined and bored tunnels wherever practical to minimize environmental, community, and businesses impacts of construction. Cut and cover constructions, however, can ease disruption through the use of temporary decking over the excavation to restore traffic as soon as possible while having a utility-support function at the same time.

Tunnel portals are often in soil or in weak and/or weathered rock. They are shallow and relatively easily accessible for geotechnical investigations. In the past, they were often insufficiently explored, and this resulted in a high percentage of portal failures. For example, in 1967, on the 5 mile Oso Tunnel, in New Mexico, excavations had begun well in hard shale, but 1000 ft in from the portal, a buried valley filled the excavation with water-bearing sand and gravel, which buried the excavator and caused a 6-month delay. Similar adverse subsurface conditions have caused numerous costly delays and constructability issues on tunnel projects over the years.

12.5 Tunnel Design

Tunnel design takes place during the planning stage where multidisciplinary assessments are performed and design standards and guidelines followed in accordance with state and local ordinances as well as generally accepted standards of tunnel design and construction practice. Such standards are generally coordinated with other elements of the projects, at grade or aboveground.

The primary consideration for the tunnel design is the extent and type of support required to temporarily secure the surrounding ground safely during excavation and later during the tunnel's permanent service. Determining the tunnel support includes a detailed analysis of the type of support, its capacity to support the ground, and the timing of its installation during and after the excavation process. One of the key factors in timing is the so-called stand-up time, which signifies the time period during which the ground can safely support itself at the tunnel "heading," or face, while the excavation support is installed. In soft ground, stand-up time varies; in loose sand, it can be seconds, while in cohesive clays, this time can be hours. Stand-up time does not exist in flowing ground below the water table; any water seepage would cause the ground to collapse and flow into the tunnel. In rock stand-up times can also be very short, often minutes for weathered and closely fractured rock, up to days in moderately jointed rock, and centuries in nearly intact rock (where the distance between joints in the rock equals or exceeds size of the tunnel) requiring no support. The saying "good ground makes good tunnellers" is partially true: while tunneling in rock is generally preferable to soft ground, major flaws within the rock mass can quickly result in conditions that require the use of the soft-ground support methods and the consideration of a very short stand-up time.

Tunneling causes a transfer of the ground "load" to the sides of the excavated opening through an "arching" effect. This effect at the tunnel heading is usually three-dimensional and provides a ground "dome" effect where the arching happens on the sides of the tunnel but also forward and back. If the arching effect is in place, then the stand-up time is permanent; however, this effect tends to weaken over time while increasing the "load" on the support. The total ground "load" therefore is shared between the support and the ground arch; relative to their stiffness, this is called structure-ground interaction. Often, prompt installation of proper support (depending on nature of the ground being excavated) tends to arrest "loosening" of the ground while maintaining the stability of the excavation – primarily by preserving the strength of the ground arch as the strongest load-bearing element of the tunnel system. Excavation proceeds in stages, and larger openings are excavated in a series of smaller drifts using the same principle.

The support of tunnels has developed over time; previously, timber was used for temporary support and followed by brick or stone masonry as the permanent support (tunnel lining). Subsequently, steel became available for use as temporary support; it is usually encased in concrete for corrosion and fire protection purposes. Steel-rib support with timber blocking, to ensure close contact with the ground, was widely used in rock tunnels historically; inadequate timber blocking was often a cause of significant failures. Modern tunneling engages ground support capabilities actively through the use of rock bolts and dowels, as well as sprayed concrete or shotcrete; this is concrete placed pneumatically and "sprayed" over the excavated area. It is used as temporary support (lining) prior to the installation of a final lining, if required, to locally support the tunnel excavation and stabilize it. The shotcrete is placed in layers and reinforced with welded wire mesh and/or with steel fibers incorporated into the concrete mix; its inside surface can have a smooth finish and is sometimes used as a final lining. These days, tunnels are more often lined with reinforced concrete and sometimes have finished surfaces.

The consideration of tunnel configuration and spatial planning, to house all necessary internal facilities and systems, greatly impacts the size and shape of a tunnel, be it circular, rectangular, or horseshoe (curvilinear) shaped. For hard rock tunnels, the use of a horseshoe shape is common (in all but the weakest of rock), as the flat invert reduces the quantity of rock that needs to be excavated and promptly facilitates the removal of the excavated spoil or "muck." However, the most efficient shape for a tunnel is circular, as this readily supports greater loads, especially in soft ground. The tunnel shape greatly depends on the construction method used as well as ground conditions. For example, cut and cover methods generally lead to rectangular tunnels, as do immersed tube or jacked box methods; tunnels bored by TBMs are generally of circular shape (although noncircular TBMs do exist); and those constructed by drill and blast methods in hard rock, or by mechanical excavation (and SEM construction) in softer formations, can either be horseshoe shaped or circular.

In the course of the tunnel design, it is important to take the operational requirements of the facility into consideration, as well as maintenance and safety requirements. Housing vehicle clearance requirements, for example, must be considered, as well as lighting, ventilation and fire life safety elements, drainage, tracks or roadways, high- and low-voltage conduits, power and water lines, sumps and pumps, emergency egress and maintenance pathways, signals and communication equipment, and others, as required. In addition, tunnels are often protected from the effects of fires or explosions, as per design criteria, that may include elements of active or passive protection or methods of deterrence or preemption, including survey and security measures.

Bored tunnels constructed by TBMs, in soft ground and in rock, usually utilize a precast concrete segmental lining of rings that are erected within the tail shield of the machine as it proceeds. Segments can also sometimes be made of cast iron, steel, or a composite when design or service durability requires, for example, when crossing an active seismic fault. The segments are usually fabricated with gaskets, which make the final tunnel structure waterproof, and are connected together using special bolts and dowels. Some precast segmental linings are treated as a temporary lining within which a cast in place final lining is provided.

Some tunnels, such as road or station/platform tunnels that are visible to the public, are fitted with interior finishes that are mounted or adhered to the final lining and consist of ceramic tiles, epoxy-coated metal panels, porcelain-enameled metal panels, or various coatings. They have the purpose of enhancing lighting and visibility, attenuating noise, enhancing fire durability, and providing an easy-to-clean surface.

12.6   Risk and Cost Considerations

Risk and cost considerations have become a critical part of every tunnel project. Formalized risk assessments inform the cost of the project and assumed cost contingencies directly; they are implemented for every project option considered during the project development stage and are continuously reviewed and updated. Risk evaluations consider subsurface exploration results as they become available including related hazards and subsequent consequences, all presented as formalized risk assessments for each project option identified.

As project options are evaluated for their alignment, as they relate to subsurface conditions encountered, tunnelling methodology, and operational considerations including maintenance, safety, environmental, third-party, and others, the risk assessments assign the risks to a party best equipped to handling them from "cradle to grave" on a particular project. Also, they establish a method of achieving certainty of the project costs by establishing risk-informed project decisions and related cost contingencies. These directly correspond to the project levels of design completion and subsurface exploration as well as the overall hazards facing the project, relating to market, labor, and material availabilities and prices; bonding; insurance availability; and other conditions.

The risk-informed decision process is often implemented on the project to deal with the largest and most impactful project risks in the order of their severity and order of appearance; this process often brings most value when continuously implemented and all the risks tracked and systematically addressed. Moreover, program "sensitivity tests" are often established and undertaken to determine projected cost and schedule overruns in relation to specific risks; there, confidence levels are appropriately assigned considering suitable mitigation measures to prevent or minimize identified risks. If this method is implemented for every credible project alternative, it is highly likely that a technically viable option, with the highest level of programmatic confidence in terms of its execution, would be deemed a preferred option.

In the course of a formalized risk assessment, a risk register is prepared for the preferred project option that includes the associated hazards and perceived risks as well as potential mitigation measures, annotated with detailed explanations for their basis (as determined by previous project studies). The risk register is usually included as reference document in the project's contract procurement stage used to select a preferred constructor.

Examples of major construction risk elements that may be identified early in the program are: underground collapse and associated surface subsidence resulting in damage to existing overlying structures and elements of infrastructure (transportation facilities, utilities); total or partial collapse of the tunnel face due to adverse geology; ground movement from construction resulting in damage to adjacent buildings; construction equipment operations resulting in third-party claims; temporary or permanent tunnel support failure; delays in material or equipment delivery; delay in temporary power or right-of-way acquisition; market saturation driving the bid prices; raising labor or material prices; uncertainties in obtaining required insurance or bonding the project, and others. A proper risk management encompasses planning, engineering, procurement, and construction aspects from safety, security, contractual, and financial standpoints. The risk management process is shown in Fig. 12.21.

Risk assessment involves assessing each risk and providing a response strategy to avoid, control, retain, or transfer the risk elements. New risks are evaluated as soon as they are identified; responsibilities are updated (they belong to owner, designer, or contractor); the risk program is monitored and status reported, and periodic audits are performed. Through risk identification, likely events affecting a project, or its work element, are tabled and their main features documented (cause, description, and effect of each risk) so that they can be addressed early in the design

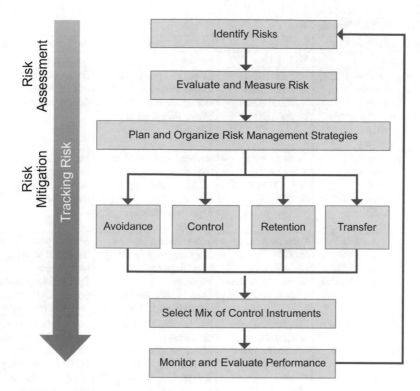

Fig. 12.21 Typical risk management flowchart

Table 12.1 Risk Likelihood Score

Score	Descriptor	Description
1	Non-credible	Judged not likely to occur
2	Improbable	About 1 in 1000
3	Remote	About 1 in 100
4	Occasional	About 1 in 10
5	Probable	More likely to happen than not
6	Frequent	Likely to happen several times
0	Risk not scored	Insufficient information to rank

stage of a tunnel project. The risk evaluation process determines the likelihood of risk events occurring and the associated range of probable outcomes or risk severity. A possible Risk Likelihood Score can be seen in Table 12.1.

The Risk Severity Score, as noted in Table 12.2, is an example of a method used to evaluate the severity of the outcome should the risk event occur.

Table 12.2 Risk Severity Score

Score	Health and safety	Public liability	Railroad operations	Project cost/schedule
0	No impact	No liability	No impact	No impact
1	Reportable injury to worker Multiple non-reportable injuries to workers	Public relations embarrassment Nuisance claims (under $5K)	Minor revenue losses Public relations embarrassment Minor delays up to 1 hour	Extra cost including cost of avoiding delays (under $20K)
2	Potential for major worker injury Minor injuries to public	Property damage under $850K Temp. closure of street	Reroute trains to other locations Medium delays up to 2 hours Significant revenue losses	Moderate extra cost including cost of avoiding delays (under $200K)
3	Worker fatality Potential for public fatality Multiple injuries to public	Property damage $500K Disruption of traffic greater than 4 hours	Minor effects resulting from closure of railroad operations for up to 24 hours Major public alert at Grand Central Terminal	Several-week delay in project (critical path affected) Cost required to get project back on schedule (under $2M)
4	Multiple worker fatalities Public fatality	Property damage under $5M Closure of street for more than 1 day Major effects to structure and building	Major effects to railroad operation Major effects from service outage up to 2 days	Several month delay in project (increase in cost of $5M) Major restructuring of project required Major redesign required including major costs (under $20M)
5	Catastrophic worker and public fatalities	Building collapse Street closures longer than 1 week	Major effect to railroad Longer service outage (3 days)	Potential to close down project

Table 12.3 Risk level assessment matrix

Likelihood Score	Severity Score					
	0	1	2	3	4	5
6	0	6	12	18	24	30
5	0	5	10	15	20	25
4	0	4	8	12	16	20
3	0	3	6	9	12	15
2	0	2	4	6	8	10
1	0	1	2	3	4	5

Legend: | Low | Medium | High |

Combining the likelihood of a risk occurrence with its severity – usually a matrix is provided that defines risk level assessment values (Table 12.3) – forms the basis for establishing a path toward risk management for a particular project.

When creating the matrix, a particular risk nature and origin as well as project impacts should be identified, such as scope, safety, performance, cost, or schedule. The evaluation considers the probability and severity of each risk occurrence and produces an action level in terms of health and safety, public liability, operational aspects, and cost and schedule. Therefore, each risk is assigned an appropriate "risk level" and carries a specific strategy of mitigation measures. Those might include preventive measures, reducing risk exposure by adopting design alternatives, transferring risk to others through contractual arrangements, employing financial protection mechanisms such as bonds and insurance, including risk allowances in engineering cost estimates and schedules, or simply adding contingency allowances to the project budget.

Generally, risks are handled by avoidance, control, retention, and transfer. Using the risk management process, the risks are handled based on probability and severity of risk occurrences. Low-level risks are usually handled by their retention or transfer; medium- and high-level risks should be avoided or controlled through appropriate planning and engineering as well as via selection of appropriate construction methodologies. Risk management is an ongoing process and encompasses the entire project life, through the design, construction, and start-up. Generally, it is not possible to eliminate all risks from a project, so an appropriate risk management process should be established in a timely way to control uncertainties while ensuring successful project completion.

The cost estimate for a tunnel project depends greatly on the risk assessment and management, and at the same time, is a quantitative representation of all resources expected by a contractor to complete the project work as represented by the design documents. These resources include direct trade expenditures specific to project construction activities such as labor, materials, equipment, supervision, overhead, and profit associated with contracted services. They also include the soft costs outside of a typical construction contract that are usually borne by the owner such as right-of-way, rolling stock, insurance costs, and others.

The estimates always reflect an assessment of probable costs of the project, for the given and documented scope prepared using qualified and experienced personnel in accordance with accepted cost estimating principles and practices. The estimated costs often signify a point on a probability distribution curve representative of a range of potential cost outcomes that reflect variability in labor, material, and equipment prices; market; contractor's strategy or negotiation techniques, and other factors.

The estimate is based on a bottom-up, deterministic estimating approach to the construction scope described in the design and contract documents. Pricing is developed for construction items and activities and includes labor, material, equipment, and subcontract cost elements with markups for contractor indirect costs, overhead, and profit. The scope of each construction cost item is determined by an estimator review of the subject design information in which discrete construction elements are identified and evaluated including any and all construction and support activities reasonably associated with each particular element. Pricing for each construction element is based on labor costs, material costs, and equipment costs associated with the construction activity and applied against a direct measurement of quantities. Where design development did not allow for the deterministic pricing of construction elements, a lump sum allowance is usually used.

For each construction activity of a tunnel or underground project, it is critical that the estimator establishes an appropriate crew mix inclusive of the labor (craftsmen, journeymen, foreman) and construction equipment needed to perform the work. Each crew mix offers the labor and equipment costs to be incurred during a work shift and provides a cost per man-hour. The cost per man-hour is then applied against the quantity of materials associated with each work activity divided by the estimated production rate of work that can be achieved by the selected crew. Production factors are always based on historical data available from other similar projects (in the area, if any) with both location and work similarities. In the United States, federal agencies retain credible databases of historically incurred project costs for each state and project location.

12.7 Tunneling in the United States

Today, generally, more people live in urban than rural areas. In the 1950s, 30% of the world's population was urban; it has been projected that by 2050, this percentage will rise to 68%. Cities are becoming dense and overpopulated, yet they continue to grow. This progress is much more pronounced in the United States as the nation's urban population growth began to outpace overall growth more than a decade ago. In 2015, about 83% of the total population of the United States resided in urban areas. It is projected that by 2050 over 87% of the population will live in cities. This trend has been somewhat stalled by the Covid-19 pandemic, with people starting to gravitate toward less-urban areas; however, this drift is expected to be temporary.

Preparing US cities for growing urban populations and improving the effectiveness of transportation infrastructure are of great concern especially when the use of surface space is at a premium. Well-organized infrastructure and transportation solutions have become an absolute necessity, and the sustainable planned use of both surface and underground space is a "must." To improve existing and develop new infrastructure – allowing for economic growth and connectivity – cities, metropolitan areas, and states shall invest; a planned federal infrastructure investment of over one trillion dollars is greatly anticipated in 2021. A large portion of this investment would include public transit, rail, water and wastewater tunnels, and public stations and other complex underground facilities and structures. The diversity of these projects is remarkable across the country; however, the need to make them more affordable and timelier is a common issue.

A recent study (completed in 2021) compares costs and timelines of the projects in the US cities to similar metropolitan areas in Western Europe and Canada. The study finds that the US pays an almost 50% premium (on a per-mile basis) when

Table 12.4 Average construction costs per mile (USD)

Percent tunneled	Non-United States	United States	Difference (US premium)
0–20%	$81M	$118M	46%
20–80%	$286M	$323M	13%
80–100%	$346M	$1.2B ($511M excl. NYC)	247% (48% excl. NYC)

Source: Eno Capital Cost Database

constructing transit projects both at grade and underground (see Table 12.4). The premium for tunnel projects reaches 250% if New York City projects are included. It is notable that these tunnel transit projects are less expensive and more common internationally. In the United States, only 12% of rail and transit projects are located underground, while internationally these projects comprise 37% of all projects in the same category. Moreover, many international tunnel projects cost the same as at-grade projects in the United States. According to the Eno Center study, Metro Line B in Toulouse, France, is 9.3 miles long and was built entirely underground at a cost of about $176 million per mile; Houston Metro's 3.2 mile Green Line is all at grade and costs $223 million per mile.

In addition, international tunnel transit projects are sometimes more complex than those in the United States; they often run through very dense and historic city centers and share the street with light rail vehicles and cars. The US cities such as Seattle, Boston, Los Angeles, and New York have used tunnels to connect densely developed urban areas, and this is being a very complex undertaking. Based on the study, however, whether connecting well- or less-developed dense urban areas, projects in the United States (even those with minimal tunneling) take 6 months longer to construct, on average, than similar projects internationally. Projects that are completely underground take almost a year and a half longer to build than those abroad. It appears that the aphorism "time is money" could be used to somewhat justify the excessive costs that the US underground construction industry is facing.

In fact, there are many drivers impacting transit construction costs and timelines, and these can be grouped into three major categories: governance, processes, and standards. The compounding effects of these underlying factors translate into inefficiencies, and those result in generally fewer projects, shorter transit lines, and underutilized project alignment decisions. Necessary actions and concerted efforts are required at all levels – federal, state, and local – to reverse this trend.

Sometimes municipalities turn away from tunnel projects, considering them too expensive, and eliminate them from consideration at an early stage of project planning. In many cases, the public is supportive of transit investment; when it tends to show opposition, however, demands are usually related to mitigating negative construction impacts, rather than requesting faster construction schedules. There is also a heightened willingness to sue or delay projects, contributing to skyrocketing costs.

Internationally, support for transit and tunneling projects, especially, is much greater; this was unambiguously expressed in the successful completion of a complex subway expansion in Madrid (with the first use of a single bore tunnel for transit that housed stacked station platforms within the tunnel) or the recent expansion of the Paris Metro, which gained the support of much of the metropolitan area.

Often US public agencies delivering a major transit construction project, especially when the project is mainly underground, need support from local agencies and authorities to acquire rights-of-way, obtain local permits to close streets, and relocate utilities. They also need the flexibility to hire talent and provide adequate training to manage projects, construction staff, and consultants, especially for projects that are federally funded (federal projects need to meet federal guidelines and obtain certainty of the project scope, cost, and schedule early to be included in the federal budget).

In addition, the US labor costs are higher; often benefits that make up 20–30% of the labor costs are incorporated into the direct costs of a project, as opposed to other countries, where healthcare and pension plans are nationalized and paid for through taxes. High labor costs are due to federal and state prevailing-wage laws combined with specific work rules negotiated by each specific union. In many Western countries, healthcare costs are covered by government healthcare and retirement plans easing such cost issues.

In fact, the cost of construction, including labor, is over 50% of the overall cost for tunneling projects in the United States, while third-party commitments and soft costs amount to 45% of the cost and are an order of magnitude higher compared to similar projects internationally. Moreover, the labor cost consumes 40–50% of the construction cost alone, and these costs are often driven by labor union rules. The

Fig. 12.22 A new section of the Second Avenue Subway, in NYC, opened on January 1, 2017, at a cost of $2.5Bn per mile. (Source: cc. charleylhasa.com)

Fig. 12.23 The Elizabeth line, a modern London rail line, was built by Crossrail at a cost of $500M per mile. (Source: Crossrail Ltd.)

soft cost may account for 35% of the overall costs for the tunnel project and include owner, program and construction manager, designer, insurance, legal, bonding, and other costs. Right-of-way acquisitions alone are much higher in urban areas where projects encounter site availability restrictions; for instance, New York, San Francisco, and Los Angeles are especially challenging in this respect. In addition, government approval processes, federal, state, local, and third-party approvals and permits, including environmental impacts assessments and community inputs, add sizable cost and schedule increases to projects (See Figs. 12.22 and 12.23 for costs of subway tunnels in New York City and London, respectively).

Other factors that affect the cost and schedule of complex tunnel projects in the United States are procurement methods and payment provisions. Costs on lump sum projects that have a limited level of design and subsurface investigation completed have contingencies that are higher than those encountered on projects with more progressive forms of procurement where the design is advanced to fit the contractor's constructability approach and risks are better understood and shared prior to arriving at a bid price. Consequently, contractors assuming higher risks assume higher contingencies and provide higher bid prices; owners therefore end up paying for the risk in advance, whether or not it materializes.

It is often said that design-build projects should be performed in a highly collaborative manner and contract terms exposed to a shared interpretation; experienced project leaders from the owner, contractor, and designer play conciliatory roles and often form the foundation for project success. Similarly, it has been found that collaboration among all parties and an "open book" approach are extremely useful when progressive procurement methods are implemented (design is advanced prior to facing the price proposal), and more often than not, this yields fruitful results. Public-private partnerships (P3) in the United States are of particular concern as the owners often see these projects a risk-transfer mechanism to a P3 entity; this approach had met major challenges.

Unanticipated subsurface conditions are a cause for many change orders; the process for handling these is often constrained, and it usually takes too long to arrive at a resolution, with multiple parties involved. A stronger design review and risk management process and early investment in underground explorations could help curb these issues.

Market situation, availability of opportunities in the area, increased competition, availability of qualified local labor and qualified contractors, constrained supply chains and a shortage of materials can also play a major role in determining the costs of a tunnel project in the United States. A table indicating comparative analysis of a few cost influencers in the US regions, versus in Europe, India, China, and Southeast Asia, is indicated in Fig. 12.24.

In conclusion, it appears that the improvement of control processes is needed in order to improve successful outcomes for tunneling projects, especially where cost

Cost drivers by region	New York	Rest of US	Europe	Asia
Geology and geotechnical challenges	H	H	H	M
Labor cost and requirements	VH	H	M	VL
Materials and plant	H	H	H	M
Construction risks and contingencies	VH	H	M	L
Soft costs	VH	H	M	L
Safety and environmental issues	VH	VH	M	L
Market structure and situation	VH	H	M	L
Government approvals and permitting	VH	VH	M	L
Stakeholders and community issues	VH	H	M	VL
Project procurement type	VH	H	M	L
Client sophistication and knowledge in underground construction	H	H	M	M
Owner's risk sharing	VH	H	M	M
Labor and other laws	H	H	H	L
Political influence on infrastructure	H	H	L	L

Influence of drivers on the cost
VH = very high; H = high, M = medium; L = low; VL = very low

Fig. 12.24 Comparison of the influence of cost drivers per region. (Source: Tunnel Business Magazine, August 2020)

and timelines are concerned. Controlling soft costs and streamlining the environmental and approval processes would greatly help especially in this age of increased environmental consciousness and a growing focus on socioeconomic inequality. Establishing and following equitable risk sharing between owners and contractors would also bring improvements, along with fair and impartial practices when dealing with changes, claims, and disputes. In addition, unified contractual terms and conditions for underground works should be established as well as equitable project labor agreements that would review labor laws and union rates and regulations.

It appears that a global call toward cutting greenhouse gas emissions is gaining great support among US lawmakers whose political backing for public infrastructure, where tunnels play a major role, is clear. The US tunneling industry has extended this call toward changing approaches to project delivery and improving and implementing best practices to secure more favorable outcomes and serve growing communities better.

Bibliography

1. ASCE, American Society of Civil Engineers, About Civil Engineering, History and Heritage, Historic Landmarks
2. ASCE, American Society of Civil Engineers, ascelibrary.org
3. Eno Center of Transportation, A Blueprint for Building Transit Better, Report https://projectdelivery.enotrans.org/wp-content/uploads/2021/07/Saving-Time-and-Making-Cents-A-Blueprint-for-Building-Transit-Better.pdf
4. Kuesel, Thomas R., King Elwyn H., Bickel John O., Tunnel Engineering Handbook, 2011
5. M.F. Roach, C.A. Lawrence, D. R. Klug, History of Tunneling in The United States, The Society for Mining, Metallurgy and Exploration (SME), 2007
6. Martin Herrenknecht and Karin Bäppler, Tunnel Boring Machine Development, North American Tunneling Conference 2008
7. Martin Herrenknecht and Karin Bäppler, State of the Art in TBM Tunneling, North American Tunneling Conference 2006
8. The International Tunnelling Insurance Group, A Code of Practice For Risk Management Of Tunnel Works, 2nd Edition, May 2012
9. USDOT FHWA, United States Department of Transportation Federal Highway Administration, Technical manual for Design and Construction of Road Tunnels – Civil Elements, Publication No. FHWA-NHI-10-034, December 2009

Sanja Zlatanic, P.E., graduated from the School of Civil Engineering at the University of Belgrade, in former Yugoslavia, in 1988, at the top of her class. Her academic standing led to a job offer prior to graduation at one of the country's most prominent engineering firms, Energoprojekt, which went on to endure the tests of the country's political challenges and economic hardships during the 1990s and still thrives today.

Sanja began her career at Energoprojekt working on international projects and continued in this domain following a move to the United States with her husband, in 1991, shortly before the start of the civil war in Yugoslavia. Within a few short years, the bloody regional conflict led to the complete dismantling of the country and the creation of new states. With her parents trapped in the region until the end of the war, Zlatanic raised her two sons in New York and pursued her carrier with great resolve. Her enthusiasm for engineering, especially complex underground structures, overlapped with her appreciation of being a part of the "American dream," where personal growth is achieved through hard work, persistence, continuous self-improvement, and love, empathy, and the care of others.

In New York, Sanja joined a well-known tunneling company where she exercised all the "tools of the trade" in terms of tunnel design, construction, and a sophisticated approach to risk-based decision-making; she shared these experiences with many prominent national and international experts engaged on the largest tunnel projects in the United States, primarily for transportation.

Over the past 30 years, Sanja has been responsible for managing all phases of major multi-billion-dollar projects, including extensive multidisciplinary joint venture staff, from feasibility and conceptual engineering to final design and construction. Her superb results in project management and multidisciplinary coordination and integration of complex underground structures and tunnels have been witnessed and appreciated by clients and major transit agencies nationally and internationally. Her ability to bring forward state-of-the-art innovative solutions through collaboration with top industry experts had brought value to many mega-transit programs.

As an active member of various tunneling and underground societies, she is well recognized in the profession and has published numerous articles, chaired conference sessions, and made numerous presentations on the design of construction of tunnels and underground facilities at national and international tunneling conferences. She received a Technical Excellence Award and had been recognized as a fellow, for extraordinary carrier-long accomplishments, practicing technical excellence, and championing innovative approaches to solving underground engineering issues, especially in relation to minimizing the impacts of tunneling on densely populated urban environments, communities, and businesses. She is an elected board member and secretary general of Associated Research Centers for Urban Underground Space (ACUUS), an international, nongovernmental organization dedicated to partnerships among experts who research, plan, design, construct, and decide upon the best use of urban underground space.

Since 2016, Sanja has been chair of HNTB's National Tunnel Practice and has led and mentored dozens of tunnel consultants bringing value to multi-billion-dollar tunnel projects, including the independent design verification of the Istanbul Strait Road Crossing Tunnel project, in Turkey; overseeing design and construction issues for the Alaskan Way SR-99 tunnel project, in Seattle, WA; and developing a novel large-diameter single-bore tunnel option for transit in the United States. Her projects have won many industry awards.

Sanja firmly believes in the important role women perform in the tunnel industry; the teams who benefit from diverse participation, especially when solving challenges and exploring innovation, are generally more productive. A few decades ago, when Sanja first chose her career, there were just a handful of female professionals in this realm; today, many young women are interested in the field of tunneling and underground engineering, and they generally find the industry supportive and rewarding. Having never met a woman who expressed a regret about being in tunneling industry, Sanja trusts it is a "happy" career choice as well.

Tunneling and underground projects are among the riskiest engineering practice areas. Sanja trusts solid engineering judgment and practical solutions that always have safety as a primary concern. Throughout the years, she has learned the only way to successfully conquer great challenges is to rely on team contribution as well as having the courage to pursue one's own vision and convictions. Often it is not easy; however, in practicing the perseverance, respect, and camaraderie that is typical of the tunneling industry, it is possible. Courage is also a big component – one should speak their mind, especially when it comes to ideas or solutions that can move a project forward. The tunneling and mining industry is a very warm and gratifying environment and a very conducive atmosphere for women engineers to thrive. This originates from a long-developed culture of caring – the lives of miners are often in the hands of their teammates. This culture has transferred into the consulting industry as well, and a feeling of camaraderie and mutual respect is ever present. "Occasionally, early in my career, I would find myself needing to work harder to 'break the ice' in terms of obtaining a team's trust or having to prove a point – in retrospect, I am very grateful for those instances, as they made me a fast learner, gave me courage to think 'outside the box' and propelled me to develop and put forward innovative solutions," Sanja notes.

Part III
Controlling Water

Chapter 13
Dams

Beth Matzek Boaz

Abstract This chapter introduces the types of dams, including how they are constructed and who owns and maintains them. It explains how the inherent risks are defined and mitigated. Finally, it discusses the future of dams and dam safety, including what individuals can do to ensure that they remain safe and that dams continue to provide their vital benefits as safely as possible.

Keywords Dam · Safety · Flood · Spillway · Hydroelectric power · Irrigation · Hazard · Risk · Levee

13.1 Introduction

"Life as we know it would not be possible without dams." That message was articulated by Denise Bunte-Bisnett, water resource engineer and 2019–2021 president of the board of the United States Society on Dams (USSD). Lori Spragens, executive director of the Association of State Dam Safety Officials (ASDSO), was quick to add "but they have to be safe." While dams may not be your first thought when you think of infrastructure, you probably receive benefits from one or more of America's 91,000 dams. They may contribute to your water supply, protect you from floods, provide you with renewable electricity, or give you a place to camp, fish, or boat. While you enjoy dams' benefits, you may also be concerned for the risks they pose to the natural environment and to those who live nearby. This chapter introduces you to the types of dams, including how they are constructed and who owns and maintains them. It explains how the inherent risks are defined and mitigated. Finally, it discusses the future of dams and dam safety, including what you can do to ensure that you remain safe and that dams continue to provide their vital benefits as safely as possible.

B. M. Boaz (✉)
U.S. Bureau of Reclamation (Retired), Thornton, CO, USA
e-mail: Beth@bethboaz.com

© The Author(s), under exclusive license to Springer Nature Switzerland AG 2022
P. Layne, J. S. Tietjen (eds.), *Women in Infrastructure*, Women in Engineering and Science, https://doi.org/10.1007/978-3-030-92821-6_13

305

13.2   What Do You Think About Dams?

While dam professionals like Bunte-Bisnett and Spragens believe that dams are essential, you may not be entirely convinced. I recently conducted a (entirely unscientific) poll asking people their opinions about dams: are they good, bad, safe, or dangerous? Some responded by extolling the multiple benefits of dams, while others cited their environmental and safety risks. The truth is that dams have both benefits and risks, and each dam must be carefully examined to ensure that its benefits always outweigh its risks.

Those speaking of dams' benefits often related personal stories: reminiscences of growing up on a dryland farm, which was irrigated with water stored behind a dam, or of having electricity at their farm only because of the power produced by a dam. They spoke of witnessing historic floods that could have been prevented by dams. Some expressed their awe at these amazing feats of engineering.

Others noted the massive impacts on the environment, such as fish that can no longer migrate up the traditional streams. As several pointed out, even The Walt Disney Company has portrayed dams as environmental villains: in the 2019 movie *Frozen II*, nature could only return to normal once the dam had been destroyed.

Many mentioned the danger dams can pose to those living downstream, both the immediate threat to human lives and the damage inflicted to health and property.

No matter what your opinions on dams, I hope you'll be convinced that they are vital to sustaining our lifestyles and that with proper safeguards, the benefits far outweigh the risks.

13.3   What Are the Benefits of Dams?

If dams are potentially risky, why are there so many of them? Dams have been constructed since ancient times. The Mesopotamians, the Romans, and the Chinese all built dams and reservoirs to control flooding and ensure a reliable source of water for irrigation and household use.

Today, as shown in Fig. 13.1, dams and reservoirs continue to provide flood control and water supply benefits and now produce valuable recreation, hydroelectric power, and other benefits. Many dams serve multiple purposes.

13.3.1   Recreation

Nearly one-third of the dams listed in the US Army Corps of Engineers' National Inventory of Dams list "recreation" as their sole purpose, and more than 42% list it as one of multiple purposes. There is no question that dams provide prime recreational facilities throughout the world. Boating, waterskiing, camping, picnic areas,

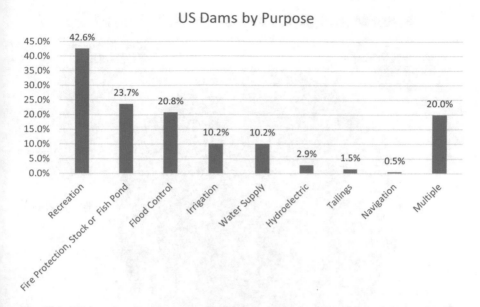

Fig. 13.1 US dams by purpose. (Note: percentages are based on the number of dams and not by storage capacity. Because some dams serve more than one purpose, percentages total greater than 100) [1]

and fishing are all supported by dams and their associated reservoirs, from the giants like Lake Powell and Lake Mead to the smallest neighborhood pond.

13.3.2 Water Supply

According to the United States Geological Survey (USGS) [2] and as shown in Fig. 13.2, in 2015, water users in the United States withdrew 322 billion gallons each day from their water supplies. That's over 1000 gallons for every woman, man, and child. That seems like an impossibly high number until you realize that over 41% of that total is used to generate electricity with steam-driven turbine generators and an additional 39.7% is devoted to agricultural use.

13.3.2.1 Agriculture (Irrigation and Livestock)

Irrigation is critical to the production of livestock and many crops, and dams are a major source of the water used for irrigation. In 2017, the US Department of Agriculture (USDA) Economic Research Service [3] reported that roughly 58 million acres (18.1%) of all harvested US cropland were irrigated, producing crops including corn, soybeans, vegetables, orchard crops, cotton, wheat, and rice, and

Water Withdrawals by Use in the US, 2015

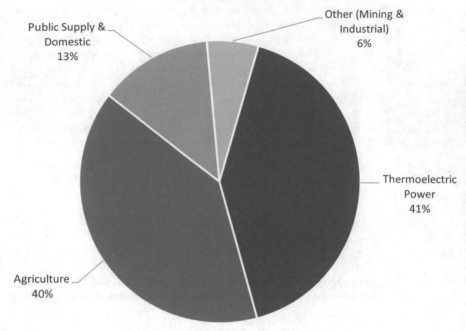

Fig. 13.2 Water withdrawals by use in the United States (2015) [2]

that nearly three-quarters of irrigated acres were in the 17 westernmost contiguous states. According to the Association of State Dam Safety Officials, 10% of American cropland is irrigated using water stored behind dams [4].

13.3.2.2 Public Supply and Domestic Use

Reliable sources of water are also critical to our public water supplies. By storing water during times of high flow, dams ensure a reliable water supply for use when natural flows are reduced, such as in the hot summer months. Of the 13.1% of water withdrawn for public supply and domestic purposes, the majority is for household use. Other uses include public services, such as public pools, parks, firefighting, water and wastewater treatment, and municipal buildings.

13.3.3 Flood Protection

River and stream flooding cause deaths and economic damage every year in the United States. While it's hard to quantify the total impacts of dams in preventing those deaths and economic losses, statistics from just two federal agencies indicate that it is significant.

Since 1948, USDA Natural Resources Conservation Service (NRCS) has assisted local sponsors in constructing 11,845 dams and estimates that these projects provide an estimated $2.2 billion in annual benefits in reduced flooding and erosion damages, recreation, water supplies, and wildlife habitat [5].

The Tennessee Valley Authority (TVA) estimates that its 49 dams prevented $1 billion in flood damages in fiscal year 2020, when the area received 150% of its average rainfall [6].

These examples document that just a fraction of US dams results in billions of dollars of flood damage reduction in a given year; the total benefits are clearly many times higher.

13.3.3.1 How Is a Dam Different from a Levee?

Levees also play a role in protecting the public from stream and river flooding, as described in "Chapter 15: Levees." The basic difference is that dams run perpendicular to the flow of the stream and usually have water behind them; levees run parallel to the stream and are usually dry. Phoebe Percell-Taureau, chief of Dam and Levee Safety for the US Army Corps of Engineers (USACE), explains that "In the Corps, we're trying to pull the worlds of dams and levees closer together, because they're both part of the infrastructure that helps us manage flood risks. What it's all about, ultimately, is finding the right solution for the nation to provide flood risk management." She shared the graphic shown in Fig. 13.3, illustrating how dams and levees both contribute to overall flood risk management. We'll look at the programs it references later in this chapter.

13.3.4 Hydroelectric Power Generation

Hydroelectric power generation (or hydropower) may be the original renewable source of electricity. Humans have harnessed water to perform work for thousands of years: grinding wheat into flour, sawing wood, and powering textile mills and manufacturing plants. Beginning in the 1880s, industries began using flowing water to generate electricity, and by the early twentieth century, US federal agencies began developing power plants that used the water stored behind dams to turn turbines and generate electricity.

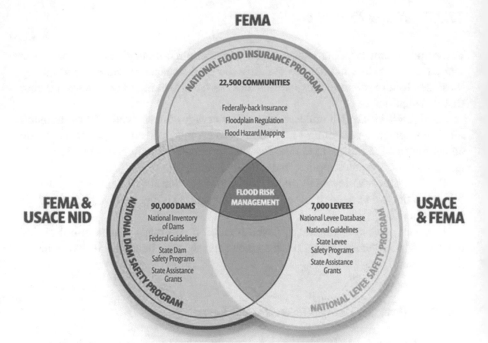

Fig. 13.3 Components of flood risk management

In 2020, the nearly 2,500 hydropower plants in the United States accounted for about 7.3% of total US utility-scale electricity generation and 37% of total utility-scale renewable electricity generation [7]. While federal agencies still operate 49% of the hydropower capacity, public owners (public utility districts, irrigation districts, states, and rural cooperatives) and private owners (individuals, homeowners' associations, farmers, investor-owned utilities, and industrial companies) share the remaining 51% [8].

At present, less than 3% of dams include hydroelectric generation. It is estimated that adding power plants to some of the dams that do not currently generate electricity could increase existing conventional hydropower by 15%. These 12 gigawatts (GW) of new renewable capacity would be enough to power up to 4.8 million homes [9].

In short, dams are an important component of our renewable energy portfolio and could become an even larger component.

13.3.5 Other Benefits

Dams provide benefits beyond those listed above, from storing mine tailings to providing a stable system for river navigation.

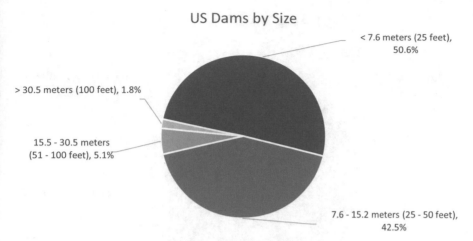

Fig. 13.4 US dams by size. (National Inventory of Dams, 2018)

13.3.6 Multipurpose

Analysis of the data in the National Inventory of Dams finds that one in five US dams serves multiple purposes. For example, a single dam may provide water for irrigation, livestock, and fire protection and also create a pond for recreation and fish habitat.

13.3.7 Who Owns All these Dams?

When we think of dams, we often picture giant federal structures like the Hoover Dam, which stands 221.3 meters high and spans the Colorado River between Arizona and Nevada. The truth (as shown in Figs. 13.4 and 13.5) is that over half of the dams in the United States are under 7.6 meters tall and nearly two-thirds are privately owned. Private owners include irrigation companies and power companies.

13.4 Dam Basics

Stated most simply, a dam is an artificial barrier built across a stream or river to allow storage of water in the resulting artificial lake. The water can be stored in times of high river flow to prevent flooding and released as it is needed. Understanding a few key terms will add to your understanding.

Figure 13.6 shows a typical dam, on a stream that originally flowed from the upper right to the lower left. The dam has blocked the stream from one side of the

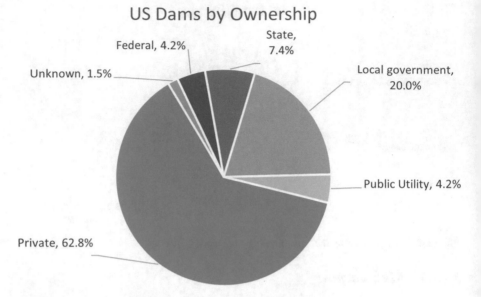

Fig. 13.5 US dams by ownership. (National Inventory of Dams, 2018)

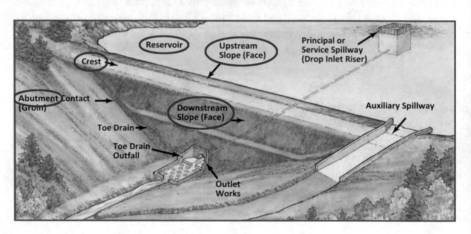

Fig. 13.6 Dam basics. (Adapted from FEMA (2016) [10])

valley to the other to create a *reservoir*. Those sides are referred to as *abutments*. The *upstream slope* of the dam is in contact with the reservoir, while the *downstream slope* is on the dry side of the dam. The top of the dam is called the *crest* and determines the maximum depth of the reservoir.

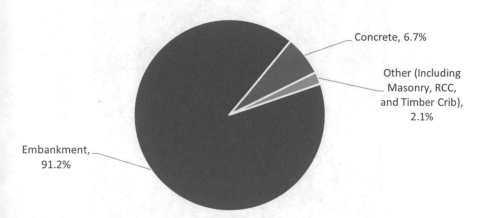

Fig. 13.7 US dams by primary type (USACE, 2018) [1]

13.5 How Are Dams Built?

The purpose of a dam is to block a river or stream, creating a reservoir until the stored water can be released to continue downstream. To perform this purpose successfully, each dam must resist the force exerted by the water stored in the reservoir, which is trying to push it downstream. The deeper the water in the reservoir, the stronger the force it exerts. Dams must be designed so that the shape and materials are able to resist that constant force, as well as to minimize and control water seeping through or under the dam. All dams must also include mechanisms for water to be released to prevent overfilling and for downstream water use. Design of a dam depends on factors including its purpose, the site geography, the foundation conditions, and what materials are readily available.

The oldest known human-constructed dams were built of the natural materials at hand: earth, stone, and rock. This method continues to be the most common today. Over 91% of dams in the United States today are classified as embankment dams, as shown in Fig. 13.7.

13.5.1 Types of Dams

13.5.1.1 Embankment

Embankment dams are well suited for wide valleys; can be built from a combination of soil, sand, and rock; and must have a water-resistant core to control the amount of water seeping through the dam. Embankment dams use the weight of the dam to

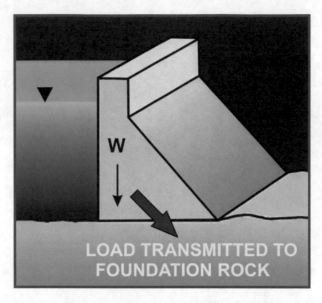

Fig. 13.8 Cross section of a concrete gravity dam (FEMA, 2016) [10]

resist the downstream force by transmitting the force to the foundation. The resistant core is often constructed from clay but can also be made from asphalt, concrete, or other materials.

13.5.1.2 Concrete

Concrete dams are further categorized by their shape, including gravity, arch, and buttress.

13.5.1.2.1 Gravity

Gravity dams, like embankment dams, resist the downstream force of the water by transmitting the weight of the concrete to the foundation. To attain sufficient weight, they require a large amount of concrete (Fig. 13.8).

13.5.1.2.2 Buttress

In a buttress dam, a watertight upstream surface is supported at intervals on the downstream side by a series of buttresses. As in a gravity dam, the weight of the structure transfers the load to the foundation. The addition of downstream buttresses adds an upstream force to keep the dam from toppling over (Fig. 13.9).

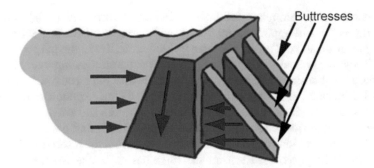

Fig. 13.9 Cross section of a buttress dam. (Adapted from Public Broadcasting Service (2000) [11])

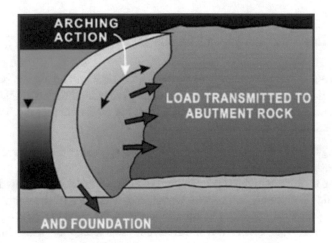

Fig. 13.10 Cross section of an arch dam (FEMA, 2016) [10]

13.5.1.2.3   Arch

Arch dams are curved and transmit the force of the water into the abutments, as well as into the foundation (Fig. 13.10).

Some dams combine more than one type, such as having a central segment that is concrete, flanked by embankment portions.

13.5.2   Spillways and Outlet Works

Since the purpose of a dam is to store water for later use, it must have a mechanism to release water from the reservoir. It must also have a way to ensure that water doesn't reach the crest of the dam. As we'll see later, water overtopping the dam crest is a dangerous cause of dam failure. Water can be released through spillways and outlet works. By definition, the two seem identical. USSD defines a spillway as "a structure over or through which flow is discharged from a reservoir," and an outlet works as "a dam appurtenance that provides release of water (generally controlled) from a reservoir" [12]. The primary difference lies in the purpose.

Spillways are typically designed to prevent water from flowing over parts of the dam that are not designed to convey water, such as over the crest. Spillways may be controlled by gates, or flow may be regulated only by the elevation of the spillway. Spillways can often handle larger flows than outlet works and are designed to regulate the amount of water stored during flood or non-flood conditions.

Outlet works provide normal, controlled releases from the reservoir, allowing the dam to fulfill its purposes. Each outlet works has an intake structure; a conduit through, around, or under the dam; and a series of regulating gates or valves.

13.6   What's the Impact of Dams on the Environment?

The nonprofit, nongovernmental environmental, and human rights organization International Rivers has cataloged a number of issues that the construction of dams, and particularly large dams, can have on the natural environment [13]. The negative impacts cited include the following:

- Blocking fish migrations, in some cases completely separating spawning habitats from rearing habitats.
- Trapping sediments, which previously replenished the habitat provided by deltas, fertile floodplains, and coastal wetlands downstream of the dam.
- Changing the characteristics of the downstream flows by altering natural water temperatures, water chemistry, and river flow characteristics, sometimes making it more suitable to non-native and invasive species than to the native aquatic plants and animals.
- Increasing erosion in the downstream channel.
- Transforming the waters upstream of the dam from a free-flowing river ecosystem to an artificial reservoir habitat, changing characteristics such as temperature and chemical composition, and tipping the balance from native aquatic plants and animals to non-native and invasive species.

13.7   What About the Threat to Human Safety?

The pool of water stored behind a dam can cause massive damage if released suddenly in an uncontrolled manner, and the history of US dams is marked by a number of deadly failures. The deadliest occurred on the afternoon of Friday, May 31, 1889, when a wall of water 11 meters high slammed into Johnstown, PA, killing 2209 residents. The cause of the flood? The failure of South Fork Dam, 14 km upstream on the Little Conemaugh River [14]. The failure was the result of several days of unprecedented rainfall, coupled with several dangerous modifications to the dam [15].

Can such a disaster happen today? We can be reassured by the significant evolution of technical knowledge and protective legislation since 1889. The National Dam Safety Program, administered by the Federal Emergency Management Agency (FEMA), works to "reduce the risks to life and property from dam failure in the United States." In addition, 49 states have implemented dam safety programs. (As of 2021, Alabama had yet to pass dam safety legislation [16].)

Yet dams continue to have issues. The 2021 ASCE Report Card for America's Infrastructure gives the nation's 91,000 dams a grade of D. This rating is largely the result of a lack of specific funding programs to help private dam owners maintain and upgrade their aging dams [17]. Except for a D+ in 2005, this rating has not changed since dams were added to the report card in 1998 [18].

What *has* changed is that when dams do fail, there is less chance of loss of human life. In 2017, the near failure of the spillway of Oroville Dam in California resulted in the evacuation of nearly 200,000 people and required $1 billion in damage repairs but caused no fatalities. Likewise, the 2020 failure of the Edenville Dam and partial breach of the Sanford Dam in Michigan resulted in thousands being evacuated and an estimated $200 million in damages but no deaths.

13.8   How Risky Are Dams?

We live in a world full of risk, and we do what we can to protect ourselves. Dina Hunt, chief seismic hazard engineer with Gannett Fleming and 2021–2022 secretary treasurer of the United States Society on Dams, explains it well: "We live in a world where we're constantly exposed to risks. Every time you wake up in the morning you take risks, and we do things to mitigate those risks; You're driving in a car, you put on your seat belt."

To understand how dams contribute to our daily risk, we first need to understand how "risk" is defined.

13.8.1   Definition of Risk

The US Bureau of Reclamation defines "risk" as the probability of adverse consequences" [19]. It can be defined mathematically as follows:

(The probability of an adverse event) × (The expected consequences of that event)

As an example, your risk of dying by being struck by lightning over your lifetime is as follows:

(The probability that you will be hit by lightning over your lifetime) × (The likelihood that you will die by if you are struck by lightning)

According to the National Weather Service [20], your chance of being struck by lightning in your lifetime is 1 in 15,300, and only 10% of those struck by lightning are killed by the strike, so your risk of dying by a lightning strike are as follows:

$(1/15,300) \times (1/10)$ or 1 in 153,000

In the case of the risks of dam failure, the equation becomes the probability of a dam failing multiplied by the expected consequences. Those consequences could be economic or environmental damage or lives lost. Federal and state dam safety programs aim to reduce the risks posed by dams by reducing both factors: decreasing the probability of failure and minimizing the consequences should a failure occur.

13.8.2 Hazard Potential

As you research dams in the United States, you will see them classified by their hazard potential: high, significant, or low. Seeing a dam listed as "high-hazard potential" can be frightening, until you realize that the hazard potential reflects only the "consequences" part of the risk equation. It means that there are people and structures downstream which would be impacted if the dam failed or was mis-operated. Be assured that receiving a "high-hazard potential" dam classification does *not* mean that the dam is in bad condition and likely to fail.

While specific definitions vary slightly from authority to authority, the classifications typically agree with the descriptions from the ASDSO [21]:

High Hazard: Dams assigned the high hazard potential classification are those where failure or mis-operation will probably cause loss of human life.

Significant Hazard: Dams assigned the significant hazard potential classification are those dams where failure or mis-operation results in no probable loss of human life but can cause economic loss, environmental damage, disruption of lifeline facilities, or can impact other concerns. Significant hazard potential classification dams are often located in predominantly rural or agricultural areas but could be located in areas with population and significant infrastructure.

Low Hazard: Dams assigned the low hazard potential classification are those where failure or mis-operation results in no probable loss of human life and low economic and/or environmental losses. Losses are principally limited to the owner's property.

Once again, it is important to emphasize that a dam's hazard potential reflects the expected consequences of a dam failure and does *not* indicate how likely a dam is to fail. It is also reassuring to note that as shown in Fig. 13.11, more than 65% of all US dams fall into the "low hazard potential" category.

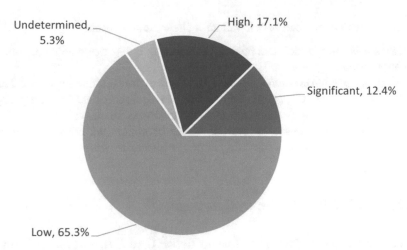

US Dams by Hazard Potential Classification

Undetermined, 5.3%

High, 17.1%

Significant, 12.4%

Low, 65.3%

Fig. 13.11 Hazard potential classification of US dams (National Inventory of Dams, 2018) [1]

13.9 What Causes Dams to Fail?

The ASDSO maintains a database of dam failures and incidents (episodes that, without intervention, would likely have resulted in dam failure). While not all failures or incidents in the United States are included, the database shows 306 dam failures and 579 "incidents" between 2006 and 2019, including events in all 50 states [22].

13.9.1 Mechanisms of Dam Failures

The ASDSO database gives us a good idea of what causes dams to fail. The cause of failure was unknown or under investigation for nearly a quarter of the dams that failed between 2006 and 2019, but for those which a failure mechanism was listed, nearly 70% failed by *overtopping*, that is, by water spilling over the top of a dam.

"Piping" is another common failure mechanism. It occurs when seepage through an embankment dam removes soil particles and progresses to form sink holes in the dam. Seepage often occurs around pipes and spillways, through animal burrows, around roots of woody vegetation, and through cracks in dams, dam appurtenances, and dam foundations.

Other causes of failure include problems with the *spillway*, which may be too small or not strong enough to handle the flows encountered, and *foundation issues* which can cause the dam to settle or crack (Fig. 13.12).

13.9.2   Triggering Conditions

Conditions which can lead to dam failure include the following:

1. *Extreme inflows*, which can lead to the spillway being overwhelmed, overtopping the dam
2. *Seismic (earthquake) activity*, which may deform the structure or its foundation
3. *Inadequate design or construction*, which can also contribute to overtopping if the spillway is undersized or can cause foundation deficiencies
4. *Inadequate maintenance*, where the failure to notice and correct developing issues can allow piping or cracking to progress to a dangerous extent

13.9.3   What Makes Failure More Likely?

As reported by ASCE's 2021 Infrastructure Report Card [17], the average age of the 91,000 dams in the United States is 57 years. According to Amanda Sutter, Dam Safety Program Manager for the USACE's South Atlantic Division and a board member with the USSD, "Aging infrastructure is only a problem if we don't maintain it." So, while age itself does not make a dam more likely to fail, older dams may not be designed to modern standards and should be more carefully inspected and maintained. This is exacerbated by the fact that the owners, the majority of whom are private entities, may not have the expertise or the resources to provide the level of care warranted by older dams (Fig. 13.13).

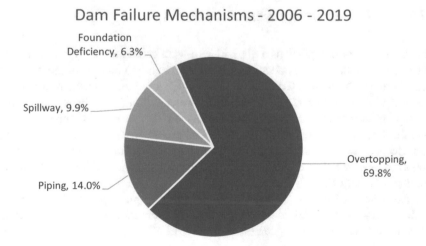

Fig. 13.12 Dam failure mechanisms (2006–2019, ASDSO) [22]

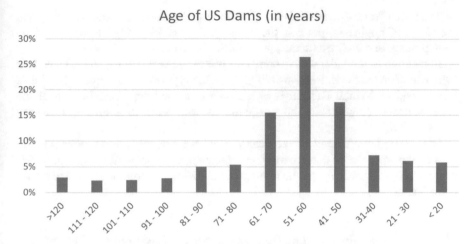

Fig. 13.13 Age of US dams (National Inventory of Dams, 2018) [1]

13.10   How Can We Reduce the Risks Posed by Dams?

This section will discuss the measures that are in place or could be instituted to reduce both the environmental and safety risks posed by dams.

13.10.1   Who's Responsible for Protecting the Environment?

The good news is that environmental regulations control the construction, modification, and removal of dams. The bad news is the regulations vary by the location, purpose, and ownership of the dam.

You may be familiar with the National Environmental Policy Act (NEPA), which requires government agencies to review proposed projects to discover any significant environmental and public health impacts before a decision is made or construction begins [23]. It provides the public with ample opportunities to provide input into the proposed project. More details on the public input process of NEPA can be found on the Environmental Protection Agency's website [24] or by downloading the Council on Environmental Quality's "A Citizen's Guide to NEPA: Having Your Voice Heard" [25]. Federal agencies also can use the NEPA process to comply with other environmental requirements like the Endangered Species Act, the National Historic Preservation Act, the Environmental Justice Executive Order, and other federal, state, tribal, and local laws and regulations.

While NEPA was originally intended to apply only to "major federal actions," projects being conducted by a federal agency, interpretation has since expanded so that it now applies to any project requiring a federal license or permit. For example, every nonfederal hydroelectric project is required to obtain and maintain a license from the Federal Energy Regulatory Commission (FERC).

The Clean Water Act, administered by the Environmental Protection Agency and the Corps of Engineers, protects the water quality of the navigable waters of the United States and their tributaries, through Sections 401, 402, and 404.

In addition to the federal environmental protection regulations, states have their own sets of regulations. Responsibility for dams typically falls within the state's Department of Natural Resources or a similarly named department or under the state engineer. A guide to the programs in each state is available from the ASDSO [26].

Obviously, I can't describe the details of all the programs and regulations. Be assured that approval of construction, modification, or removal of a dam, no matter who owns it, will involve a public input process, giving you the chance to make your voice heard.

13.10.1.1 What More Can Be Done to Reduce the Environmental Risks of Dams?

Our first thought in minimizing the impacts of dams might be to remove them. In cases where a dam is no longer providing the intended benefits, dam removal may be a viable alternative. The USGS maintains a database of dams which have been removed or are being studied for removal. It shows that from 1986 to 2015, at least 126 US dams were removed [27].

But – as we've seen – many dams provide critical benefits that would be lost if the dam was removed, and dam owners can take steps short of removal to minimize the impacts of existing dams. Fish passage and ladder devices can be constructed to help migrating species (such as salmon, steelhead, and shad) move through river systems, while a number of best practices and technologies help preserve the aquatic environment for in-river native fish species. For example, a 2018 blog post by *Scientific American* describes how releases from large dams can be timed to meet agricultural water needs, benefit native fish species, and deter invasive fish species [28].

13.10.2 Who Is Responsible for Keeping Us Safe?

The safety of each dam is the responsibility of its owner. As we saw in Fig. 13.5, the great majority of dams are privately owned; fortunately, even those dams are subject to federal and state regulations. Let's look back at the history of the programs before examining the current status.

13.10.2.1   Federal Dam Safety Programs

13.10.2.1.1   History of Federal Dam Safety Legislation

The 1970s were marked by a string of deadly dam failures: In February 1972, a coal-waste impoundment failed in West Virginia, taking 125 lives and causing more than $400 million in damages, including destruction of over 500 homes. Only a few months later, the failure of the Canyon Lake Dam near Rapid City, SD, took an undetermined number of lives (estimates range from 33 to 237), destroyed 1335 homes, and caused more than $60 million in damages. The 1976 failure of Teton Dam in Idaho caused the loss of 11 lives and an unprecedented amount of property damage totaling more than $1 billion. 1977 brought failures in Pennsylvania and Georgia, each of which was deadly and costly.

In response, President Jimmy Carter directed the US Army Corps of Engineers to inspect the nation's nonfederal high-hazard dams. This "Phase I Inspection Program" lasted from 1978 to 1981. The findings of the inspection program were responsible for the establishment of dam safety programs in most states and, ultimately, the creation of the National Dam Safety Program, which supports dam safety programs in 49 states.

13.10.2.1.2   Current Federal Dam Safety Programs

At present, there are two federal programs in place to reduce dam safety risks.

13.10.2.1.2.1   The National Dam Safety Program

The Federal Emergency Management Agency (FEMA) describes the National Dam Safety Program (NDSP) as "a partnership of states, federal agencies and other stakeholders to encourage and promote the establishment and maintenance of effective federal and state dam safety programs to reduce the risk to human life, property, and the environment from dam related hazards" [30]. The NDSP cannot *require* states to establish dam safety programs, but it establishes guidelines and criteria for state programs. In 1987, FEMA partnered with the Association of State Dam Safety Officials to develop a "Model State Dam Safety Program." The most recent update was in 2007 [31].

Two *committees*, the Interagency Committee on Dams and the National Dam Safety Review Board, advise FEMA on establishing dam safety guidelines and setting priorities.

FEMA administers two *grant programs*. One supports state efforts in the areas of dam safety training, inspections, identification of dams needing repair or removal, and public awareness. The other program provides grants to support technical assistance, planning, design, and construction for rehabilitation of eligible high-hazard potential dams. Grants are only available to states with approved dam safety programs.

FEMA also coordinates dam safety training, publications, and technical assistance. They collaborate closely with the US Army Corps of Engineers to maintain the National Inventory of Dams (NID), a publicly accessible database with information on more than 91,000 dams in the United States [1].

13.10.2.1.2.2   National Flood Insurance Program

The Federal Emergency Management Agency (FEMA) administers the National Flood Insurance Program (NFIP). Its two purposes are to:

1. Enable homeowners, business owners, and renters in participating communities to purchase federally backed flood insurance
2. Encourage participating communities to adopt and enforce floodplain management ordinances to reduce future flood damage

More information on the NFIP is available from FEMA [32].

13.10.2.2   State Dam Safety Programs

As mentioned above, every state except Alabama has established a dam safety program. Research by ASDSO measures how well state programs are aligned with the components of the Model State Dam Safety Program mentioned above. Their latest analysis shows that alignment has increased from 59% in 1989 to 80% in 2020 [33].

13.10.3   How Do State Programs Reduce Risks?

As mentioned previously, dam safety regulations vary from state to state, but most are based on the National Dam Safety Program's "Model State Dam Safety Program" [31]. It provides guidance on reducing both factors in the risk equation, by making it less likely that a dam will fail and less likely that any failure will cause loss of life.

13.10.3.1   Reducing the Probability of Dam Failure

Provisions which reduce the probability of dam failure include the following:

- *Requiring permitting* for the construction, modification, abandonment or removal, or change in ownership of a dam. This allows the state to ensure that dams are designed and constructed to acceptable standards and that plans are in place for ongoing maintenance.

- *Implementing a program of periodic inspection and safety evaluation* for all regulated dams. The program ensures that all dams have a correct hazard potential classification and prioritizes inspections accordingly.
- *Establishing a mechanism for enforcing the requirements.*
- *Ensuring appropriate staffing levels,* with proper qualifications and ongoing training.

13.10.3.2 Minimizing the Consequences of a Dam Failure

The consequences of failure are minimized by defining and practicing the emergency response in advance of any incident. The primary mechanism for accomplishing this is through the development and "exercising" of *emergency action plans* (EAPs) for all high- and significant-hazard potential dams.

A good EAP includes the information that is necessary to detect a potential incident and respond appropriately. It will include the following:

- *Emergency detection, evaluation, and action procedures*, detailing what steps are to be taken to assess the situation
- An *emergency notification flowchart* detailing who is notified by whom, so they can take appropriate actions
- *Inundation maps*, indicating what areas are likely to be flooded in case of a dam incident

EAPs are tested – or "exercised" – regularly, to ensure that all the entities involved, dam owners; local, county, and state emergency response teams; and schools, prisons, and other facilities, have considered potential impacts and practiced how they will work together to respond. It is much more productive to discover a glitch during an exercise than during the stress of an actual event.

Emergency preparedness is improving, with the percentage of state-regulated high-hazard potential dams with an emergency action plan (EAP) increasing from 35% in 1999 to 81% in 2018 [34].

13.11 What Is the Future of Dams and Dam Safety?

In May 2021, I had conversations with six leaders in the dam safety field to get their input on the major challenges and opportunities facing dams today. Their insights were most valuable.

13.11.1 What Are the Most Serious Challenges Facing US Dams?

- *Hazard creep:* Many dams were built in remote rural areas, only to have the downstream population grow, perhaps drawn by the dam itself, increasing the hazard potential from low to significant or even to high. An increase in hazard potential triggers greater responsibilities on the part of dam owners.
- Money: All the dam safety professionals interviewed cited lack of funding as a key concern. Dina Hunt, of the USSD, explains, "Maintaining a dam is expensive and the things we're doing right now to make them safer require even more money." The major rehabilitation required to address deterioration or a change in hazard potential can be well beyond the budget of the many dam owners who are small entities without a steady income. Lori Spragens, of ASDSO, says of this situation, "I don't know how we're going to fix it without some federal involvement." Federal funding would help dam owners meet their obligations to monitor, maintain, and rehabilitate their dams safely, but federal infrastructure funding often overlooks the needs of dams for more visible infrastructure, like roads and bridges. Hunt agrees that when it comes to the safety of our dams, "society as a whole has to take responsibility for it, even though it's not in their back yard."
- *Attracting and maintaining talented engineers:* In the era of booming dam construction, it was easy to hire and retain technical professionals, but as Phoebe Percell-Taureau, of the USACE, points out, "Maintaining dams is just not as sexy as building them. We haven't been able to attract qualified, motivated people to this industry in the same way that we used to."
- *Internal communication:* Within organizations that own, manage, and maintain dams, cultures and personalities can impede communication between disciplines and discourage personnel from voicing critical concerns. There are often no mechanisms in place to ensure that appropriate staff learn the lessons of dam failures and incidents.
- *Public perception of dams:* Entertainment (like the movie *Frozen II*) and the media often paint dams as villainous, making it too easy for us to ignore our responsibility to ensure the continued benefits while minimizing the environmental and safety risks. Alternatively, dams are portrayed as the foolproof solution to all our water-related problems, distracting us from important issues such as climate change and water conservation.
- *Difficulty communicating the risks:* Explaining the risks of dams is a delicate balance. We all need to be aware of the risks posed by dams and need to put them into perspective with other risks we face voluntarily: from driving in traffic to boarding an airplane. Karen Knight, director of Dam Safety and Infrastructure for the US Bureau of Reclamation adds, "The challenge we have is being able to communicate the risk in a way that the owners, responsible parties or decision makers can understand it, without simplifying it so much that we aren't properly framing the risk."

- *Security:* The high-profile 2021 ransomware attacks on infrastructure raise concerns for the vulnerability of dams and their related water and power distribution systems. Dam safety professionals categorize the security risk as real, but not increasing. Since 2005, the Department of Homeland Security has identified dams as one of 17 critical infrastructure and key resources sectors. They established councils to foster communication, coordination, sector-specific expertise, and advice between the government and private owner/operators during protection, response, and recovery activities [35]. Entities are working together to protect dams against physical threats as well as cyberattacks.

13.11.2   What New Developments Are Improving the Safety of US Dams?

- Dam safety organizations encourage networking and a collegial atmosphere. Two of the organizations mentioned in this chapter are the ASDSO and USSD:
  - ASDSO has as its mission: "Improve the condition and safety of dams through education, support for state dam safety programs and fostering a unified dam safety community." They accomplish this mission by working not just with state officials but also with federal dam safety professionals, dam owners and operators, engineering consultants, emergency managers, manufacturers, suppliers, academia, contractors, and others interested in improving dam safety. ASDSO serves as an educator and an advocate for dam safety.
  - USSD is the US member of the International Commission on Large Dams and has as its mission "Empower professionals to advance sustainable benefits of dams and levees for society." Like ASDSO, USSD provides education and advocacy in support of dams and their safety.
  - Organizations like these allow dam safety professionals to network with their peers to share best practices and lessons learned. Sutter, of the USACE and USSD, says of dam safety professionals, "Because we're a small group and we all have very similar concerns and interests, we've developed a camaraderie. Most of the people in the field really care about what they're doing."
- There is increasing support for safe removal of those dams that are obsolete and have outlived their intended purposes. Even traditional supporters of dams are becoming more receptive to dam removal. Spragens of ASDSO has said, "We want the public to be safer. If we can remove thousands of dams that don't need to be there, then let's do it."
- The dam industry is moving toward a "risk management" mindset, so that limited resources are focused on those dams posing the highest risks. Knight, of Reclamation, says, "in the Bureau of Reclamation we've seen a lot of benefit in introducing risk management practices. They're helping us really focus on where to spend the money."

- As mentioned earlier, EAPs are in place for most high-hazard potential dams.
- New technologies allow us to manage dams more efficiently, using computerized systems to measure structural changes and drones to inspect dams and developing three-dimensional virtual models.

13.12   What Can I Do to Help?

13.12.1   Be Aware

- Learn what dams are near you and what risks they pose. Hunt of Gannett Fleming and USSD advises, "Don't put your safety in other people's hands. Educate yourself and have an understanding." A good place to start is with the USACE's NID, a database of 91,000 US dams. You can access the NID at https://nid.sec. usace.army.mil/ and filter it to show all dams in your state and county, even your specific city. The NID will show you each dam's hazard potential, owner, height, and purpose. Clicking on the links, you can see the dam's location and find out when it was constructed, whether an emergency action plan is in place, when it was last inspected, what agency regulates the dam, and other details. At the time of this writing, a new version of the NID was scheduled to be released in November 2021 and will include additional information on USACE dams. USACE will be sharing information regarding the risks associated with its dams as well as measures in place to mitigate those risks. A second component of this risk communication for USACE dams is providing the flood inundation map details. Site users will be able to use the new modernized search interface to easily locate dams. For the USACE dams, the map viewers will also include multiple flood inundation scenarios.
- If there is a dam you regularly encounter near home or while recreating, know what's normal, and if you see something, say something. In the words of the late Dr. Ralph Peck, "An instrument too often overlooked in our technical world is a human eye connected to the brain of an intelligent human being" [36]. Know who owns the dam and how to contact them. Sutter tells of being on a dam inspection in a small town when a citizen called the police to report "trespassers" on the dam. Rather than resenting their interference, Sutter reports, "I really appreciated that somebody thought 'this might be a problem; I'm going to get someone to check it out.'"
- Observe warning signs at dams. "Keep out" signs are posted to keep you away from potentially dangerous areas. Flow conditions at and downstream of dams can change suddenly, so preserve your own safety by obeying all posted warnings.
- In the words of the CBS public service campaign of the 1980s and 1990s, "read more about it," by diving into any of the references cited in this chapter.

13.12.2 Encourage Young People to Pursue Careers in Dam Safety and Those in the Profession to Remain

- As a parent, encourage your children – both girls and boys – to embrace math and science from an early age. Help them avoid arriving at college unprepared to pursue a technical degree.
- As an educator, do your best to identify and encourage promising students at all levels and to make the engineering curriculum interesting and engaging.
- As a professional, provide employees with challenging assignments, mentoring, and a welcoming, flexible workplace.

13.13 What's the Bottom Line?

This chapter has explained the many benefits of dams, the basics of their design and construction, the environmental and human safety risks they pose, and how those risks are mitigated. I hope it has motivated you to take what actions you can to ensure your own safety and that of others and led you to conclude that in most cases, the benefits of dams far outweigh the risks.

Perhaps Hunt said it best: "In today's world, dams are essential; together as a society we must recognize their importance to keep the benefits high and risks low."

References

1. US Army Corps of Engineers (USACE) (2018) National Inventory of Dams. http://nid.usace.army.mil/ Accessed 14 June 2021
2. Dieter CA, Maupin MA, Caldwell RR, Harris MA, Ivahnenko TI, Lovelace JK, Barber NL, Linsey KS (2018) Estimated use of water in the United States in 2015: U.S. Geological Survey Circular 1441 https://doi.org/10.3133/cir1441
3. US Department of Agriculture (USDA) National Agricultural Statistics Service (2017) Agricultural Census https://www.nass.usda.gov/Publications/AgCensus/2017/Full_Report/Volume_1,_Chapter_1_US/st99_1_0009_0010.pdf
4. Association of State Dam Safety Officials (ASDSO) Dams 101 https://damsafety.org/dams101 Accessed 14 June 2021
5. USDA Natural Resources Conservation Service (2019) Watershed Rehabilitation Progress Report https://www.nrcs.usda.gov/wps/portal/nrcs/main/national/programs/landscape/wr/
6. Tennessee Valley Authority (2020) River Management Fact Sheet for FY 2020 https://tva-azr-eastus-cdn-ep-tvawcm-prd.azureedge.net/cdn-tvawcma/docs/default-source/annual-report/fact-sheets/river-management3415427b-214e-47a5-ab2e-e7d82bbcbadc.pdf?sfvrsn=4b324a8a_3
7. US Energy Information Administration (2020) Hydropower Explained https://www.eia.gov/energyexplained/hydropower/#:~:text=Hydropower%20is%20energy%20in%20moving%20water&text=In%202019%2C%20hydroelectricity%20accounted%20for,utility%2Dscale%20renewable%20electricity%20generation

8. National Hydropower Association https://www.hydro.org/waterpower/hydropower/# Accessed 14 June 2021
9. Hadjerioua B, Wei Y, Kao S-C (2012) US Department of Energy An Assessment of Energy Potential at Non-Powered Dams in the United States https://www.energy.gov/sites/prod/files/2013/12/f5/npd_report_0.pdf
10. Federal Emergency Management Agency (FEMA) (2016) Pocket Safety Guide for Dams and Impoundments
11. Public Broadcasting Service (PBS) (2000) Building Big, Dam Basics https://www.pbs.org/wgbh/buildingbig/dam/buttress_forces.html
12. United States Society on Dams (USSD) Glossary https://www.ussdams.org/resource-center/glossary/ Accessed 14 June 2021
13. International Rivers Environmental Impacts of Dams https://archive.internationalrivers.org/environmental-impacts-of-dams Accessed 14 June 2021
14. Association of State Dam Safety Officials (ASDSO) South Fork (aka Johnstown Dam) - PA00000-102 https://damsafety.org/incidents/pa00000-102 Accessed 14 June 2021
15. ASDSO Case Study: South Fork Dam (Pennsylvania, 1889) https://damfailures.org/case-study/south-fork-dam-pennsylvania-1889/ Accessed 14 June 2021
16. ASDSO Alabama Does Not Have a Dam Safety Program https://damsafety.org/alabama Accessed 14 June 2021
17. American Society of Civil Engineers (ASCE) (2021) Report Card for America's Infrastructure, Overview of Dams https://infrastructurereportcard.org/cat-item/dams/ Accessed 14 June 2021
18. ASCE (2021) Report Card History https://infrastructurereportcard.org/making-the-grade/report-card-history/ Accessed 14 June 2021
19. US Bureau of Reclamation (2011) Interim Dam Safety Public Protection Guidelines https://www.usbr.gov/ssle/damsafety/documents/PPG201108.pdf
20. National Weather Service How Dangerous is Lightning, https://www.weather.gov/safety/lightning-odds Accessed 14 June 2021
21. ASDSO Frequently Asked Questions https://damsafety.org/media/faq Accessed 14 June 2021
22. ASDSO Dam Incident Database Search, https://damsafety.org/incidents Accessed 14 June 2021
23. Environmental Protection Agency (EPA) Summary of the National Environmental Protection Act https://www.epa.gov/laws-regulations/summary-national-environmental-policy-act Accessed 14 June 2021
24. EPA How Citizens can Comment and Participate in the National Environmental Policy Act Process, https://www.epa.gov/nepa/how-citizens-can-comment-and-participate-national-environmental-policy-act-process Accessed 14 June 2021
25. Council on Environmental Quality (2021) A Citizen's Guide to NEPA: Having Your Voice Heard https://ceq.doe.gov/docs/get-involved/citizens-guide-to-nepa-2021.pdf
26. ASDSO (2020) Summary of State Laws and Regulations on Dam Safety https://damsafety-prod.s3.amazonaws.com/s3fs-public/files/FINAL%20-%202020%20Update%20State%20Laws%20and%20Regulations%20Summary_0.pdf
27. USGS Dam Removal Information Portal (DRIP) https://www.sciencebase.gov/drip/ Accessed 14 June 2021
28. Chen W (2018) We Can Make Large Dams More Friendly to the Environment. Scientific American https://blogs.scientificamerican.com/observations/we-can-make-large-dams-more-friendly-to-the-environment/
29. ASDSO Learning from the Past: A Snapshot of Historic US Dam Failures https://damsafety.org/dam-failures Accessed 14 June 2021
30. FEMA Dam Safety https://www.fema.gov/emergency-managers/risk-management/dam-safety#:~:text=The%20National%20Dam%20Safety%20Program%20is%20a%20partnership,property%2C%20and%20the%20environment%20from%20dam%20related%20hazards Accessed 14 June 2021

31. The National Dam Safety Program (2007) Model State Dam Safety Programhttps://damsafety. s3.amazonaws.com/s3fs-public/files/ModelStateDamSafetyProgram_July2007_All.pdf
32. FEMA Flood Insurance https://www.fema.gov/flood-insurance Accessed 14 June 2021
33. ASDSO (2019) Dam Safety State Performance Reports https://damsafety-prod.s3.amazonaws. com/s3fs-public/files/State%20Performance%20Reports%20Complete%20-%202019_ Reduced.pdf
34. ASDSO State Performance and Current Issues https://dev.damsafety.org/state-performance Accessed 14 June 2021
35. Department of Homeland Security (DHS) National Infrastructure Protection Plan, Dams Sector https://www.dhs.gov/xlibrary/assets/nipp_snapshot_dams.pdf Accessed 14 June 2021
36. Professor Ralph Peck's Legacy Website https://peck.geoengineer.org/resources/words-of-wisdom Accessed 14 June 2021

Beth Matzek Boaz, P.E., P.M.P., entered Augustana College, Rock Island, Illinois, in 1973, intending to save the environment by becoming an environmental lawyer. As she perused the catalog, she thought the prelaw classes looked tedious. Her astute advisor (the head of the Mathematics Department) looked at her high school transcript and her advanced placement credits in calculus and suggested that she might be better suited to saving the environment as an environmental *engineer*. Before classes started, Beth had become the first female student in Augustana's cooperative program with the University of Illinois. In 1978, she graduated with a BA cum laude in mathematics from Augustana and a BS in civil engineering from University of Illinois. In 1986, after years of night classes, Beth earned an MS in civil engineering from the University of Colorado – Denver.

Beth spent her entire career with the US Bureau of Reclamation. She embraced the variety of roles available within a large organization, starting as a design engineer in the Spillways and Outlet Works Branch. Beth took full advantage of Reclamation's rotation engineer program and worked in several Reclamation disciplines, including spending 5 months inspecting construction of the Twin Lakes Dam and Mt. Elbert pump-generating power plant.

Her other positions included performing designs for the Yuma Desalting Plant and heading the engineering division for the Eastern Colorado Area Office (ECAO) in Loveland, Colorado.

Having been a member of her high school and college speech and debate teams, Beth continued to maintain and refine her communication and leadership skills throughout her career through active participation in Toastmasters International. Those skills came into play when ECAO undertook high-profile dam rehabilitation projects of the Pueblo Dam and the four dams at Horsetooth Reservoir. Beth became the project manager/spokesperson for the projects, attending public meetings, appearing in television interviews, and negotiating agreements with the projects' many public and private stakeholders.

Involvement in the two dam safety projects led Beth to become a program manager in Reclamation's Dam Safety Office. Her final Reclamation assignment was as program manager for the Department of the Interior's Working Group on Dam Safety and Security, striving to establish consistent dam safety policies and practices across six interior bureaus.

Beth has been a licensed professional engineer in Colorado since 1982 and a project management professional since 2009. She discovered the Society of Women Engineers while at the University of Illinois and has remained a member ever since, supporting their vision of "a world with gender parity and equality in engineering and technology." Now retired, Beth remains active in Toastmasters, plays the viola in a community orchestra, and makes presentations on project management and communication.

Chapter 14
Managing Levees in the Modern Age

Tammy L. Conforti and Janey Camp

Abstract Floodplains have served important functions in human livelihoods for millennia. Early settlements in the United States frequently occurred along waterways due to the many benefits offered such as navigation for transport of goods and supplies, fertile soils in the low-lying floodplain areas for growing crops, and simply access to water for irrigation and household or industrial purposes. Over time, it became commonplace to try and control the nuisance flooding that occurred in these areas through the use of physical barriers such as floodwalls or levees. Today, there are approximately 7000 levee systems identified throughout the United States, and communities continue to become more reliant on levees as an important tool for reducing risk to life and property from flooding (U.S. Army Corps of Engineers (2016) National Levee Database, https://levees.sec.usace.army.mil). However, many of those benefiting from these levees are unaware of their flood risks, either from levee failure or overtopping. Challenges also exist with increased development behind levees and a desire to continue to build higher levees – a phenomenon known as the levee effect (i.e., putting more reliance on the levees' ability to perform and protect the public, with a potential unintended consequence of transferring flood risk elsewhere). There is an emerging current opportunity to make forward progress toward a unified, national approach to better manage levees and improve how decisions about levees are made while recognizing other floodplain considerations through a National Levee Safety Program.

Keywords Levees · Floodplain · Risk · Flooding · Floodwall · Hazard · Mitigation · Breach · Resilience · Waterways

T. L. Conforti (✉)
U.S. Army Corps of Engineers, Washington, DC, USA
e-mail: Tammy.Conforti@usace.army.mil

J. Camp
Department of Civil and Environmental Engineering, Vanderbilt University, Nashville, TN, USA

P. Layne, J. S. Tietjen (eds.), *Women in Infrastructure*, Women in Engineering and Science, https://doi.org/10.1007/978-3-030-92821-6_14

333

14.1   The Floodplain and Modern Society's Development

Floodplains have served important functions in human livelihoods throughout human history. Floodplains are the lowlands adjoining the channels of rivers, streams, or other watercourses or shorelines of other bodies of water. They are the lands that have been or may be inundated when the volume of water exceeds what is normally confined within the banks or normal course of the waterbody which is referred to as a flood [2]. Early settlements in the United States frequently occurred along waterways due to the many benefits offered such as navigation for transport of goods and supplies, fertile soils in the low-lying floodplain areas for growing crops, and simply access to water for irrigation and household or industrial purposes. Settling on and developing lands near waterbodies had many benefits but also came with risks. Over time, residents began to realize that flooding occurred, sometimes frequently, in low-lying areas along rivers and their tributaries which would sometimes destroy crops, damage infrastructure, and create additional problems for communities including fatalities. As a result, it became commonplace to try and control the flooding through use of physical barriers such as floodwalls or levees.

Human beings continue to struggle with the risks and benefits of living and working near waterbodies. At times, the approach is combative, trying to use engineered solutions like dams or levees to either control flooding or eliminate it. According to the US Federal Emergency Management Agency (FEMA), 99% of US counties were impacted by a flooding event between 1996 and 2019 [3]. Additionally, flooding continues to be a regular hazard faced by many communities, and it is seen as an increasing threat due to climatic changes.

On the other hand, it is necessary to coexist with flooding in a balanced way. Flooding offers benefits such as groundwater recharge, improved water quality, and replenishment/distribution of nutrients for habitats along waterways and at times assistance in clearing sedimentation for improved water flow. In reality, water must go somewhere, and humans will always occupy some floodplains. Striking a balance of coexistence is challenging because there will always be trade-offs that need to be collectively considered and accepted. Flood mitigation measures are usually tailored and implemented based on localized conditions to obtain localized benefits but could have much more far-reaching effects and impacts. Gilbert F. White, known as a pioneer in floodplain management, stated in his dissertation in 1945, "Just as one engineering project on a stream system may set in motion a series of readjustments affecting reaches above and below it, so flood-abatement measures are likely to have profound effects upon the agricultural economy of the upstream drainage areas" [4]. This chapter describes the use of levees for flood control in the United States.

14.2 Levee Basics

Levee systems, or just "levees" for short, are a structural tool that can reduce the frequency of flooding, but no levee system can eliminate all flooding. A levee is generally designed to prevent water from entering a specific area under certain flood events. Basic characteristics of a levee include the following:

- A man-made barrier along a watercourse (not across a watercourse like a dam). Levees may be built along rivers, tributaries, coastlines, canals, or other waterways.
- Excludes flooding from encroaching on a limited area for a range of flood events (the levee's height and design specifications are determined by the flood level that it is intended to withstand).
- Typically composed either of earthen embankments, as shown in Fig. 14.1, or concrete floodwalls or a combination of both.
- Can have other features such as pedestrian gates, traffic closures, and pump stations.
- Generally, ties into high ground (elevated land that is taller than the floodplain and less likely to flood) on either end, but some levees do exist that are open-ended.
- May be composed of other man-made structures in a landscape which may divert or exclude flooding, but were not designed specifically for that purpose such as roadway and railroad embankments and dredging disposals. Yet, sometimes, these other structures' purpose transforms to a flood- prevention purpose over time. Sometimes levees are linked to these other man-made structures.
- May be linked to dam-related structures and coastal barriers, which can also be integral to a levee system or can function like a levee.

Fig. 14.1 Example of a typical earthen levee embankment, Upper Wood River Levee System, Illinois. (Source: USACE)

If a larger flood occurs than what the levee was designed to withstand, floodwaters may exceed the height of the levee and flow over the top of the levee, referred to as *overtopping*. Sometimes levee systems can have designed overtopping locations for potential overtopping to occur in less- vulnerable areas. Flooding also can damage levees, allowing floodwaters to flow through an opening, or *breach*, in the levee. In either case, the area behind the levee, or the *leveed area*, could be flooded.

14.3 Use of Levees to Control Flooding and Protect People and Property

Historically, the frequency of flooding would dictate the land use – type of farming, ability to live there (i.e., development), access to water, etc. The Swamp Land Acts passed in the mid-1800s transferred from the federal government lands deemed as "swamp and overflowed lands" to certain states for reclamation, including the ability to use levees and drain and fill these lands to use for other purposes [5]. As a result, millions of acres of flood-prone wetlands were converted to other uses for economic benefit. By the late 1800s, settlers had cleared and drained many of these lands for agriculture purposes. Settlements were established along rivers for close access to water for drinking, irrigation, and transportation. Then, the problem of flooding risks to life and property began to become more prevalent as the occupancy and use of floodplains began to increase.

Significant historic flood events have led to widespread adoption of the use of levees and legislation focused on protecting people and property from flooding. As an example, in 1913, a flood in the Ohio River Valley killed more than 400 people with property damages exceeding $200 million (in 1913 dollars). This basin-wide flood resulted in more public interest in flood control and the creation of organized groups to focus on flooding issues [6].

The US Army Corps of Engineers was established as a permanent military branch in 1802 and began to be viewed as the nation's builders to serve both military and public purposes [7]. The Flood Control Act of 1917 was the first legislation that included involvement of the federal government in looking at more comprehensive solutions to flooding along the lower Mississippi and Sacramento rivers. In addition, it included a provision for the US Army Corps of Engineers to start looking at flood management measures for navigation and hydropower purposes. Prior to this, flooding was considered a local problem.

Perhaps one of the most important pieces of landmark legislation related to flooding was the 1936 Flood Control Act [8]. It was the first legislation that explicitly stated that the federal government should take an interest in flooding as shown below:

SECTION 1 It is hereby recognized that destructive floods upon the rivers of the United States, upsetting orderly processes and causing loss of life and property, including the erosion of lands, and impairing and obstructing navigation, highways, railroads, and other channels of commerce between the States, constitute a menace to national welfare; that it is the sense of Congress that flood control on navigable waters or their tributaries is a proper activity of the Federal Government in cooperation with States, their political subdivisions, and localities thereof; that investigations and improvements of rivers and other waterways, including watersheds thereof, for flood-control purposes are in the interest of the general welfare; that the Federal Government should improve or participate in the improvement of navigable waters or their tributaries, including watersheds thereof, for flood-control purposes if the benefits to whomsoever they may accrue are in excess of the estimated costs, and if the lives and social security of people are otherwise adversely affected.

As stated by Lieutenant General E. R. Heiberg III, the chief of engineer from 1984 to 1988, "The hundreds of reservoir, levee, and channelization projects that resulted from the 1936 act and subsequent amendments have literally changed the face of the nation. The projects have contributed to both the growth of towns and the protection of rural farmlands" [8]. Controlling flooding had become commonplace and provided increased opportunities for growth and development across the nation across multiple sectors – energy, transportation, agriculture, and more.

14.4   Evolution of Levee Design and Construction

Engineering practices for levee design and construction have widely varied across the nation. Levees in the United States are built by various governmental agencies or by private property owners, often using different standards, materials, and flood scenarios to inform their design (see Fig. 14.2). Until the 1930s, levees were constructed by farmers and local and regional entities without the benefit of engineering practice and often using readily available soil materials [9]. As geotechnical, hydraulic, and hydrologic understanding and practices began to evolve, levee design and construction practices predominantly continued to be based on local or regional experiences and available materials.

Today, we have the benefit of learning from history. Engineering practices for levee design and construction continue to evolve and progress through improved understanding of levee performance, new technologies, and improved accessibility to materials that are not just local. Significant events, including Hurricane Katrina (2005), Hurricane Sandy (2012), and flooding on the Mississippi and Missouri rivers at various times, have provided a more detailed understanding of factors associated with levee performance and potential failure mechanisms. Processes today allow for application of critical thinking, modeling, and even use of a risk-based approach to consider unique conditions that may be likely to happen. Therefore, engineers can now make better informed decisions as to when designs should be adapted or enhanced. As an example, a design may include adding armoring on the landside portion of the levee to prevent erosion due to overtopping. Fig. 14.3 shows

Fig. 14.2 Early levee construction along the Mississippi River. (Source: USACE)

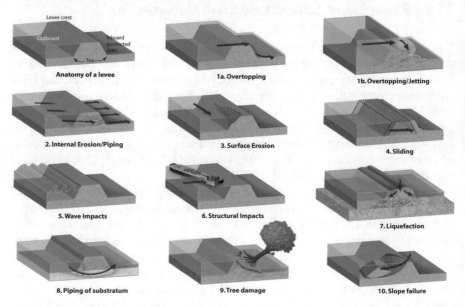

Fig. 14.3 Schematics of levee performance considerations. (Source: Zina Deretsky, National Science Foundation)

examples of modern performance considerations (i.e., potential failure mechanisms to prepare for) during design.

Fig. 14.4 A view of a terraced floodwall that also serves as a walking path along the Napa River in California. (Source: USACE)

Because levees are visible within communities, sometimes, other purposes are added around or incorporated into the levee (see Fig. 14.4), such as recreational features like bike paths, docks, boat ramps, or ecological enhancements; thus, designing levees has increased in complexity due to the addition of multiple purposes.

14.5 The Levee Effect

Many individuals do not realize how prevalent levees are in our nation and local communities. More than 17 million people live and work behind the many miles of levees across the nation – including $2.3 trillion in property value and many of the nation's critical infrastructure assets. There are approximately 7000 levees (totaling 25,000 miles in length) of identified levees in the United States [1]. Communities continue to become more reliant on levee systems as an important tool for reducing risk to life and property from flooding. According to the US Army Corps of Engineers (USACE), in 2019, levees under the organization's jurisdiction alone prevented $300 billion in potential flood damages. There are communities in every state and the District of Columbia, Guam, and the Virgin Islands that rely on levees. About two-thirds of the US population live in a county with a levee [1].

Fig. 14.5 A levee improvement project constrained by riverine vegetation and development, Sacramento region, California. (Source: USACE)

Floods occur every year in the United States. Some of these floods severely test the performance of these levees and in some cases overwhelm levees causing catastrophic flooding. Levee owners and communities work hard to minimize the impacts from these events but often struggle due to limited resources to effectively manage these levees or ensure communities are resilient to flooding. Levees are often viewed as a community's primary and often only measure against flooding, mainly because levees provide a cost-effective method of reducing the chance of flooding for existing development while also providing continued opportunities for new development in the floodplain. In some areas of the country, there are levees along both sides of the watercourse (see Fig. 14.5), meaning that increasing the height of the levees (referred to as a levee raise) must include consideration of both levees. However, continuing to raise the same levees over time may not be a sustainable nor feasible solution due to potential induced flooding impacts that may be transferred upstream or downstream to areas less protected or lack of real estate to expand and raise levee systems, as shown in Fig. 14.5 where development is adjacent to the levees.

14.6   Unified Recommendations

Since the Flood Control Act of 1936, there have been numerous studies by task forces or committees to investigate national approaches to policy and processes related to flooding issues, some specifically looking at levees. Even though specific recommendations and effectiveness of associated efforts have changed over time, all the concepts that have been suggested, debated, or attempted fall into four basic overall themes, which are each described below. To date, there may not yet be a unified approach to floodplain management for this country, but over the past 75 years, the recommendations have been consistent and interrelated. In general, the basic actions on what should be done are recognized; the challenges reside with finding the best ways on "how" to make progress.

14.6.1 Theme 1: Develop and Implement a National Approach to Flood Management

The first theme is the need to develop and implement a consistent national approach for planning for and managing impacts due to flooding while protecting and restoring natural benefits of floodplains. Regular assessment and recommendations for a national approach were first championed by President Lyndon Johnson as recommendations for "A Unified National Program for Managing Flood Losses" by the Task Force on Federal Flood Control Policy in 1966. These recommendations evolved into the formation of the Federal Interagency Floodplain Management Task Force (Task Force) in 1975 to make progress toward a unified program. That Task Force still exists today [10].

Periodic assessments of the Task Force continue to advocate for integrated policies with clear, measurable goals and understanding of trade-offs between construction and maintenance costs, environmental impacts, and impacts on upstream and downstream residents and industry, for example, to achieve the appropriate balance. However, the complex governmental structure of the United States with the federal government, states, and local organizations all having varying authorities and priorities has made developing an agreed-upon national approach extremely challenging.

Specific to levees, a national approach is usually promoted due to a lack of common standards and policies that would improve the consistent understanding of the predictability of levee performance across the country. In addition, it is important to understand how one or more levees interact within the broader watershed. As recommended in a 2013 report by the National Research Council of the National Academy of Sciences, "There is a clear need for a comprehensive, tailored approach to flood risk management behind levees that (1) is designed and implemented at the local level; (2) involves federal and state agencies, communities, and households; (3) takes into account possible future conditions; and (4) relies on an effective portfolio of structural measures, nonstructural measures, and insurance to reduce the risk to those behind levees" [11].

14.6.2   Theme 2: Collect and Utilize Data to Inform Decisions

The second theme is related to the early recognition that to be able to make decisions related to the floodplain and levees, there must be consolidated, timely, and accurate data available. In the Flood Control Act of 1917, there was a provision to undertake surveys for flood control improvement. In the Rivers and Harbors Act of 1927, there was funding allocated for surveys of 180 rivers. The 1960 Flood Control Act authorized the collection and dissemination of flood information to states and local bodies. The 1972 National Dam Safety Inspection Act required an inventory and inspection of dams across the nation. The National Levee Safety Act of 2007, as amended, also requires a national inventory and assessment of all levees to be conducted and information made available to all. Often associated with recommendations related to information collection is the statement that this is a role for the federal government.

Having best available relevant data is the foundational basis for being able to make informed decisions, not just on a localized basis but for understanding impacts on a broader sense as well, such as impacts constructing a levee may have on the frequency of flooding upstream and downstream of the levee as well as impacts to the environment. Quality, consistency, credibility, and availability of the data are also important factors. With modern technology, fast access to data is not only achievable; it is expected. With more access to data also comes challenges with credibility of the data and in understanding which sources to use for the proper purposes.

14.6.3   Theme 3: Increased Public Education and Awareness

The third theme focuses on improving public awareness and understanding of flood risk. This recommendation is prefaced by the assumption that if people understand their flood risk, then they will make the right decisions to protect themselves. One challenge is the need to have a common understanding of the types of behaviors to encourage through policies, programs, and authorities. In other words, what does success look like for improved public awareness and understanding of flood risk?

Examples may include having/requiring flood insurance in areas not typically required to have it, reducing development behind levees and floodwalls, and elevating structures or other critical assets to reduce damages if there is a breach. Another challenge is with the fact that some information associated with flood risk and levees is typically very technical and difficult to explain in a simple way, for example, the misconceptions about the probability of occurrence and the design levee height related to 100-year and 500-year storm events (i.e., 1% chance of exceedance in any given year, not that level of flood would only happen once in 100 years). To compound this challenge, there is a lack of common terminology. The 2012 report entitled "Dam and Levee Safety and Community Resilience: A Vision for Future

Practice by the National Academies" recognizes "Without a clear understanding of the limitations of flood mitigation infrastructure, community members and stakeholders are likely to be ill-prepared for emergencies that might place lives and livelihoods at risk" [12].

14.6.4   Theme 4: Develop Clearly Defined Management Roles and Responsibilities

The fourth theme of the recommendations is related to understanding and articulating the roles and responsibilities at each level of government to ensure decisions are made within a commonly understood framework and authorities are used in a manner to be complementary toward common goals. This is probably the most challenging of the recommendations and is directly related to the need for a common national approach.

Flooding does not care about political boundaries. Jurisdictional authorities do. Clarification of roles and responsibilities is needed; however, so is consistent enforcement and implementation of policies, programs, and authorities. The 2006 report entitled the "National Levee Challenge: Levees and the FEMA Flood Map Modernization Initiative" recommends that federal agencies recognize state and local roles in floodplain management and create support mechanisms to assist states and local bodies in effectively carrying out those responsibilities [13]. There continues to be debate as to whether more public assistance, such as flood insurance, afforded by the federal government encourages continued development and occupation of floodplains and discourages redundant actions and other behaviors. In 2002, the Task Force on the Natural and Beneficial Functions of the Floodplain observed that some government programs seem to subsidize or encourage development in floodplains regardless of flood risk or consideration of preserving beneficial floodplain functions [14].

14.7   Influence of the National Flood Insurance Program

As part of the national discussions related to flooding and levees, the role of flood insurance as a way to help mitigate for flood losses was interwoven throughout (see Fig. 14.6). For decades, the typical response to flood events has been through the use of structural means, such as dams and levees. This left a gap in helping with individual economic flood losses and encouraging wise floodplain management. To add to the problem, private insurance that would cover flood losses was generally not available. In 1968, Congress passed the National Flood Insurance Act creating the National Flood Insurance Program to provide affordable flood insurance and promote sound floodplain management practices.

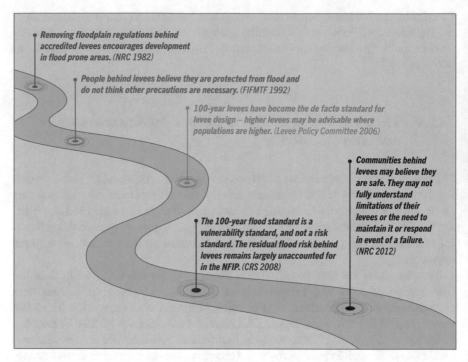

Fig. 14.6 Timeline of example reports with observations related to the National Flood Insurance Program and public awareness [2, 12, 13, 15, 16]

As a result of lack of community participation in the program along with continued flood events, Congress later enacted the Flood Disaster Protection Act of 1973 to establish mandatory flood insurance for buildings with federally backed mortgages in areas identified as Special Flood Hazard Areas. Special Flood Hazard Areas are areas that have a 1% annual chance of flooding also known as the base flood or 100-year flood. The 1% annual chance of flooding means that the area has a one in 100 chance of a given level of flooding being equaled or exceeded in any given year. Within the Special Flood Hazard Areas, the National Flood Insurance Program also requires all new or substantially improved structures to be constructed at or above the elevation of the 1% annual chance of flooding.

How levees are treated under the National Flood Insurance Program continues to be debated. Early on it was discussed that the flood reduction afforded by levees should be given some kind of recognition under the program. This led to interim procedures being issued in 1981 and eventually codified in 1986 as Title 44, Chap. 1, Section 65.10 (44 CFR 65.10), *Mapping of Areas Protected by Levee Systems.* Communities or parties seeking accreditation of a levee system for the National Flood Insurance Program must provide data and documentation in accordance with program requirements, detailed in 44 CFR 65.10. Once criteria for 44 CFR 65.10 are met and "accredited" under the program, areas behind an accredited levee system are considered outside of the Special Flood Hazard Area, so there are no development requirements in the leveed area and no mandatory insurance purchase requirement for buildings in the leveed area with federally backed mortgages [17].

As of January 2021, there are about 1200 levee systems accredited under the National Flood Insurance Program with a population of more than ten million people behind those levees [1]. This means that the population behind these levees is not required to purchase flood insurance, but can do so voluntarily. However, many people behind accredited levees do not purchase flood insurance because they equate not being required to purchase flood insurance with meaning they are safe from being inundated by floods.

Even when the program was in its infancy, it was recognized that how levees were treated under the National Flood Insurance Program could have the unintended consequence of incentivizing communities to seek levees designed for the 1% annual chance of flood to avoid mandatory flood insurance and floodplain development restrictions instead of considering appropriate measures based on site-specific flood risk. With a majority of the population behind "accredited" levees and an increase of levees designed solely for the 1% annual chance of flood event (see Fig. 14.7), these unintended consequences became reality.

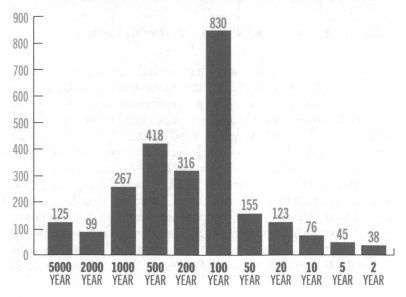

Fig. 14.7 Levees in the National Levee Database with known overtopping frequencies (design level) [1]

14.8   Progress Being Made: What We Know About Levees in the United States

For many years, one of the key gaps in knowledge for levees was simply identifying the location of all levees in the nation and understanding their basic condition. Without this information, understanding the scope and scale of issues associated with levees and making decisions related to prioritization of potential solutions to those issues could not be accomplished.

The National Levee Database, found at https://levees.sec.usace.army.mil, is intended to serve as a dynamic, searchable inventory of information about all known levee systems in the nation and to be a key resource for supporting decisions and actions affecting levees. The National Levee Database is a robust data system that contains thousands of levee systems and has the ability to track hundreds of data points on each system. Today, the USACE is responsible for operating and maintaining the database itself, but data is managed through a collection of stakeholders who hold data management responsibilities for the infrastructure records. FEMA was the largest contributor of levee locations to the National Levee Database that was captured through its production of flood hazard maps as part of the National Flood Insurance Program. It is assumed the majority of the levees across the nation and the consequences associated with their potential failure have been located and identified. The data quality and amount of information available for each levee varies; however, data continues to be refined and improved. Figs. 14.8 and 14.9 show example levee information from the National Levee Database.

14.9   Understanding Risks Associated with Levees

There has been a lot of progress toward looking at levees in a risk-informed way. Through historic performance events, it became clear that various levee systems are composed of different features and good performance requires that all the features work together. It was also observed that levees that met all relevant design standards still could breach or have serious defects. Until the late 1990s when risk-informed approaches started to evolve, there was no framework to evaluate the interrelation of all these features or consider the entire levee system relative to risk.

Risk is a measure of the probability (or likelihood) and consequence of uncertain future events. If there is no chance of an event occurring, then there is no risk. If there are no consequences resulting from an event occurring, then there is no risk. There could be two situations that seemingly have identical risk, but what is driving the risk for each of the two situations can be extremely different. To understand flood risk in the context of levees, the following three components are considered (see Fig. 14.10):

Fig. 14.8 Example information about all known levees and what is in their leveed areas

1. Hazard: This is an event that has the potential to cause an adverse consequence. For levee systems, the typical hazards are flood, seismic activity, and security (intrusions, attacks, or effects of natural or man-made disasters).
2. Performance: This involves analyzing how the levee has functioned before or is anticipated to function during the occurrence of the specified hazards.

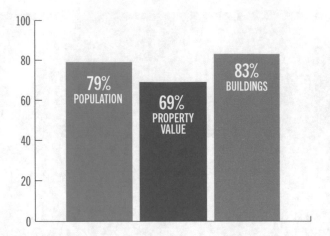

Fig. 14.9 What is behind 25% of the levees in the National Levee Database

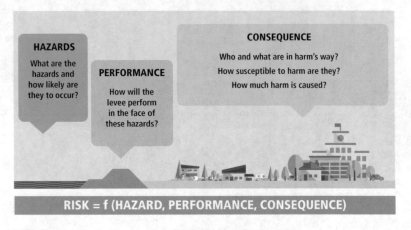

Fig. 14.10 The three main components of risk associated with levees

3. Consequence: This is the effect, result, or outcome from different scenarios considering the combination of the hazard and levee performance. Consequences can be reflected as potential loss of life, economic losses, adverse environmental impacts, or other direct or indirect losses.

The process used to analyze the three components of risk for a levee is called a risk assessment. Risk assessments provide a systematic, evidence-based approach for estimating and describing the existing and future risk associated with levee systems. Risk assessments consider what can go wrong, how it can happen, the consequences if it happens, and how likely it is to happen. Typically, for levee systems, a risk assessment explicitly addresses the likelihood of various flood loadings on the levee system, the response of the levee system to these flood loadings, and the consequences of the combination of loading with the response of the levee system. Risk assessments can range from qualitative, to semiquantitative, to quantitative, incrementally increasing in level of detail and confidence to reduce uncertainty in the results. In the National Levee Database, about 1800 levees have had risk assessments completed on them, with about 5% characterized as having very high to high risk and 95% characterized as having moderate to low risk. Residual flood risk is the risk of flooding in a community that remains at any point in time after accounting for the flood risk reduction contributed by the levee system.

Risk-informed decision-making uses qualitative or quantitative information about flood risk in conjunction with other considerations to lead to more comprehensive, transparent, and informed decisions about flood risk management. Flood risk management is the activity in which measures are identified, evaluated, implemented, and monitored with the goal of managing and reducing flood risk.

The chance of flooding as well as the condition of levees is dynamic over time. Considerations related to levees are a component of overall flood risk management with an intent to continuously understand, monitor, and manage levees in the broader community context of flood risk management activities, priorities, and goals. The benefits of a risk-informed approach include the following:

- Allows for scalability of level of effort to be based on scope and intent of decisions that need to be made.
- Allows for consideration of historic levee performance in addition to deterministic standards.
- Provides a way to acknowledge uncertainty and whether or not reducing that uncertainty would change the decision.
- Helps in prioritizing and optimizing resource decisions (i.e., supports a risk-informed "fix-the-worst-first" approach).
- Informs design and construction decisions to decide where more or less robustness and resiliency are warranted.
- Allows for recognition of different perspectives and how that may lead to different choices being preferred by different entities.
- Helps identify which risk management measures can influence which of the three components of risk (Fig. 14.11).
- Improves understanding of the purpose and benefits of the levee within the community.

14.10 Now Is the Time for a National Levee Safety Program

Year after year, periodic flood events continue to shine the light on the importance of levees and the continued need for a consistent national approach to better predict their performance and manage them in the broader context. To answer this need, Congress enacted the National Levee Safety Program, codified in 33 U.S.C. Chap. 46.

Why does the nation need a National Levee Safety Program? Responsibility for levees is currently distributed across all levels of government. Levee owners and

Fig. 14.11 Example risk management measures to address the different components of risk. (O&M stands for operation and maintenance activities)

operators work diligently every year to maintain levees with limited resources. However,

- Much of the levee infrastructure is decades old and was built without the benefit of modern engineering practices.
- Levees are designed, constructed, and managed by various entities, utilizing different processes and standards.
- Development continues to intensify behind levees, putting more reliance on the levees' ability to perform and the consideration of other means, such as evacuation and land-use planning, for managing flood risk.
- Much of the public remains unaware of their flood risks, either from levee breach or overtopping, or the actions they can take to reduce those risks.

Moving toward a National Levee Safety Program is an important investment for the nation. It complements other national goals, such as reducing disaster relief costs, identifying levee-related needs and resources, prioritizing resources, reducing loss of life and damage to property, and increasing local resilience against flooding. The purposes enumerated by Congress [18] for the National Levee Safety Program are as follows:

1. *To ensure that human lives and property that are protected by new and existing levees are safe*
2. *To encourage the use of appropriate engineering policies, procedures, and technical practices for levee site investigation, design, construction, operation and maintenance, inspection, assessment, and emergency preparedness*
3. *To develop and support public education and awareness projects to increase public acceptance and support of levee safety programs and provide information*
4. *To build public awareness of the residual risks associated with living in levee-protected areas*

5. *To develop technical assistance materials, seminars, and guidelines to improve the security of levees of the United States*
6. *To encourage the establishment of effective state, regional, and tribal levee safety programs*

14.10.1   A Consistent National Approach

To answer the need for consistent and comprehensive nationwide levee practices and a common framework for the management of levees, under the National Levee Safety Program, National Levee Safety Guidelines are intended to be developed to serve as a national resource that will be continuously updated to reflect the best consistent modern engineering and flood risk management practices applicable to life cycle of levees (such as decision-making related to the planning, designing, constructing, and managing of levees). These practices are intended to reflect the most current and advanced technical information available to promote the safe performance and management of levees and flood risk management for communities behind levees. The guidelines will also be able to serve as a reference document providing fundamental information about levees and a common set of terminology to be used. As more strains on floodplains continue and there is more recognition and understanding of climate considerations, the guidelines also provide an opportunity to reinforce practices that address decisions about levees in a larger watershed and floodplain system context.

14.10.2   Timely and Accurate Data

The National Levee Database was made publicly available in October 2011. Efforts under the National Levee Safety Program will continue to seek ways to combine the data in the database with other datasets like the National Inventory of Dams (https:// nid.sec.usace.army.mil/), improve how the data is assessed and displayed to assist in decision-making, and develop training materials that will help users best utilize and understand the information [19]. The goals of the National Levee Database include being the nationally recognized resource of the highest quality and most complete data record for all levees in the nation and contributing information to support science and technology advancements.

The National Levee Database is the delivery mechanism for data, but there is also an opportunity under the National Levee Safety Program to collect valuable information, similar to the collection of information about the nation's dams that helped to build the National Inventory of Dams. This information includes levee location, identification of the different levee features, and assessment of the levee condition. This levee information may help identify and inform previously unknown risks, repair and rehabilitation needs, partners for flood risk management, investments, flood fighting and emergency management activities, and the ability to describe "what is at stake" to residents and businesses behind the levee.

14.10.3 Improved Public Awareness

A provision of the National Levee Safety Program is a public education effort to educate communities behind levees about their risk and promote consistent communication of information at the different levels of government. The National Levee Safety Guidelines in combination with consolidated readily available information help provide the tools to improve public awareness and understanding of both levees and flood risk in general (Fig. 14.12).

Fig. 14.12 Areas that can be improved due to increased education and awareness

Fig. 14.13 Example roles and responsibilities

14.10.4 Understanding Roles and Responsibilities

The governmental structure in the United States is very complex. There are numerous agencies with differing authorities – some own levees, some oversee management of levees, and some have programs that provide assistance to support either floodplain management or levee-related activities (Fig. 14.13). In some cases, there may be overlap, for example, with different kinds of plans required by different agencies intended to promote the same kinds of behaviors or activities (e.g., hazard mitigation plans, floodplain management plans, or emergency plans). The overlap in responsibilities between organizations can cause inefficiencies, confusion, conflicts, and duplication of effort. Thus, the best outcome is for their actions to be coordinated. Knowing what each entity's role is in flood risk management and levees is the only way to move forward in identifying where processes can be streamlined or changed to remove conflicts or disincentives. It is envisioned that the National Levee Safety Program will provide improved clarity about roles and responsibilities.

14.11 Summary

Levees are a key part of resiliency for many communities across the United States (Fig. 14.14). Efforts have been made to document and evaluate many of the levee systems in the United States in recent years through the National Levee Database. Throughout our country's history, progress has been made through laws, policies, and programs to improve management of these critical infrastructure systems. Several questions remain unanswered – how safe is safe? Who is responsible for what? Who gets the benefits? This is an exciting point in time, in which there is more emphasis on infrastructure in general. The path forward likely includes

Fig. 14.14 Levee systems at work in a community being integrated into the landscape and intersecting with multiple infrastructure and landscape features. (Source: USACE)

additional work to consider climatic changes, natural and beneficial functions of the ecosystems around levees, and improving understanding of the risks as we continue to collect more data about the levee systems we have. With a convergence of experience, technology, and information, there is an opportunity to significantly evolve and put into practice the approaches that have been discussed and considered for more than 75 years.

References

1. U.S. Army Corps of Engineers (2016) National Levee Database, https://levees.sec.usace.army.mil
2. Federal Interagency Floodplain Management Task Force (1992), Floodplain Management in the United States: An Assessment Report, Volume I
3. Federal Emergency Management Agency (2021), Historical Flood Risk and Costs, https://www.fema.gov/data-visualization/historical-flood-risk-and-costs
4. White G. (1945) Human Adjustment to Floods. Dissertation, University of Chicago
5. 43 U.S.C. Chapter 23, Grants of Swamp and Overflowed Lands

6. American Institutes for Research (2005), A Chronology of Major Events Affecting The National Levee Safety Program. https://www.fema.gov/sites/default/files/2020-07/fema_nfip_eval_chronology.pdf. Accessed 11 Nov 2020
7. US Army Corps of Engineers (2013), Engineer Pamphlet 870-1-68, The U.S. Army Corps of Engineers: A History. https://www.publications.usace.army.mil/USACE-Publications/Engineer-Pamphlets.
8. Arnold J (1988), The Evolution of the 1936 Flood Control Act. Office of History United States Army Corps of Engineers. Available via https://www.publications.usace.army.mil/USACE-Publications/Engineer-Pamphlets/u43545q/4550203837302D312D. Accessed 16 Nov 2020
9. US Army Corps of Engineers (2018), Levee Portfolio Report. Available via https://www.usace.army.mil/Missions/Civil-Works/Levee-Safety-Program. Accessed 10 May 2018
10. Task Force on Federal Flood Control Policy (1966), A Unified National Program for Managing Flood Losses. Washington, DC: US Government Printing Office
11. National Research Council (2013), Levees and the National Flood Insurance Program. Washington, DC: The National Academies Press. http://nap.edu/18309
12. National Research Council (2012), Dam and Levee Safety and Community Resilience: A Vision for Future Practice. Washington, DC: The National Academies Press. https://doi.org/10.17226/13393
13. Interagency Levee Policy Review Committee (2006), The National Levee Challenge – Levees and the FEMA Flood Map Modernization Initiative. Washington, DC: FEMA
14. The Task Force on the Natural and Beneficial Functions of the Floodplain (2002), The Natural Beneficial Functions of Floodplains – Reducing Flood Losses by Protecting and Restoring the Floodplain Environment
15. National Research Council (1982), A Levee Policy for the National Flood Insurance Program. Washington, DC: The National Academies Press. https://doi.org/10.17226/19600
16. Congressional Research Service (2008), Flood Risk Management and Levees: A Federal Primer. https://crsreports.congress.gov, RL33129
17. Title 44, Chapter 1, Section 65.10 (44 CFR 65.10), *Mapping of Areas Protected by Levee Systems*
18. 33 U.S.C. Chapter 46, National Levee Safety Program
19. US Army Corps of Engineers (2020), National Inventory of Dams, https://nid.sec.usace.army.mil

Tammy L. Conforti, - P.E., CFM, grew up in Huntington, West Virginia, and always loved math and science. Right out of high school, she got a job with the US Army Corps of Engineers (USACE) in Huntington as a co-op student working in their structural engineering office. As a co-op student, she alternated working full time one semester and attending college one semester. This provided real-world experience that helped inform her decision to major in civil engineering. In 1995, she graduated summa cum laude from Virginia Tech with a bachelor of science degree in civil engineering. Upon graduation, she accepted a full-time position as a geotechnical engineer with the Corps of Engineers and returned back to Huntington. After working for 4 years, she became a registered professional engineer in West Virginia and the lead planner for her first levee project. Soon afterward, she also became a certified floodplain manager.

In 2005, she was offered a position to create a new agency-wide program for the Corps of Engineers. The program goal was for USACE and the Federal Emergency Management Agency (FEMA) along with other federal agencies to create an interagency team at the state level to develop and implement solutions to state natural hazard priorities. She accepted this position and relocated to California. After creating and leading the first interagency team for this program with the State of Ohio and demonstrating its success, USACE permanently established the program, which still exists today.

Three years later, in 2008, Tammy accepted a position with the US Army Corps of Engineers headquarters to create the agency's new levee safety program and relocated to Washington, DC. In this role, she led the development and issuance of program policies, served as technical advisor on levee safety aspects and issues, prioritized the agency's program activities, and fostered partnerships by serving on related national committees and task forces. She also became the first woman to chair the agency's Levee Senior Oversight Group, which is a group of diverse members representing key disciplines in levee safety with the purpose of reviewing all agency risk assessments for levees and providing recommendations on levee safety matters.

In 2014, in addition to her continued work with levees, she was tasked to establish a new agency permitting program related to requests by others to temporarily or permanently alter existing USACE projects. This program was permanently established in 2016, and she continued to lead both the levee safety program and the permit program until 2019. In 2015, she served as one of the champions in support of a new online graduate program focused on risk management. She participated in the first class and graduated with a master of science in risk management from Notre Dame of Maryland University. In 2019, upon receipt of funding from Congress, Tammy became the lead for developing the complex strategic approach for the development of the National Levee Safety Program, a new national-level federal program. The program aims to provide access to best available information; develop and maintain national guidelines covering key activities for new and existing levees; encourage federal agencies, states, and tribes to establish effective levee safety programs; and align existing USACE and FEMA programs to support these objectives. Currently, Tammy resides in Alexandria, Virginia, with her husband and enjoys running, reading, and fine dining.

Janey Camp, PhD, P.E., GISP, CFM, grew up on a small farm in rural middle Tennessee. Puzzles and problem-solving were favorite pastimes as well as being outdoors – everything from gardening and trying to create a nature center in the backyard treehouse. Throughout high school, she became more interested in hiking and exploring nature beyond the family farm and small wooded acreage at the back. She was enamored with the thought of being an astronaut like Sally Ride or an F-16 fighter pilot, but the fear of being shot in war turned her attention to aeronautics. Being the oldest of five children and a first-generation college student, Janey worked hard to do well in school and secure scholarships for college. When the time came, being close to home to help with the family led her to first attend Motlow State Community College. There, realizing that any courses over a "full load" was essentially free, Janey signed up for extra courses such as psychology and criminal justice which have actually been useful in interdisciplinary projects that she's involved in now due to a better understanding of the social sciences.

Janey obtained an associate degrees in preengineering and psychology and transferred to Tennessee Technological University (TTU) with an intent to study mechanical engineering and focus on aerospace engineering. However, she accidentally stumbled into a presentation about civil engineering at the transfer student visit day, and it changed everything. Civil engineering combined her love of helping people with the outdoors through environmental engineering and seemed like the perfect fit. She immediately changed majors with a focus on environmental engineering but continued taking some "extra" courses in mechanical engineering for "fun" (i.e., thermodynamics and heat transfer) causing some confusion as to what her major really was for faculty. Thanks to the "extra" courses at Motlow, she was off sequence and started taking graduate courses to round out the schedule. Upon graduation with her BS in '02, she took a position with the TTU water center to work on advanced oxidation of organic matter in water to improve drinking water treatment and received her MS in '04. With encouragement from faculty and others, she decided to take a plunge and keep going for the PhD which she obtained from Vanderbilt University in '09 with a focus on developing a geospatial tool for spill response. The social science courses helped in the research where experts were surveyed, and later she helped host a summit at Vandy about climate change and infrastructure systems as a postdoc. The social interactions in engineering were becoming more interesting. Janey jokes that she's not the stereotypical engineer at all and loves the soft side of the real-world challenges she works on.

At Vanderbilt, Janey transitioned from graduate student, to postdoc, to an assistant research faculty over the course of a couple of years – honing skills related to risk analysis, flood modeling, and geospatial technologies. She has continued advancing and loves working on interdisciplinary projects that are unique with social science aspects that also are more applied with potential to directly impact communities and individuals in the near term. Much of her work involves using data analysis and visualization (through geospatial tools and models) to inform decisions related to community resilience, infrastructure management, natural hazards, and climatic change. Recent projects that she loves to discuss include evaluating opportunities for transportation investment to address the opioid epidemic in Tennessee, developing a rubric to assess smart mobility readiness of communities, a diversity and inclusion study of a state government agency, and work to develop K–12 risk education curriculum. These are quite different from the original interest in aeronautics but sill involve her real passions of exploring new things, helping people, and solving problems – just at a less-risky elevation.

Part IV
Cleaning Up

Chapter 15
Contaminated Sites

Rebecca Lance Svatos

Abstract Soil, groundwater, surface water, and air at contaminated sites may contain chemicals that could cause harm to human health or the environment. Spills, leaks, or other mismanagement of chemicals cause this contamination. Once a site is contaminated, it is very difficult, expensive, and sometimes impossible to clean up. During the modern environmental movement in the 1960s, awareness of the risks from contaminated sites increased. After the first high profile contaminated site, Love Canal, resulted in emergency declarations by President Jimmy Carter in 1978 and 1980, the federal Comprehensive Environmental Response Compensation and Liability Act, or Superfund, was passed to address the nation's highest risk contaminated sites. Since then, the technical expertise required for discovering, understanding, investigating, cleaning up, and redeveloping contaminated sites has advanced considerably. A case history of the Woolfolk Chemical Works Superfund site demonstrates the importance of community involvement and consideration of environmental justice during the investigation and cleanup of contaminated sites. Although continued technical advances are needed to improve the investigation and cleanup of contaminated sites, more emphasis is needed on predicting the future environmental impact of current innovations to prevent the continued creation of new contaminated sites.

Keywords Contamination · Superfund · Cleanup · Groundwater · Soil · Remediation

R. L. Svatos (✉)
Stanley Consultants, Iowa City, IA, USA

P. Layne, J. S. Tietjen (eds.), *Women in Infrastructure*, Women in Engineering
and Science, https://doi.org/10.1007/978-3-030-92821-6_15

15.1 Introduction

Chemicals are present at contaminated sites in the soil, groundwater, surface water, and/or air at levels that could cause harm to human health or the environment. Contaminated sites such as industrial facilities or the corner gas station become contaminated from spills, leaks, or other mismanagement of chemicals. Although contaminated sites have been present in the United States since the Industrial Revolution, their risks to human health and the environment did not become evident until the modern environmental movement began in the 1960s. This movement raised awareness of environmental issues and led to the passage of key federal environmental regulations in the 1970s and 1980s. After the first high profile contaminated site, Love Canal, resulted in emergency declarations by President Jimmy Carter in 1978 and 1980, the federal Comprehensive Environmental Response Compensation and Liability Act, or Superfund, was passed to address the nation's highest risk contaminated sites. Since then, the technical expertise required for discovering, understanding, investigating, cleaning up, and redeveloping contaminated sites has advanced considerably. Community involvement during the investigation and cleanup of contaminated sites as well as consideration of environmental justice issues are critically important, yet often overlooked, aspects of successfully addressing contaminated sites. The importance of these aspects is illustrated in a case history of the Woolfolk Chemical Works Superfund site along with the technical aspects of the site discovery, understanding, investigation, cleanup, and redevelopment.

Contaminated sites are often created because the future environmental impact of current practices is unknown; our ability to innovate far exceeds our ability to predict the environmental impact of our innovations. Even with the significant technological advances that have been made in the last 50 years, once a site is contaminated, it is still very difficult, expensive, and sometimes impossible to restore to its previously uncontaminated condition. Technical advances are needed not just to improve the investigation and cleanup of contaminated sites but also to better predict the environmental impact of new chemicals and products and prevent contamination from occurring.

15.2 Social Backdrop

The origins of the modern environmental movement are often traced to Rachel Carson [1]. As a biologist who spent 15 years working for the United States Fish and Wildlife Service, she had a deep understanding and love of the natural world. As a writer, she communicated the beauty of nature to a broad audience. When her third book, *Silent Spring*, was published in 1962, she warned about the long-term effects of the expanded use of synthetic chemical pesticides. She documented places in the United States where the use of pesticides, such as DDT, resulted in a spring without birds singing. Most importantly, she said that humans are a vulnerable part of the natural world subject to the same damage as the rest of the ecosystem [21].

Environmental conditions in the United States were threatening human health; leaded gasoline was releasing such high levels of lead into the air that the average preschooler had four times the currently allowable blood lead level, smog obscured the sun in many large cities, and waterways were so polluted that they could catch on fire.

After reading or hearing about these conditions and the grim future ahead if the country stayed on its current path, many Americans became concerned about the environment for the first time. Beginning with Rachel Carson's dire warning, the 1960s became a decade of environmental awakening that ended with the first Earth Day in 1970 [22]. Senator Gaylord Nelson [24] of Wisconsin proposed the first Earth Day and wrote:

> When April 22, 1970, dawned, literally millions of Americans of all ages and from all walks of life participated in Earth Day celebrations from coast to coast. It was on that day that Americans made it clear that they understood and were deeply concerned over the deterioration of our environment and the mindless dissipation of our resources. That day left a permanent impact on the politics of America. It forcibly thrust the issue of environmental quality and resources conservation into the political dialogue of the Nation. That was the important objective and achievement of Earth Day. It showed the political and opinion leadership of the country that the people cared, that they were ready for political action, that the politicians had better get ready, too. In short, Earth Day launched the Environmental decade with a bang.

The environmental movement of the 1960s led to the passage of environmental regulations that have been protecting our air, water, and land ever since.

15.3 Environmental Regulation

The creation of the United States Environmental Protection Agency (EPA) in 1970 was the beginning of a new emphasis on the environment that included passage of the following foundational environmental laws (Fig. 15.1):

- *National Environmental Policy Act (NEPA)* – Passed in 1969, this was one of the first laws that established the broad national framework for protecting the environment. NEPA requires all branches of the federal government to consider the environment prior to undertaking any major federal projects.
- *Clean Air Act* – In 1970 this comprehensive federal law was passed to regulate air emissions and authorize EPA to establish National Ambient Air Quality Standards to protect public health.
- *Clean Water Act* – Although the Federal Water Pollution Control Act of 1948 established basic water pollution control requirements, significant reorganization and expansion in 1972 created the current framework for regulating discharges of pollutants into waterways. Under the Clean Water Act, as it is now known,

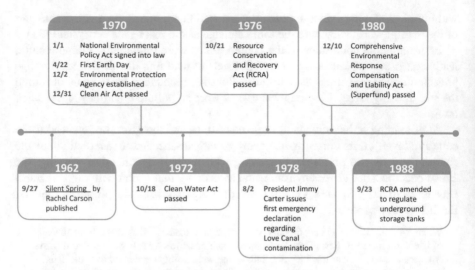

Fig. 15.1 Timeline of environmental milestones

water quality standards are established with the goal of attaining fishable, swimmable waters throughout the United States.

Following the establishment of these key environmental laws, attention turned to hazardous and solid waste. In 1976, Congress amended the Solid Waste Disposal Act of 1965 with the Resource Conservation and Recovery Act (RCRA) to address problems from a growing volume of municipal and industrial wastes. RCRA created "cradle to grave" hazardous waste management by authorizing EPA to control hazardous waste generation, transportation, treatment, storage, and disposal. RCRA also created a framework for the management of non-hazardous solid wastes in municipal landfills. In 1988, RCRA was amended to enable EPA to address environmental problems from underground storage tanks (USTs) storing petroleum and other hazardous substances because of widespread groundwater contamination from leaking USTs.

The severity of contamination at the infamous Love Canal site in Niagara Falls, New York, led to passage of the Comprehensive Environmental Response, Compensation, and Liability Act (CERCLA), commonly known as Superfund, in 1980. This law created a tax on the chemical and petroleum industries for a trust fund (Superfund) for cleaning up abandoned or uncontrolled hazardous waste sites. It also provided broad federal authority to respond to releases of hazardous substances that could endanger public health or the environment. The Superfund enforcement program identifies companies responsible for contamination at a site, called Potentially Responsible Parties (PRPs), and orders them to clean up the site or pay for the cleanup to be completed by the EPA, a state, or another responsible party. CERCLA also created a National Priorities List (NPL), a national list of the highest risk sites that have priority for cleanup. Love Canal was the first Superfund site on that list [14].

15.4 Love Canal

In 1894, William T. Love began building a shipping canal in New York that would bypass Niagara Falls to provide inexpensive hydroelectric power for industrial development. However, only one mile of the canal was actually built before the project was abandoned. The excavation was partially filled with water and initially used for recreation until the 1920s when the City of Niagara Falls used the abandoned canal as a municipal landfill.

In the 1940s, Hooker Chemical was given permission to place industrial waste in the abandoned canal. After lining it with clay, Hooker Chemical disposed of more than 21,000 tons of drums containing chemical wastes over a 10-year period. The wastes included caustics, alkalines, fatty acids, and chlorinated hydrocarbons resulting from the manufacturing of dyes, perfumes, and solvents for rubber and synthetic resins. In 1953, the landfill was covered with clay and leased to the Niagara Falls Board of Education. An elementary school and many residential properties were built over the disposal area. These construction activities breached the clay cap and allowed rainwater to leach the chemicals.

During the 1960s, complaints about odors and residues were first reported at the Love Canal site. In the 1970s, complaints from residents living adjacent to the Love Canal landfill increased as rising groundwater levels brought contaminants to the surface. Various federal and New York State studies indicated that numerous toxic chemicals, including dioxin, had migrated through existing sewers and, ultimately, drained into nearby creeks. The EPA and New York State began investigating the Love Canal groundwater along with indoor air and sump water contamination in the various residences.

In 1978, after a dramatic increase in skin rashes, miscarriages, and birth defects among residents in the area, President Jimmy Carter issued the first of two emergency declarations regarding the Love Canal site. The first declaration provided federal funding for remedial work to contain the chemical wastes and to assist New York State in the relocation of some of the people living around the Love Canal landfill. This was the first time in American history that emergency funds were used for a situation other than a natural disaster.

In 1980, President Carter issued a second emergency declaration, establishing the 350-acre Love Canal Emergency Declaration Area surrounding the landfill. The second declaration authorized federal funds to purchase homes and relocate approximately 950 families under the management of the Federal Emergency Management Agency (FEMA) and New York State. The homes were demolished, and the demolition debris placed under the Love Canal landfill cap.

In 2004, after more than 20 years of work, EPA deleted the Love Canal Superfund site from the NPL. EPA, together with New York State, contained and secured the wastes disposed in the abandoned canal so that they are no longer leaking into surrounding soils and groundwater. The secured 70-acre site includes the original 16-acre hazardous waste landfill and a 40-acre disposal area covered by a synthetic liner with a clay cap and surrounded by a barrier drainage system with a leachate collection and treatment system.

Today, the Love Canal area is once again a flourishing community. Neighborhoods to the west and north of the canal have been revitalized, with more than 200 formerly boarded-up homes renovated and sold to new owners and ten newly constructed apartment buildings. The area east of the canal has been sold for light industrial and commercial redevelopment [15].

15.5   Contaminated Site Investigation/Remediation

15.5.1   Contaminated Sites

At contaminated sites, like Love Canal, chemicals are present in the soil, groundwater, surface water, and/or air at levels that could cause harm to human health or the environment. The chemicals could be toxins that cause cancer or naturally occurring chemicals that are present in unnaturally high concentrations due to human activities. These contaminated sites could be industrial facilities or the corner gas station. Sites become contaminated from spills, leaks, or other mismanagement of chemicals.

Contaminated sites are often created because the future environmental impact of current practices is unknown. For example, it was previously believed that disposing of waste in the ground was safe because the soil would filter out contaminants. Because of this mistaken belief, landfills and industrial wastewater lagoons were previously constructed without liners. Because of the soil and groundwater contamination that resulted from many unlined landfills and lagoons, current regulations require landfills and lagoons to be lined. Another common example of past practices that caused site contamination is gas stations. USTs containing gasoline at many gas stations have corroded and leaked fuel that caused soil and groundwater contamination that sometimes also resulted in hazardous vapors migrating into nearby buildings. This widespread problem resulted in stronger regulation and monitoring of underground fuel tanks. Regulations are just now being developed to address per- and polyfluoroalkyl substances (PFAS) manufactured and widely used for decades in non-stick coatings, fire-fighting foam, and many other products. Even though PFAS are now ubiquitous in the environment and the human body, research is still in progress to establish levels that are protective of human health and the environment. There are certainly other currently accepted practices that will cause future contaminated sites because our ability to innovate far exceeds our ability to predict the environmental impact of our innovations.

Not only are humans inadvertently causing contamination, but the ability to completely clean up contaminated sites is limited. Once a site is contaminated, it is very difficult – and often impossible – to completely restore the site to its previously uncontaminated condition. The most contaminated sites can require decades of significant effort costing millions of dollars to investigate and remediate. Even after

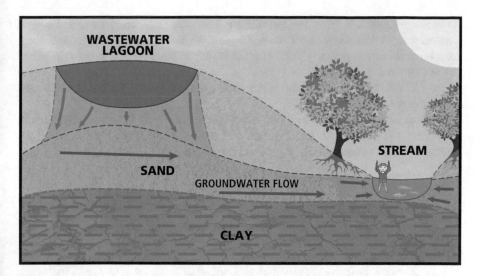

Fig. 15.2 Contaminated site schematic

that long-term commitment of time and money, there may still be residual contamination present requiring long-term monitoring and management. Rather than investing these resources in addressing a problem after it has been created, it would be preferable to invest time and money into preventing contamination from occurring in the first place. Figure 15.2 presents a schematic of a contaminated site, and the following paragraphs provide an overview of the complex process required to investigate and remediate contaminated sites.

15.5.2 Discovery

Contaminated sites are often identified after complaints are made to state or federal agencies regarding conditions such as unusual odors, discolored streams, problems with drinking water, leaking drums, dead fish, and/or health problems in nearby residents. However, there are many other ways that contaminated sites can be identified. Permitted landfills, USTs, and other facilities that manage chemicals are required to conduct routine monitoring that may detect contamination. A facility that manages chemicals may accidentally spill or leak chemicals into the environment. Sometimes excavation during construction activities reveals odors, stained soil, and/or contaminated groundwater from a previously undiscovered contaminated site.

Due diligence prior to purchasing a property is now a common way that contaminated sites are identified. The owner of a property can be liable under CERCLA for cleaning up environmental contamination – even if a previous owner caused the contamination – unless they exercise due diligence in evaluating potential environmental issues prior to purchasing the property. Therefore, a Phase I Environmental Site Assessment (ESA) is almost always completed before a commercial or industrial property purchase to evaluate whether environmental problems are present on the property that could cause future liabilities. Sometimes these Phase I ESAs indicate potential contamination, and additional testing is conducted that confirms contamination.

When contamination is identified or suspected, state or federal environmental agencies must be notified. Investigation and remediation efforts will then be conducted under the oversight of the governmental agency.

15.5.3 Investigation

15.5.3.1 Source

Once a contaminated site is identified, investigation is required to define the scope of the problem beginning with the source of the contamination. Sometimes the investigation starts with a reported leak or spill so the contaminant source is known. In other cases, such as a contaminant detected in a drinking water well, investigation is needed to determine the source. In all cases, any available information regarding the volume of chemical(s) that were released into the environment as well as the timing and duration of the release are helpful for understanding the potential environmental impacts.

If the contaminant source is unknown, an evaluation of available information on the chemicals detected and their common sources, nearby uses of those chemicals, and other relevant facts is necessary to determine the source(s) of the contamination. Sometimes, significant investigation, including environmental testing and interviews with property owners and nearby residents, is required to identify the source of a contaminated site.

Once the source is identified, immediate action must be taken to control or eliminate ongoing contamination, if possible, because cleanup is ineffective if there is an ongoing source. If an unlined pond is leaking and contaminating groundwater, it should be taken out of service or a liner system installed to stop the leakage (see Fig. 15.3). However, some sources cannot be immediately stopped. Leaching from an unlined landfill will continue to cause groundwater contamination unless all the buried waste is removed and placed in a lined landfill, or an impervious cover is built on top of the landfill to prevent infiltration of rainwater.

15.5.3.2 Site Understanding

Once the source is identified, the site investigation is focused on determining the scope of the contamination. In order to evaluate potential impacts to human health and the environment from the contaminated site, the setting must be understood.

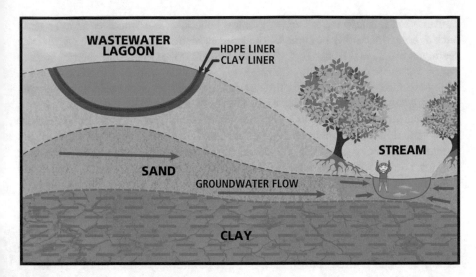

Fig. 15.3 Contaminated site – pond with liner

The site setting includes the built and natural environment where the contamination occurred and where the contamination could migrate. The natural environment includes the land (surface soils and subsurface geologic formations), water (streams, lakes, and groundwater), and air. In addition to investigating impacts to the natural environment, including the flora and fauna inhabiting that environment, impacts to the human environment and human health must be investigated. The human environment includes places where people live and work, sources of drinking water, and places where people recreate. Particular attention must be paid to places where children could be exposed to contaminants because they are more susceptible to health impacts.

If contaminants were released on the land, an understanding of the local geology will reveal how contaminants might infiltrate vertically into different geologic layers and in which direction contaminated groundwater would flow within a water-bearing geologic layer or aquifer. If subsurface soils consist mostly of coarse sand or gravel, contaminated groundwater can move freely and quickly in the subsurface. However, if subsurface soils are mostly tight clays, groundwater movement will be restricted. There are a wide range of geologic conditions that can be present, and they can vary considerably with depth. Groundwater typically flows into nearby surface waters so groundwater contamination can also cause surface water contamination. Contaminants released onto the ground surface could also be washed into nearby surface waters even if they don't migrate through groundwater to surface water. If contaminants were released directly into a waterway, the investigation will need to consider flow to other downstream waterways.

In addition to understanding the site setting, the nature of the chemicals that were released must be understood to evaluate their movement and transformation in the environment. Some chemicals readily volatilize and can cause airborne contamination. These vapors are of greatest concern if they enter enclosed spaces such as homes and other buildings. Some contaminants do not volatilize but migrate readily in water, so they can be transported long distances in groundwater or surface water. Yet other contaminants are relatively immobile and tend to stick to soils. Many chemicals will degrade in the environment into other chemicals; sometimes the degradation products are less toxic, and sometimes they are more toxic.

The current uses of the site by humans must be understood in order to evaluate the potential harm to human health. People could be directly exposed to chemicals in surface soil. For example, if contaminated soil is present in someone's yard, their children could accidentally ingest soil when they are playing outside. If someone is gardening in contaminated soil, they could absorb contaminants through their skin, or contaminants could be present in the produce. If construction workers are digging in contaminated soil, they can inhale dust containing contaminants. If contaminated soil is present beneath a house, vapors from that contamination could seep into the basement or other lower levels of the house. If contaminated groundwater or surface water is used for drinking or other purposes, people can be exposed to contaminants.

The variety of information needed to understand the site illustrates the interdisciplinary nature of contaminated site investigations. Hydrogeologists are critical to the investigation of sites with subsurface contamination because of their expertise in understanding subsurface geology and the movement of groundwater in the subsurface. If surface water is impacted, a biologist provides expertise in evaluating impacts to fish and other aquatic biota. A risk assessor is a scientist with expertise in calculating risks to human health from various contaminants and exposure scenarios. A chemist can provide expertise in the ways that chemicals are degraded and transformed under different conditions. In particular, a geochemist provides specialized expertise in the behavior of chemicals in the subsurface geology. With a focus on solving complex problems, engineers often investigate contaminated sites on their own or as part of a team of specialists. Engineers also may serve as project managers for multi-disciplinary teams.

15.5.3.3 Site Investigation

Once the source has been determined and the site is better understood, a plan is developed to define the areal and vertical extent of contamination in soil, groundwater, surface water, and/or air. Samples of potentially impacted media will be collected and sent to a laboratory to be tested for a list of chemicals that are associated with the source of the contamination. Because contamination is often present in the subsurface, site investigations often involve drilling borings to collect underlying soil and groundwater samples. Monitoring wells are often installed to test groundwater and determine groundwater flow direction. Triangulation of groundwater elevations surveyed in a minimum of three monitoring wells is required to determine groundwater flow direction. However, site investigations can require the installation and testing of many monitoring wells in multiple geologic layers (see Fig. 15.4).

If the contaminant source is a fuel or solvent, it may be present in the subsurface as a pure product rather than as residual chemical concentrations dissolved in groundwater or attached to soil. During the investigation phase, this "free product" may be observed in monitoring wells and/or soil samples. Free product can be lighter than water and float on groundwater, or it can be heavier than water and sink through groundwater. Contamination with free product typically occurs when the source is a fuel or solvent that has spilled or leaked from a tank, drum, or pond. The pure fuel or solvent migrates through the unsaturated surface soils and settles on top of the groundwater or, if it is heavier than water, migrates vertically through groundwater until it settles on an underlying geologic layer with limited permeability. Wherever the free product settles, it acts as a continual source by leaching contaminants into groundwater. During the investigation phase, it is critical to define the extent of free product in the subsurface environment.

A simple site investigation can be completed in a matter of days, while extensive investigations of complex contaminated sites can require years of effort. Complex site investigations are often conducted in phases with each phase providing information that guides the next phase of investigation.

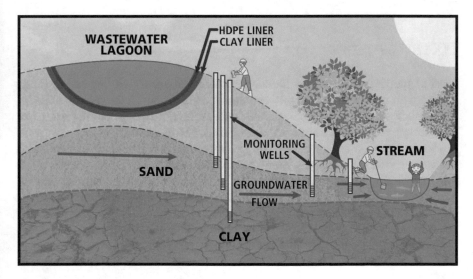

Fig. 15.4 Contaminated site with monitoring wells

15.5.4   Cleanup

After the site investigation is complete, the areal and vertical extent of contamination in groundwater, surface water, soil, and/or air is known. The next step is to evaluate whether the contaminants could cause harm to human health or the environment now or in the future. This evaluation must consider the current and future site uses. For example, acceptable chemical concentrations at residential sites are lower because the potential human health risk is higher. People generally spend more time in their homes, and more vulnerable people, such as children and the elderly, are exposed to chemicals in homes. Acceptable chemical concentrations at industrial or commercial sites are higher because people generally spend less time at these sites, and only working-age adults are exposed to chemicals at these sites. If all chemical concentrations are found to be safe for current and future site uses, cleanup is not required. However, if chemical concentrations could cause harm to human health or the environment now or in the future, corrective action (remediation) is needed.

It may not be technologically feasible or cost-effective to clean up the contamination. In these cases, action must be taken to prevent contact with the contaminated media. For example, contaminated surface soil could be paved over and the site used as a parking lot. If people at a residence were drinking contaminated well water, their well can be closed and their home connected to a municipal water supply that comes from another source. In some cases, a deed restriction or other legal document can be used to prevent future use of contaminated groundwater for drinking or future residential development of a site even if the property is sold.

If contamination can be cleaned up, there are many technologies to consider. The following paragraphs summarize commonly used cleanup technologies for soil and groundwater.

15.5.4.1   Soil Cleanup

Most of the risks associated with contaminated soil are usually related to direct contact with the contaminated soil, contaminants in the soil leaching into underlying groundwater, or vapors from volatile contaminants in the soil entering buildings. Soil cleanup technologies must reduce or eliminate these risks.

One of the most common corrective actions is the excavation of contaminated soil and disposal in a lined and permitted landfill to contain the contaminants. When full, the landfill will be covered to prevent direct contact with contaminants and to minimize exposure of the contaminated soil to rainwater that could cause contaminants to leach into groundwater. The liner minimizes the risk of leachate reaching groundwater, while the landfill is being filled as well as after it is full. Access to permitted landfills is restricted by fencing, and the landfills are monitored and maintained in perpetuity to prevent the release of contaminants to the environment in the future.

In some cases, contaminated soils can be left in place if they are not impacting groundwater and are covered with a cap that prevents direct contact with the soils and minimizes the potential for leaching to groundwater. The cap typically consists of low permeability clay and/or an impermeable synthetic liner that is typically constructed of high-density polyethylene (HDPE). Access to the capped soils must be restricted by fencing and deed restrictions. Long-term monitoring and maintenance are also required.

Some contaminated soils can be treated in situ by mixing with materials that solidify and/or stabilize the contaminants in place. These admixtures often include cement, lime kiln dust, fly ash, bentonite, and/or grout although a wide range of admixtures can be considered to react with a variety of contaminants. Solidification/ stabilization typically works best with soils that are contaminated with heavy metals although use of this technology to treat a wide range of contaminants is expanding. Treatability testing is required in advance to determine the specific admixture formula that will be effective on the specific contaminated soils present at a site. In order to treat contaminated soils in situ, an auger is used to bore holes at regular intervals throughout the contaminated area and mix the stabilization/solidification admixture into the soils. Treated soils can be left in place without a cap because they are stable and will not leach contaminants to groundwater at concentrations of concern.

Some organic contaminants can biodegrade under favorable conditions into less toxic or benign chemicals. Some organics biodegrade best under aerobic conditions (oxygen is present), while others biodegrade best under anaerobic conditions (little to no oxygen is present). Some soils contaminated with organic contaminants can be excavated and treated by landfarming to accelerate natural aerobic biodegradation of the organic soil contaminants. Another option for treating soils contaminated with organic contaminants may be thermal treatment with or without incineration. The contaminated soils are excavated and thermally treated to remove or destroy the contaminants. Soil testing is required following landfarming, thermal treatment, or incineration to confirm that organic contaminants have been removed or reduced to acceptable levels. The treated soils can then be placed back in the excavation, and no further controls are required to limit direct contact with the soils or to reduce leaching to groundwater. These types of soil remediation technologies are often used to treat soils contaminated with petroleum products or solvents.

Soils contaminated with organics may also have significant levels of organic vapors present within the soil matrix. These organic vapors pose a threat as they rise to the surface where they can accumulate in overlying structures or be released to the atmosphere where nearby people or animals can be exposed. Organic vapors can be removed from the unsaturated soils above the groundwater table by applying a vacuum to the in situ soils or by introducing air to the soils through sparging or venting. The extracted organic vapors often require treatment before they can be safely vented to the atmosphere. It may take months or years before testing confirms that these vapor removal technologies have reduced organic vapors in contaminated soils to acceptable levels.

15.5.4.2 Groundwater Cleanup

When groundwater has become contaminated, it is very hard to restore it to natural background concentrations because the groundwater is present underground within the small pore spaces of the soils and rock. The only way to access this subsurface contaminated groundwater is through wells that are drilled vertically or horizontally into the contaminated groundwater or by installing trenches either into the contaminated groundwater or downgradient of the plume at a location that will intercept the groundwater flowing in that direction. Groundwater remediation technologies typically consist of pumping out contaminated groundwater (see Fig. 15.5), treating it in situ, or monitoring natural biodegradation already occurring in the subsurface.

The characteristics of the geologic environment often complicate cleanup. For example, fractured bedrock is a rock matrix that has many fractures running in variable and often poorly understood directions. The groundwater will flow predominantly in these fractures so the direction and magnitude of groundwater flow may not be easily predicted. Even if groundwater is present in a highly transmissive, uniform sand or gravel matrix with a known flow direction, it is not possible to quickly pump all of the contaminated groundwater out of the subsurface. Only a portion of the contaminated groundwater flows from the soil matrix into a pumping well. Continued pumping through a network of recovery wells over a period of years is required to remove the contaminated groundwater and allow clean groundwater to flow into the area. If the soil matrix is a low permeability clay or other matrix that holds the groundwater more tightly than sands or gravels, it is even harder to pump out the contaminated groundwater. Continuous pumping over a period of many years is often required to remove enough contaminated groundwater to reduce concentrations to safe levels. All of the extracted groundwater must be

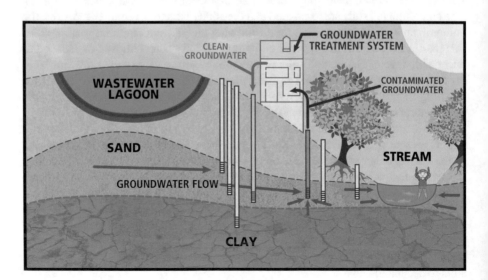

Fig. 15.5 Contaminated site with groundwater remediation

treated to remove the contaminants and the treated water either discharged to surface water or returned to the groundwater to accelerate flushing of contaminated groundwater toward the pumping wells.

In addition to the complications posed by the geologic matrix that holds the groundwater, the contaminants themselves may not be amenable to cleanup. Many contaminants are quite stable in the subsurface, while others are naturally biodegradable under the right conditions. Groundwater contaminants such as heavy metals do not biodegrade at all although they may be attenuated under the right conditions. Attenuation could occur by sorption to soils, precipitation in the groundwater, or transformation into other compounds less likely to migrate. It is important to understand the contaminants present and the subsurface conditions (aerobic, anaerobic, pH, etc.). It may be possible to inject chemicals into the groundwater to create the right subsurface conditions to transform contaminants so they are less mobile. Alternatively, pumping groundwater to the surface for treatment is often necessary to prevent migration. In some cases, a trench can be installed in front of the groundwater plume so the contaminated groundwater flows into the trench. The trench can be filled with chemicals to treat the contaminated groundwater as it flows through the trench or to create the right conditions for natural degradation or attenuation to occur. Contaminated groundwater can also be pumped out of recovery trenches.

If a groundwater contaminant is biodegradable, remediation efforts will focus on providing the right conditions for that degradation to occur. A common example is cleanup of contamination resulting from leaking fuel tanks at gas stations. Benzene, toluene, ethyl benzene, and xylene (BTEX) are volatile organic contaminants that are biodegradable under aerobic conditions. As long as the BTEX concentrations are not so high that they overwhelm the naturally occurring bacteria, biodegradation often occurs at these sites. Groundwater monitoring can be conducted to check for evidence of natural biodegradation. If it is occurring at an acceptable rate and there are no risks to humans or the environment, it may be possible to continue to monitor the site over time to confirm that degradation continues to occur naturally. However, if free product is present or if BTEX levels are too high, the bacteria may not be able to effectively biodegrade the contaminants. In this case, there are chemicals that can be injected into the groundwater to enhance the natural biodegradation, or the contaminated groundwater and free product can be pumped out until the subsurface conditions are suitable for the natural bacteria to effectively biodegrade the contaminants.

15.5.5   Redevelopment

For most contaminated sites, the goal is to restore the site such that it can be reused. Suitable site uses are determined during the investigation phase, and the level of cleanup that is conducted must support the planned site reuse. Many sites may be cleaned up and their existing use maintained. For example, there is ongoing groundwater cleanup at many gas stations, while the site continues to operate as a gas station. Other sites may be abandoned industrial sites that are not being used at the time the investigation and cleanup occur. In this case, the goal would be to clean up the site so it can be restored to productive use. If some contaminants remain onsite, commercial or industrial use could be appropriate as long as contact with residual contamination is limited. If the site has been completely remediated, residential use might be appropriate. Some sites can be cleaned up such that the site can be used as green space. Because contaminated sites are so common in the USA, redevelopment following cleanup – or during cleanup – is common. Municipalities may own contaminated sites and seek developers who will redevelop sites consistent with site use limitations. Developers may not initially be interested in these abandoned or underutilized sites because of the perception of contamination. The federal Brownfields program provides grants for communities to investigate sites and determine whether they actually require cleanup before redevelopment.

15.5.6   Community Involvement

Contaminated sites are often present near neighborhoods. People living near, or on, a contaminated site need information to understand the contamination and the risks to them. It is frightening to think you may not be able to see, smell, or taste contamination in your drinking water or on your property, yet it could be impacting your health and the health of your family. Engineers and scientists investigating and cleaning up contaminated sites must consider not only how to protect people who may be exposed to contaminants but also how to clearly communicate the facts to them. Especially when evaluating cleanup options, the input of the impacted community must be considered and is often required by state and federal agencies. Not only will residents be concerned about potential health impacts, they will also be concerned about the impact to their property values and to their community. For example, it may be technologically sound to put a cap over contaminated soils and leave them in place with proper monitoring to be sure there are no future impacts to groundwater. However, people living in the area may not want to live near a repository of contaminated soils. They may prefer the more expensive option of digging up the contaminated soil and hauling it offsite to a secure landfill.

Presenting complex scientific information to people without a technical background can be difficult. It is important not to unconsciously bias the information that is presented or else the community members will not trust the message. For example, providing a professional opinion that a contaminant concentration is "low" is not as effective as providing the actual concentration and the relevant government standard, so community members have proof that a concentration is low. All questions should be answered with openness and honesty. If community members feel they can trust the information they are receiving, it will reduce their stress about the situation.

A common problem when providing information to community members about contaminated sites is the communication of relative risk. Scientists and engineers are often hesitant to state that a contaminated site is "safe" because the level at which there is no health risk associated with a contaminant may not be scientifically proven. The risk from contaminants is typically stated in terms of the estimated incidence of excess cancer cases or other diseases caused by contaminants. An excess cancer risk of one case in a million people exposed is considered an acceptable risk but is not necessarily "safe" because it is still possible for someone to get cancer. Because of this discomfort with saying a site is "safe," a common tactic used to communicate risk at a contaminated site is to compare the risk to some other situation that is common. For example, the risk at a contaminated site may be compared to the risk of dying in a car accident. However, this is not an appropriate way to communicate risk because people can choose whether to drive a car, but a person does not choose to drink contaminated groundwater. It is better to provide facts about the potential risks from site contaminants and the ways to reduce or eliminate those risks.

15.5.7   Environmental Justice

According to the EPA [18]:

> Environmental justice is the fair treatment and meaningful involvement of all people regardless of race, color, national origin, or income, with respect to the development, implementation, and enforcement of environmental laws, regulations, and policies. This goal will be achieved when everyone enjoys:
>
> - The same degree of protection from environmental and health hazards, and.
> - Equal access to the decision-making process to have a healthy environment in which to live, learn, and work.

Low-income and minority residents are more likely to live near industrial facilities, landfills, and other sites that often cause contamination to the air, land, and water. For this reason, particular attention must be paid to these disproportionately impacted populations when investigating and cleaning up contaminated sites. Potential health and financial impacts to nearby residents must be studied and understood. Local residents must have the opportunity for meaningful involvement in decision-making. Complicated technical information should be provided by trusted advisors in a way that is understandable to non-technical people. If appropriate, information should be provided in multiple languages. Open, accessible, and regular communication will help people understand the site investigation and cleanup process and also facilitate their involvement in the decision-making process. The Woolfolk Chemical Works Superfund Site case history (Sect. 15.7) provides an example of the investigation and remediation of a complex Superfund site and also illustrates the importance of addressing environmental justice issues.

15.6 Conclusion

Although there have been many technological advances in the investigation and cleanup of contaminated sites since the Love Canal site raised public awareness in the 1970s, there are still new chemicals and products being created that are causing new contamination issues. Once a site is contaminated, it is still very difficult, expensive, and sometimes impossible to restore it to its previously uncontaminated condition. Although continued technical advances are needed to improve the investigation and cleanup of contaminated sites, more emphasis is needed on predicting the future environmental impact of current innovations to prevent the continued creation of new contaminated sites. Our water, land, and air are precious resources necessary to sustain life. As we continue to contaminate them, we are threatening human existence on Earth.

15.7 Case History – Woolfolk Chemical Works Superfund Site

15.7.1 Background

The City of Fort Valley, approximately 100 miles south of Atlanta, is the county seat of Peach County. Fort Valley's population of approximately 9900 is 75% Black and 22% White [7, 31]. Known as the Peach Capital of the World, Fort Valley is Georgia's largest peach-producing area [20]. Fort Valley is a small, economically disadvantaged community where many of the residents are unemployed and live below the poverty level [19].

The J.W. Woolfolk Company began producing and packaging pesticides on 18 acres of land in 1910 only two blocks from the Fort Valley downtown district and in the middle of a residential neighborhood (Fig. 15.6). It was not uncommon at that time to locate factories close to worker housing; the hazards of pesticides to human health were not well known. During World War II, arsenic trichloride was reportedly produced at the facility for the War Production Board. Production expanded during the 1950s to include formulation of organic pesticides. Eventually, the facility formulated a broad range of pesticides in liquid, dust, and granular forms for the agricultural, lawn, and garden markets [5, 6, 8]. Formulation and/or packaging of pesticides, herbicides, and insecticides (including arsenic and lead-based products) at the Woolfolk Chemical Works, Inc. facility continued through 1999 under several different owners. The material handling methods over the years caused extensive contamination not only on the Woolfolk Chemical Works Site but in the surrounding residential and commercial areas [13, 16].

The earliest documented complaint associated with the site occurred in 1966 when a State of Georgia water quality inspector investigated reports from local citizens that the facility discharged waste products to a ditch which flowed into a nearby

Fig. 15.6 The facility was used for almost 90 years for formulating and/or packaging pesticides, herbicides, and insecticides

creek [16]. State records indicate numerous instances where untreated industrial waste was discharged into surface waters. During a routine inspection in 1979, EPA discovered that the facility was discharging untreated pesticide production wastewater into an onsite storm sewer. This unauthorized wastewater discharge flowed into an open ditch south of the facility and then into a creek [8]. It was not until the early 1980s that complaints to the Georgia Environmental Protection Division (EPD) resulted in investigation and cleanup actions [13]. In 1985 and 1986, the Georgia EPD detected metals and pesticides; including lead, arsenic, chlordane, DDT, lindane, and toxaphene; in onsite soil and groundwater as well as in the open ditch south of the plant [8].

15.7.2 Interim Remediation

Canadyne-Georgia Corporation (CGC) purchased the Woolfolk Chemical Works facility in 1972 and was the owner when contamination was discovered. In 1986, they began investigating and conducting cleanup at the site in consultation with the Georgia EPD [3, 13]. CGC voluntarily cleaned up some onsite contaminated soils and demolished an onsite building contaminated with arsenic. Soils that were

Fig. 15.7 Wastes in this one-acre onsite capped area used for disposal of interim cleanup wastes were later excavated and treated before being placed in an onsite lined disposal cell or disposed of offsite

considered to be hazardous waste because of high arsenic and lead concentrations were taken offsite for disposal at a permitted hazardous waste landfill. Other soils with lower concentrations of arsenic and lead were disposed onsite in a one-acre area along with some demolition debris and non-hazardous lime-sulfur sludges remaining from onsite activities. These wastes were covered with two feet of low permeability clay and a 30-mil HDPE cap (Fig. 15.7). Although no liner was placed under the wastes before disposal, the cap was designed to minimize rainwater infiltration into the buried wastes so contaminants would be unlikely to leach to groundwater present approximately 10 to 20 feet below [5, 6, 13, 16].

15.7.3   Superfund Designation

At the same time that CGC was voluntarily conducting remediation in consultation with the Georgia EPD, the EPA began investigating the Woolfolk Chemical Works Site. In 1990, EPA formally placed the site on the Superfund program's NPL because of contaminated groundwater and soils resulting from facility operations [16]. Although the majority of the contamination occurred prior to their purchase of the facility, CGC was legally liable for the cleanup and was the primary PRP that actively investigated and conducted cleanup at the site [3, 16].

15.7.4   Initial Superfund Investigations

Under EPA direction and oversight, CGC conducted investigations in 1991 and 1992 to evaluate the extent of contamination in soil, groundwater, surface water, sediments, and air [5, 6]. The results showed contamination was present well beyond the 18 acres of the Woolfolk Chemical Works Site. The designated Superfund site was eventually determined to cover an area of 31 acres and groundwater contamination migrated over a mile beyond the Superfund site boundaries. Although the initial cleanup focused on soils with high levels of arsenic and lead, additional contaminants were identified including a variety of pesticides and semi-volatile organic contaminants. As investigations continued, it was found that not only were the soils at the production facility contaminated, but the yards of surrounding residences and commercial properties also were contaminated. Further investigation revealed that airborne dust containing arsenic at unsafe levels had entered people's homes. Although groundwater was contaminated, it was not a source of drinking water, and the City of Fort Valley municipal water supply wells were not contaminated by the Woolfolk Site (Fig. 15.8).

Fig. 15.8 The author sampling waste byproducts remaining onsite

15.7.5   Legal Issues

At the same time that CGC was beginning additional site cleanup under the
Superfund program, they were sued by over 600 Fort Valley residents who were
negatively impacted by the Superfund site (Fig. 15.9). The 1993 class action lawsuit
was reportedly settled for $11,000,000 after several years of litigation [25, 28, 30].
CGC continued to investigate and remediate site contamination, while the lawsuit
was ongoing although interactions with the community members were complicated
by the active litigation. In addition to the cost of the litigation and settlement, CGC
estimated the cost for the Superfund work at approximately $50,000,000 [30]. It
became more difficult for CGC to pay these high costs, and in 1998 the EPA took
over the investigation and cleanup activities at the site using federal funds [16].
Ultimately, CGC's parent company paid EPA $5,000,000 in 2005 to settle further
claims with the US government. Eventually CGC's parent company filed for Chap.
11 bankruptcy protection in 2014 [29].

15.7.6   Contaminated Soils and House Dust

In 1993, EPA required CGC to clean up residential properties to eliminate the
immediate threat to public health from contaminated soil on residential properties
[9]. CGC purchased some properties and permanently relocated the residents, while
other residents were temporarily relocated so that contaminated soil could be excavated from their property and arsenic-laden dust removed from their homes. During

Fig. 15.9 The Woolfolk Chemical Works Superfund Site located in the middle of Fort Valley,
Georgia, had a negative impact on the community

Fig. 15.10 Some of the buildings were in poor condition and required demolition. Special handling during demolition and disposal was required for some buildings with contaminated building materials

1994 and 1995, approximately 23,000 tons of contaminated soil and debris were removed from 43 properties and five road shoulder areas and rights-of-way. An onsite wooden building was demolished because the wood was contaminated with dioxin (Fig. 15.10). EPA also required removal of sediment and soil from the storm-water ditch that drained the site [5, 6, 13].

In 1994 and 1997, the living spaces of residential houses with arsenic-contaminated house dust were cleaned by CGC using high-efficiency vacuum cleaners with special filters [2]. Following the initial cleaning, CGC retested homes to confirm that the cleanup had met EPA criteria [4]. EPA conducted additional testing in 2002 that found more arsenic contamination in residential yards and attic dust. In 2008, EPA excavated additional contaminated soil from residential properties, cleaned 63 attics that had dust-containing elevated levels of arsenic, and decontaminated a drainage pipe [13].

The contaminated soils that had been consolidated in a capped area onsite before the Woolfolk Site was listed on the NPL were investigated in 1996. It was found that the volume and depth of contaminated soils were significantly greater than originally estimated. It was determined that arsenic-contaminated soils could be treated onsite by solidification and stabilization to minimize contaminant leaching. If the solidified soils passed leaching tests, they could be disposed of onsite in new HDPE-lined containment cells to further limit the potential for contaminants to leach to groundwater [5, 6, 13].

In 1998, EPA required that CGC excavate the contaminated soils from the capped area and dispose of them offsite. However, CGC claimed they did not have sufficient funds to conduct this work because of the large volume of soils to be removed. They believed EPA's estimate of 8000 cubic yards of soil requiring excavation from the capped area was much lower than the actual volume. EPA conducted additional investigations and increased their estimate of the volume of contaminated soil to 120,000 cubic yards. Based on this higher estimated volume, EPA determined that the contaminated soils would be treated by solidification and stabilization before placing them in HDPE-lined containment cells to be constructed onsite [11].

Because of CGC's financial difficulties, the cleanup of the onsite capped area was conducted by EPA using federal funds. During the cleanup activities, the actual volume of contaminated soil was found to be even higher than EPA's second estimate. Between 2006 and 2010, approximately 500,000 cubic yards of contaminated soil were excavated to an average depth of 35 feet [10]. Batches of arsenic-contaminated soils were treated by solidification and stabilization using Portland cement, lime kiln dust, and ferric or ferrous granules or powder. A sample of each batch was tested by the Toxicity Characteristic Leaching Procedure (TCLP) to evaluate whether arsenic and other contaminants would leach from the treated soils. Those soils that passed the leaching test were placed in three HDPE-lined areas constructed onsite. Because the actual volume of contaminated soils found during the cleanup was significantly greater than originally estimated, there was not enough capacity in the onsite HDPE-lined containment areas. Once these onsite lined areas were full, treated soils that passed the leaching test were disposed of offsite in a municipal landfill. Treated soils that did not meet treatment standards were disposed of offsite in a hazardous waste landfill. After they had been filled, the lined containment cells were paved [13] (Fig. 15.10).

15.7.7 Groundwater

In 1994, EPA required that CGC remediate groundwater contaminated with arsenic and organic contaminants. Following treatability testing and design, CGC constructed a groundwater pump and treat system and began operating it in 1998. Contaminated groundwater was pumped from 24 wells to the surface, treated by precipitation, filtration, and activated carbon adsorption before being discharged to the Fort Valley municipal wastewater treatment system for further treatment prior to discharge [3]. In 2002, the groundwater pump and treat system was shut down by CGC due to inadequate performance but restarted in 2008 by EPA. The system was again shut down in 2014 because it appeared that the groundwater plume was continuing to migrate even while the system was in service. Contaminated groundwater was found more than 2.5 miles downgradient of the Woolfolk Site despite operation of the groundwater pump and treat system for more than 15 years.

From 2008 to 2019, various private wells and monitoring wells were tested, and pesticides were detected in residential and irrigation wells downgradient of the

Woolfolk Site. Groundwater beneath the Woolfolk Site also still contained high levels of arsenic and other contaminants. Higher pesticide concentrations found in private wells downgradient of the site could be due to a source of groundwater contamination other than the Woolfolk Site or pesticide contamination from the Woolfolk Site that was found in the drainage ditch leading to an offsite stream [13]. EPA continues to investigate the source of this groundwater contamination.

15.7.8 Redevelopment

During initial cleanup activities, CGC purchased and cleaned up 13 properties under EPA oversight. Deed restrictions for these properties were required to prevent residential development and the use of underlying groundwater. Following cleanup, CGC transferred the properties to the City of Fort Valley and contributed funds to support redevelopment [13]. In 2002, an EPA Brownfields pilot grant provided to the city from the Superfund Redevelopment Initiative was used for reuse planning. The city conducted a redevelopment study and sought input from local residents on potential redevelopment ideas. After extensive discussions, the community decided to build a library and literacy center on some of the remediated properties. Another property containing the Troutman House, a once-contaminated antebellum farmhouse, was converted into a Welcome Center and office space for the Fort Valley Chamber of Commerce (Fig. 15.11). Another former residence nearby was redeveloped as an adult education center. Residences and other private properties that were impacted by contamination were cleaned up and remain in use. In recognition of their proactive engagement and support of reuse, EPA presented the City of Fort Valley with an Excellence in Site Reuse Award in 2009 [12].

Much of the former Woolfolk Chemical Works site where contaminated soils were excavated, treated, and disposed in lined cells has been paved and is now being used by recreational vehicles and for school bus parking. Other areas of the site that were remediated and revegetated are used for an event space called the Fort Valley Festival Park that opened in 2012. The Fort Valley Public Works Department also built a community playground and veterans memorial on the remediated site in 2013 [17, 23].

15.7.9 Community Involvement

During the early days of the Woolfolk Superfund process, the community mistrusted the regulatory agencies and CGC. This distrust manifested itself in the perception that the entire city was contaminated and the health of Fort Valley's residents was compromised. Members of the Black community formed the Woolfolk Citizens Response Group (WCRG) to address issues related to the Woolfolk Site and potential threats to the health of the community [19].

Fig. 15.11 Following cleanup, properties like this one have been returned to productive use. The Troutman House is now a welcome center that also provides office space for the Fort Valley Chamber of Commerce

Fort Valley residents needed a way to voice their concerns and receive progress reports from EPA, other federal and state agencies, and technical experts. The Woolfolk Alliance was established in 1998 to provide a forum to discuss and reach consensus on cleanup issues at the Woolfolk Site. Woolfolk Alliance members included representatives of the local, county, and federal government; the WCRG; local businesses; any private citizens that wanted to attend; and CGC. The initial meetings, led by an outside facilitator, were tense because there was a great deal of anger, frustration, and mistrust among the citizens, the government agencies, and CGC. After the difficulty of the first few meetings, the group asked the Mayor of Fort Valley to preside. From that point on, Mayor John Stumbo effectively led the group and was able to quiet the anger, encouraging attendees to ask questions and make comments in a respectful way. He personally paid for homemade lunches for the group during each meeting because he felt that if the attendees could eat together, they would get better acquainted and talk about their families, and a community would be formed [26, 27]. At each of the Woolfolk Alliance meetings, technical presentations were made to explain ongoing investigation and cleanup activities to the group and solicit their input. Members of the Woolfolk Alliance met regularly in Fort Valley for more than 20 years [19].

In 2010, EPA awarded the Woolfolk Alliance the National Community Involvement Award for outstanding achievements in the field of environmental protection and for its dedication and commitment to the cleanup and redevelopment of the Woolfolk Site. In 2011, EPA awarded Peach County and Houston County public health representatives to the Woolfolk Alliance the Notable Achievement Award for demonstrating a sustained and thorough understanding of environmental justice concerns and assisting in providing opportunities for the community to play a meaningful role in the environmental decision-making process [16].

References

1. Carson, R (1962) Silent Spring. Houghton Mifflin Company, Cambridge
2. CGC (Winter 1997-1998) Woolfolk Update #1, Woolfolk Chemical Works Site, Fort Valley, Georgia.
3. CGC (Spring 1998a) Woolfolk Update #2, Woolfolk Chemical Works Site, Fort Valley, Georgia.
4. CGC (Summer 1998b) Woolfolk Update #3, Woolfolk Chemical Works Site, Fort Valley, Georgia.
5. CH2M HILL (1996a) Draft Feasibility Study Addendum, Operable Unit No. 3, Woolfolk Chemical Works Site, Fort Valley, Georgia. March 1996.
6. CH2M HILL (1996b) Woolfolk Chemical Works Capped Area Investigation, Fort Valley, Georgia. December 1996.
7. City of Fort Valley https://www.fortvalleyga.org/ (2021). Accessed Aril 10, 2021.
8. EPA (1988) NPL Site Narrative for Woolfolk Chemical Works, Inc., Fort Valley, Georgia. June 24, 1988.
9. EPA (2004) US Army Corps of Engineers, Savannah District, Savannah, Georgia for U.S. Environmental Protection Agency Region 4, Atlanta, Georgia. First Five-Year Review Report for Woolfolk Chemical Works Site, Fort Valley, Peach County, Georgia. August 2004.
10. EPA (2009) E^2 Inc., Charlottesville, Virginia for U.S. Environmental Protection Agency Region 4, Atlanta, Georgia. Second Five-Year Review Report for Woolfolk Chemical Works, Inc, Fort Valley, Peach County, Georgia. September 2009.
11. EPA (2014) State of Georgia, Environmental Protection Division, Land Protection Branch, Atlanta, Georgia for U.S. Environmental Protection Agency Region 4, Atlanta, Georgia. Third Five-Year Review Report for Woolfolk Chemical Works, Inc., Fort Valley, Peach County, Georgia. September 2014.
12. EPA (2018) U.S. Environmental Protection Agency Region 4, Atlanta, Georgia. Site Redevelopment Profile, Woolfolk Chemical Works, Inc. Superfund Site, Fort Valley, Georgia. December 2018.
13. EPA (2019) U.S. Environmental Protection Agency Region 4, Atlanta, Georgia. Fourth Five-Year Review Report for Woolfolk Chemical Works, Inc. Superfund Site, Peach County, Georgia. September 2019.
14. EPA (2021a) Milestones in EPA and Environmental History. https://www.epa.gov/history/milestones-epa-and-environmental-history (2021). Accessed January 25, 2021.
15. EPA (2021b) Love Canal Niagara Falls, NY Cleanup Activities https://cumulis.epa.gov/supercpad/SiteProfiles/index.cfm?fuseaction=second.Cleanup&id=0201290#bkground Accessed Jan 25, 2021.
16. EPA (2021c) Superfund Site: Woolfolk Chemical Works, Inc. Fort Valley, GA. https://cumulis.epa.gov/supercpad/SiteProfiles/index.cfm?fuseaction=second.Cleanup&id=0401315#bkground (2021). Accessed February 7, 2021.
17. EPA (2021d) Superfund Sites in Reuse in Georgia – Woolfolk Chemical Works, Inc. https://www.epa.gov/superfund-redevelopment-initiative/superfund-sites-reuse-georgia#woolfolk (2021). Accessed April 19, 2021.
18. EPA (2021e) Environmental Justice at the EPA https://www.epa.gov/environmentaljustice (2021). Accessed May 2, 2021.

19. Fort Valley (2002). Application for Brownfields Assessment Grant, Fort Valley, Georgia. December 2002.
20. Fort Valley Main Street https://www.fortvalleymainstreet.org/about-fort-valley.cfm (2021). Accessed March 31, 2021.
21. The Life and Legacy of Rachel Carson. https://www.rachelcarson.org/Bio.aspx (2021). Accessed January 25, 2021.
22. National Geographic (2020). New Challenges for Us All. April 2020.
23. 41NBC/WMGT (2012). Woolfolk Site Renamed as the "Fort Valley Festival Park" https://41nbc.com/2012/08/27/woolfolk-site-renamed-as-the-fort-valley-festival-park/. August 27, 2012.
24. Nelson G (1980) Earth Day '70: What It Meant. EPA Journal. April 1980. https://archive.epa.gov/epa/aboutepa/earth-day-70-what-it-meant.html
25. The New York Times (1993). Neighbors Say Factory Contaminated Town. https://www.nytimes.com/1993/08/15/us/neighbors-say-factory-contaminated-town.html?smid=em-share. August 15, 1993.
26. Stumbo J (2006 est.). Memorandum regarding the Woolfolk Superfund site in Fort Valley, Georgia, Prepared for the United States Senate Committee on Environment and Public Work by Dr. John E. Stumbo, Mayor of Fort Valley, Georgia. https://www.epw.senate.gov/public/_cache/files/5/2/52ce3839-b8a9-4ba1-9e0d-21d12090d91d/01AFD79733D77F24A71FEF9DAFCCB056.memorandumforsenatecommittee610.pdf
27. The Telegraph (2020). Stumbos have left indelible mark as they prepare to leave Fort Valley. https://www.macon.com/news/local/news-columns-blogs/ed-grisamore/article53894995.html. January 5, 2020.
28. The Telegraph (2021). Marvin Crafter, Sr. Obituary. https://www.legacy.com/obituaries/macon/obituary.aspx?pid=198114666. March 22, 2021.
29. United States Bankruptcy Court for the District of Delaware (2016). In re:Reichhold Holdings US, Inc., et al., Debtors. Chapter 11 Case No. 14-12237 (MFW) Jointly Administered. Settlement Agreement Between the United States and the Debtors. Filed January 12, 2016.
30. The Wall Street Journal (1997). NationsBank is Cleared of Liability in Cleanup of Toxic Site in Georgia. https://www.wsj.com/articles/SB881088704303282000?st=srhgvmcon79tmj6&reflink=article_email_share. December 3, 1997.
31. Wikipedia, Fort Valley, Georgia. https://en.wikipedia.org/wiki/Fort_Valley,_Georgia (2021). Accessed April 10, 2021.

Rebecca Lance Svatos, P.E., became an environmental engineer to try to leave the world better than she found it. As a child she spent her summers outdoors and enjoyed hiking and camping with her family. She learned to love snakes, spiders, and bats rather than fear them. While she was growing up in the 1960s, the environmental movement was raising awareness of environmental problems and she became interested in a career that would solve them. Through her father, a mechanical engineering professor at the University of Iowa, she was exposed to engineering. Because her father was the faculty advisor for the student section of the Society of Women Engineers (there were no female engineering faculty then), she met impressive women engineering students that inspired her to pursue an engineering career.

Becky enrolled in civil engineering at the University of Iowa in 1978. Although the engineering curriculum was difficult, she persevered and completed her BS in Civil Engineering in 1982. She wanted to continue her studies and obtain a graduate degree in environmental engineering but was too tired (and broke) to start graduate school immediately. While she was in college, she felt supported by the faculty and rarely encountered negative attitudes regarding women in engineering. However, this changed when she graduated and entered an engineering workforce with only 4% women. She quickly realized she was not always welcome in the engineering profession and was going to be under increased scrutiny as a woman in a male-dominated profession.

Becky joined the US Army Corps of Engineers in 1982 and was told she was only the second woman engineer in the Omaha District. After a couple of years, she realized that government work was not for her. She left her job and entered the University of Texas at Austin where she developed a computer model of water quality in a nearby reservoir that was being threatened by development. The model demonstrated that allowing the discharge of treated wastewater from the residential developments into this lake would threaten its renowned water quality. Lake Travis is still clear, clean, and beautiful because these wastewater discharges have not been allowed.

Following graduate school, Becky decided to pursue a career in consulting engineering. The fast pace and variety of problems to solve has kept her interested and engaged for 35 years. Her primary areas of expertise are the investigation and cleanup of contaminated sites, water quality permitting, and pollution prevention. She has enjoyed working for a wide range of clients throughout the USA. Much of her experience has been with private industry where she is continually fascinated by the innovative methods used in the manufacture of all kinds of products from breakfast cereals to toothbrushes and tractors. She has helped manufacturing facilities comply with myriad, confusing (and often illogical) environmental regulations as well as helped them prevent pollution by implementing improved material handling processes, training, and building better spill containment structures. Where pollution prevention failed, she investigates contamination and figures out how to clean it up. These contaminated sites have ranged from small corner gas stations and dry cleaners to sites with contamination covering hundreds of acres. Although she has worked on many contaminated sites, she has only been able to restore a few to their former uncontaminated condition, witnessing firsthand the importance of preventing pollution.

Becky also enjoys using her engineering skills to help others. For more than ten years, she has been part of a group of Iowa engineers working with a grassroots organization of small communities in El Salvador to help them obtain clean, reliable drinking water. True engineering problem-solving is needed in situations like this where resources are limited.

In addition to her technical accomplishments, Becky has held a variety of management positions and helped hundreds of engineers, and scientists achieve their full potential. She has mentored many women engineers and enjoys seeing the growth and acceptance of women in the engineering profession. Her love of engineering has led her to speak to countless students of all ages from preschool through graduate school about careers in this rewarding field. She still loves solving new problems and recommends engineering to anyone who likes to figure things out and wants to make the world a better place.

Chapter 16
Solid Waste

Jenna R. Jambeck, Eliana Mozo-Reyes, and Katherine Shayne

Abstract In this "peek behind the curtain" of what happens to solid waste after it is created, we discuss the structures around waste from its language inception to its final disposition and potential mitigation strategies. We talk about the concept of waste in different cultural contexts and the urgency to reconsider established social structures around it. We discuss solid waste management steps from home collection, hauling, etc., which significantly differs throughout the world depending on cultural, geographical, and economics factors. We delve into some of the most popular solid waste final disposition mechanisms including landfills as the most popular of the group, composting and recycling process and challenges, and briefly mention other less popular ways to dispose of solid waste. In terms of mitigation strategies, we discuss current location-bound approaches like circularity assessments and recycling apps, as well as product reimagining and governmental involvement.

Keywords Solid Waste · Culture · Entanglement · Globalization · MSW · Product Design · Collection · End-of-cycle · Leakage · Circularity

16.1 Introduction

The concept of "waste" only exists to humans. Consider the world without humans, nothing is a waste – each output is an input into another system. Nature, in all its complex and interconnected systems, has designed-out waste. The textbook definition of waste (noun) is *material that is not wanted; the unusable remains or*

J. R. Jambeck (✉)
Department of Environmental Engineering, University of Georgia, Athena, GA, USA
e-mail: jjambeck@uga.edu

E. Mozo-Reyes · K. Shayne
Athens, GA, USA

© The Author(s), under exclusive license to Springer Nature   391
Switzerland AG 2022
P. Layne, J. S. Tietjen (eds.), *Women in Infrastructure*, Women in Engineering
and Science, https://doi.org/10.1007/978-3-030-92821-6_16

byproducts of something (Google definition). However, have you ever heard the saying, "One person's trash is another person's treasure"? What you or I might consider "waste" someone else may find an extremely useful item or useful as an input into some process. The term waste is completely subjective and influenced by your history, culture, social interactions, and context [25]. Also, the word waste as we know it in the American English language doesn't even have an identical translation in some other languages. Everyone's context around the world is different. It is, then, necessary to address solid waste as an interconnected global issue that depends specifically on the human experience at the site of creation, where culture, resources, and even geography have created structures around both collective and specific experiences with waste. To illustrate these unique understandings, the authors of this chapter offer a set of personal stories.

First is Jenna's story. As an environmental engineering undergraduate student, I was told I should pick a focus on one of the following: drinking water treatment, water reclamation (wastewater treatment), air pollution control, or solid waste management. When I took my first solid waste management course, I was enthralled. The community my university was in was facing a tough choice. Their landfill was full. Where would their waste go? Would they expand the local landfill? Would they ship the waste out of town? There were a lot of heated discussions over this choice. People really cared where their waste would go. However, in reality, people didn't want the waste they created on a daily basis anywhere close to them. The city eventually decided to build a long-haul transfer station (a facility where waste is aggregated for transfer in tractor-trailer trucks). I felt that the human components – all those people who expressed their opinion on where the waste should go and what the city should do with it – were inseparable from the engineering that went into designing the integrated solid waste management system. I realized the human component is critical to any engineering design. Besides, from then on, I enjoyed my engineering courses even more, and I was completely hooked on integrated solid waste management as my focus. I went on to research and work in it for graduate school, in consulting, for my PhD, and onwards. I remain as interested in it as ever today.

Second is Eliana. The biggest encounter I had to stir me toward solid waste happened during the official switch to digital TV after I came to the USA. Having grown up in the Colombian culture, with farmer grandparents who disposed of food waste either in the backyard or as animal feed and who didn't ever consider a plastic container single-use, it was a shockingly strong point of tension to see discarded TVs around an already full of food waste and assorted plastics dumpster. When the shift happened in Colombia, we got new TV boxes that would convert the new digital signals to analog so we could still use our old TVs. We didn't consider throwing out TVs; I didn't know this was possible. We have repair shops for everything.

So, in order to understand this dichotomy, I changed my plans to get a Bioengineering MS for a Masters in Environmental Engineering, where I found my experience could make a bigger difference. After exploring solid waste in the Jambeck Research Group, I decided to look at how culture influences waste creation because in trying to encourage behavior in the USA, I ended up inadvertently encouraging consumerism [29].

Katherine also changed her mind toward environmental engineering, this time from law school, with her experience relating more to the experience of urbanized students identifying mentorship as a catalyst for understanding and applying pro-environmental behavior [26]. In her words, after starting environmental engineering – I joined the Jambeck Research Group, and Dr. Jenna Jambeck became a mentor, advisor, and role model. I loved learning about context-sensitive design, systems thinking, green engineering, environmental policy, and how to change the world through design and reducing our waste. I went from being a student, to an undergraduate researcher, to a graduate student in Dr. Jambeck's group, to, currently, running a recycling tech company based on our research. I am a product of her passion and enthusiasm for studying solid waste systems, and I want to continue designing the future to be waste-free.

These stories represent the differences that personal journeys, culture, and history (even in the form of mentorship) make in people's understanding of waste. At the same time, through their own journey, the authors found their way toward a shared point – three different paths toward the same goal of understanding how waste is created in order to properly address it.

16.2 Rethinking Waste

There are many examples of people taking solid waste and doing something useful or beautiful with it. Besides social media accounts and websites currently teeming with upcycling projects, some people have created specifically impactful projects with significant ripples. Some examples of the latter include art made entirely out of plastic by Mbongeni Buthelezi, a South African artist whose access to typical art supplies was extremely more limited than that of plastic waste [1]; the "Treasures in the Trash" exhibit at the New York Museum of Sanitation curated by Nelson Molina, a retired sanitation worker during 34 years of collecting items destined for the landfill [2]; and the beautiful music from the Landfill Harmonics orchestra in Cateura, Paraguay, whose instruments were lovingly made by their community (whose main income comes from informal recycling) using discarded items separated near the landfill where they live [3]. These examples beg the question then, when does waste become waste?

Theoretically, for humans, it happens when no person left in the world considers it a useful item for any purpose. Unfortunately, in practice, it is when, locally, an item has no use. This happens although globalization has made it such that materials that are perceived of no use on one side of the world can be transported across the globe (once they are aggregated) as will be discussed in the Waste Exporting and Importing section. If an item is no longer useful to a specific entity or community, it might be deemed waste even before all its useful components are unusable.

Consider, for instance, a cafeteria or a mall food court, like Eliana did during a case study of culture/waste entanglements in both the USA and Colombia, where food is served on trays and people carry the tray with food on it to their table. They eat – they may or may not eat all the food, and there may or may not be packaging associated with the food. However, what do most people do when they are done with food? They push the tray away. That same food that a person was just eating a few seconds ago is pushed away, and in that act, it becomes waste, although this may not stop another person from eating it or finding a second life for the discarded items.

This happens in the Colombian city that we studied [25]. Leftover foods are collected for animal feed, while recyclable items get separated at the source guaranteeing less recycling contamination, thus more post-consumer material yield. This is done, however, by a team of custodians in charge of cleaning the food court instead of the first user (term we use instead of "end user" to create an understanding of what happens later). This extra task is part of the custodians' contract, and they get paid for this work. This dynamic might explain why separation at the source is less common in the USA – no one is paid to do it.

In the food court of the US mall, the first users dispose of their own waste into the closest trash receptacle to continue with their own activities [25]. Separation at the source requires an effort that people find confusing and cumbersome, compared with the simple tasks of turning off a faucet or switching off a lightbulb which we understand to be environmentally friendly behaviors. Unfortunately, there is rarely a team of professional separators available when dealing with waste at home or on-the-go.

In our Mitigation Strategies section, we will discuss how governments, academia, nongovernmental organizations (NGO), industries, and citizens have and can develop ways to mitigate the impact of solid waste production. However, it is important that we understand that although the issue of solid waste is a global one, highly interconnected and entangled, in order to understand and address it, we need to look at the threads that tie it to history, culture, and specific circumstances like geographic location and language. This is a crucial next step toward environmental sustainability and solid waste management.

We have, so far, as engineers, done the work of tracking, collecting, and managing solid waste to the extent that people tend to forget about it. In our industrialized world, we have done an incredible job of getting the waste "away." Solid waste trucks often come door-to-door and pick up waste, and they do it with as little disturbance as possible. We love to put our waste in our trash or recycle cans, place it on the curb, and have it gone. It's not something we think about much, unless the system breaks down. People want their trash and waste out of sight and out of mind, so when the trucks miss a collection cycle and then they notice solid waste, they have to think about it. So, where is "away" when you throw something away?

Although we highly recommend visiting your local materials recovery facility (MRF – pronounced "Murf") where recyclable materials are processed and your local landfill if you have one, this chapter will allow you to peek behind the curtain of how the solid waste management works from the comfort of your own environment.

16.3 Waste Generation and Characterization

An estimated 2 billion metric tons of municipal solid waste is generated globally each year [6] as urban household waste. While we measure waste in mass, other characteristics of it are important. Materials with low bulk densities,[1] like plastic, can be a small percentage by mass but a much larger percentage by *volume. Moisture content* also adds another important characteristic to waste and that impacts mass. Characteristics of a waste stream are critical to how it will be managed. Currently, waste management is a very reactive system. Managers/engineers have to work with what kind of waste comes their way, as well as the dynamics of it – changing characteristics in materials and products and ebbs and flows of quantities.

Waste flows and characteristics are impacted, among other things, by how the economy is doing. In the 2008 recession in the USA, waste quantities decreased [7]. Moreover, historically speaking, times of great hardship and war denote a decrease in solid waste production and an increase in materials recovery efforts [4]. On the other end, we have the data pointing at high-income and middle-high-income countries as the biggest waste producers in the world [6]. Additionally, consumerism is directly related to how much and what type of waste we produce: think toys, electronic devices, furniture, and fashion. Also, in terms of availability, anything we buy is being created and shipped around the world at the fastest pace we have ever observed.

Waste generation and characterization, then, encompass quantities of waste generated around the world, characterization of waste, and changes over time in either/or both variables.

16.3.1 Municipal Solid Waste

Municipal solid waste (MSW) is what people produce every day at home in their daily lives. It is typically measured in per person (per capita) generation rates. These rates are often quoted in kg/day but can also be quoted in kg/year. Generation rates differ all around the world for various reasons and are influenced by several factors.

One of the largest factors to influence waste generation is economics (Fig. 16.1). Overall, a country's waste generation rate is correlated with their economic status, at least while the economy is developing. There is a direct correlation between

---

[1] Solid waste bulk density is calculated as the weight of the waste divided by the volume it occupies. This volume includes the volume of the waste and the volume of the space between waste items. Solid waste bulk density is typically expressed in lb./yd3 or kg/m3.

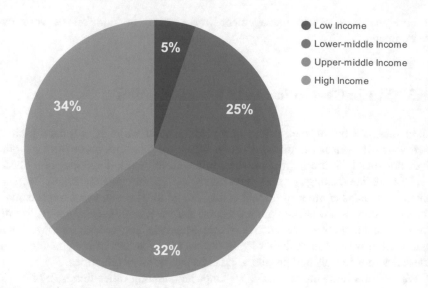

Fig. 16.1 Waste generation by income level

economic growth and increase in waste generation ([6]. P. 20). It is expected that waste generation in middle- and lower-income countries will continue to grow as the economies emerge, and their populations continue to grow. For example, the per capita waste generation in the currently low-income cities in Asia and Africa is predicted to double in the next 15–20 years. Migration trends of people moving from rural to urban areas and to the coast, as well as the number and size of cities, continue to increase [27]. By 2030, an estimated 43 megacities (population over ten million people according to the UN) will exist in 32 developing economies [8].

High- and upper-middle-income countries were estimated to generate 66% of total worldwide waste in 2016, and their per capita rates had doubled from 1970 to 2000. However, since 2005, the rates have mostly stabilized. This means that economic growth and waste generation increases are coupled, to some end point, and then may fluctuate based upon local economics.

The World Bank characterized waste according to countries' various income levels for 2016. The average rates of per person generation show low, lower-middle, and upper-middle-income countries under 1 kg/capita/day with high-income countries at around 1.5 kg/capita/day ([6]. P. 20). Characteristics of waste differ as well (Fig. 16.2). Typically organic waste is still a large fraction (more than half) of the waste stream, except in high-income countries. In higher income countries, this fraction drops to an average of 32%.

It is important to highlight, as well, that the second most common material in the waste streams is currently plastic. You may have noticed by now that this chapter does not only focus on plastic. This is because once plastic is in our waste stream, it is mixed with all the other materials there, so a plastic waste problem is an *all* waste

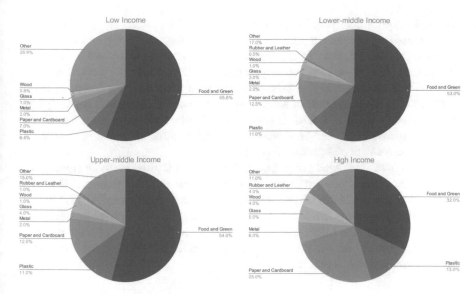

Fig. 16.2 Waste composition by income level

problem. However, plastic does have a focus because of its leakage into our environment and eventually our oceans. Plastic washes up on the shores of our beaches, like confetti in the waves (Jambeck, personal experience at an island that interrupts the ocean currents); it can be found in the depths of the ocean and at the highest peaks in the world [9, 10]. It is estimated that 11 MMT of plastic from our mismanaged solid waste enters our aquatic systems each year [18], and without further and more aggressive actions on humans' part to reduce this quantity, it could reach 20–53 MMT by 2030 [17].

16.3.2   Additional Classifications of Waste

Besides municipal solid waste, which we all generate in our homes, there are other kinds of classifications of waste that make their way into the MSW streams either directly (through contracting for disposal) or indirectly, some with significant quantities generated outside of the home or our daily activities. This is not all encompassing, but some of these more specific waste streams are described here.

16.3.2.1   Commercial Waste

Commercial waste is waste that is generated outside the home at stores, businesses, recreation areas, education facilities, entertainment facilities, sport arenas, hotels, and other similar facilities. Waste may be generated as a part of the business itself (paper from printing, waste from shipment of materials, packaging, etc.), or it may be generated by the people who visit or use a facility (concert or sporting event attendees, customers, etc.). In either case, the commercial facility needs to provide waste receptacles (trash cans and recycle bins) to allow for the internal collection of waste. Then, the waste is typically collected by a waste hauling company that is under contract. Management of solid waste at commercial entities is typically built into the cost of running the business or facility. Commercial waste entities will often have dumpsters. These can be as small as 10 cubic yards up to as big as 40 cubic yard dumpsters with compaction systems.

If a customer is charged per pickup and by weight, it provides motivation for the customer to reduce the quantity of waste generated by their business. However, historically, contracts have often had payments set for a dumpster pickup frequency, e.g., weekly, regardless if it is full or not. In this case, the customer often feels it is in their best interest to fill the dumpster. The logic is "I am paying for them to pick up the dumpster if it is full or not, so might as well fill it." This is a big disincentive for waste reduction and recycling.

16.3.2.2   Industrial Waste

Industrial waste is generated by manufacturing facilities and processing plants, and it is not easy to characterize as a whole since it is so variable. It can be anything from off-spec food products to leftover brewery mash, etc. Some of this waste can be classified as hazardous depending on its characteristics, and other times it might be a strong candidate for recycling or beneficial use since it is often a consistent type and quantity generated at one location.

16.3.2.3   Medical Waste

Medical waste is waste that is mainly generated in doctors' offices, hospitals, and clinics around the world. If it is has come into contact with bodily fluids, it is considered a biohazard, and it is specially regulated to be autoclaved[2] which kills all the bacteria and pathogens that might be on the material before it can be disposed of. It is sometimes then managed at the same facilities as regular MSW, but normally it is managed separately. The COVID pandemic saw an increase in medical waste and also personal protective equipment (PPE) use by the general public, which entered our MSW stream as well as our environment [24].

---

[2]Autoclave is like an oven; it heats up materials/items in it to high temperatures, in this case, high enough to destroy bacteria and pathogens.

16.3.2.4 Electronic Waste

Electronic waste, or e-waste, is a term that is used to describe all electronic waste that is generated. From smartphones to tablets, to computers, to televisions, all of this waste is alike in that it contains circuit boards and sometimes wiring and batteries. Circuit boards contain a lot of conducting material, copper, and solder, which can contain lead, and other hazardous materials. An older cathode ray tube (CRT) television set, like the ones mentioned in this chapter's introduction, for instance, contains high levels of lead (an average of 4 pounds) [5]. Because quantities of e-waste have been rapidly increasing in recent times as electronic devices have become ubiquitous in our daily lives, e-waste is one of the most exported wastes from high to low income countries despite the International Convention and the Basel treaty, who have defined e-waste as hazardous and have prohibited its shipment because of the high monetary value of the metals (see more in the previous chapter and the import export section on this chapter).

16.4   Waste Collection

Solid waste collection is a vital part of the management process. For most people, it is an afterthought, or at least most people don't have an opinion about it until they can't find a trash can. However, placing receptacles is not a simple task. Location, size, shape, look, and function are all critical to creating the complex system that most people don't even notice until it is needed. Beyond the bin, haulers take over to get the waste to its next destination like the transfer station, processing, recycling, or disposal. This section will focus both on bins that people interact with directly, commercial dumpsters and then the people and vehicles that haul the waste.

16.4.1 Home Bins, Receptacles, etc.

Almost everyone has a bin or place to put solid waste inside their home. In some countries, there are trash cans in nearly every room of the home – the kitchen can have a trash can, recycle bin, and a compost pail. Other rooms, bathrooms, bedrooms, and offices often have trash cans and sometimes recycling bins. In other countries, there might be only one bin or place to put trash in the entire home. All waste gets aggregated here.

From the bins inside the home, trash gets aggregated to be taken off-site. This may mean a citizen simply puts trash in a roll-cart or "toter" and pushes it to the curb, or it may mean a plastic bag gets placed outside of buildings, either on the sidewalk or hung high (on a tree branch, pole, etc.), so that animals don't get to it (Fig. 16.3). It could also mean in some cases that a citizen must take their waste to a drop-off center.

Fig. 16.3 Two methods of aggregated trash preparation for hauling

Inside the home, it is not usually a problem to get trash or recycling into these bins, but you can also look at the home as a tiny microcosm of the outside world. In a tidy home, one would not be likely to leave a wrapper or empty beverage container sitting out. However, if there were a pile of wrappers or several cans sitting on a dresser, you might just add to that since it appears that is where it goes – assuming you, or someone else, will collect that material later to properly recycle or dispose. Being a confined space, eventually household trash gets into the bins in the home. Outside of the home, the latter becomes more challenging. Then, after items make it into a bin, hauling is required.

16.4.2 Hauling

Hauling is the term for any municipal or private method (a person walking, bikes, trucks) of picking up waste to take it somewhere. There are many methods used to collect and haul waste, as well as some guidelines and requirements to make it more efficient. It can be quite a dangerous job no matter where or how it is done (e.g., working in traffic, repetitive lifting, exposure to solid waste). It is also the last step for most people for seeing their waste. Most companies try to do it with minimal disruption to citizen's daily life. In some cases, this means working in busy cities in the early morning hours or during the middle of the night. There is also a science to designing collection routes for hauling [15]. For all collection and hauling, whether on foot, bike, or with trucks, the desire is to find the most waste to be collected in the smallest area or distance traveled. This means the populated places to pick up trash on foot or where trash is already aggregated. This also means there must be enough people per mile of road to make the curbside collection with a truck economically feasible. In addition, in a densely populated high traffic city, idle time is reduced as much as possible. Collection routes are scheduled at times when traffic is less and to minimize left turns (trucks have to wait for traffic to turn left, which increases idle time). It may mean that a truck picks up both sides of the street at once or may drive down one side and back up the other to go a different direction. Technology with GPS and routing software helps collection systems to be as efficient as possible and has been available for over 10 years [16].

16.4.2.1   Foot, Cart, or Bike

Waste can be collected by foot, cart, or bike in some communities where the streets are too narrow or not designed for any other transportation. This more manual collection can be provided by a municipality or private company, but around the world, it is often a private citizen (waste picker) completing this work independently (Fig. 16.4). This waste picker may not only pickup waste that is placed outside of homes but may also pick up litter in the community – any waste that is valuable. In some areas, people visiting homes collect certain items only (e.g., all the newspapers) and citizens pay a small fee for this collection. In other cases, the trash and recyclables are set out for community collection and some of them are collected by others, just because they are valuable (but only if they are valuable). Collection on foot and with carts also occurs near, or on, the working face of landfills. These waste pickers can collect material each day to earn enough money to feed their families. The working conditions are unhealthy and dangerous. Waste pickers are exposed to anything that is in the waste, often without any protective clothing like boots, gloves, or dust masks. Working on the landfill can easily result in injury from the heavy equipment and trucks, as well as from the conditions with degrading waste and fires, which can be common in these scenarios. Organizations, like Women in the Informal

Fig. 16.4 Waste pickers transporting waste. (Credit: Jenna Jambeck)

Employment: Globalizing and Organizing (WIEGO), advocate for recognition and improved working conditions for waste pickers around the world.

16.4.2.2 Trucks

Trucks can be an efficient vehicle for solid waste collection. Trucks come in various sizes; the fuel can be compressed natural gas, or even landfill gas; and they can be equipped with the latest technology such as an on-board computer platform with large displays and DVR systems, multiple (up to 8) cameras, GPS tracking and live streaming video capability, remote vehicle system diagnostics, route management software, and integration with billing and maintenance software (McNeilus). For home collection (low-rise single family detached dwellings), crews of 1–3 people drive and pick up waste in various trucks, and citizens put their trash cans or roller carts on the curb. In multi-family dwellings and commercial settings, waste is taken out and put into a dumpster. When there is no space to store carts or cans (e.g., various urban settings regardless of size from New York City to Athens, Georgia), trash is placed out in bags, and trucks simply collect the bags manually.

16.4.2.3 Drop-Off Location

A high enough population density is needed for a municipality or private company to have curbside solid waste collection services. As an example, if only three households are along a 10-mile stretch of road, then the expense of labor and fuel for a truck is not economically favorable for collection services. In some cases, trash collection is provided in semi-rural or rural areas, but recycling is not (recycling is often picked up on a separate trip and two trips cannot be justified). In the USA and other similarly developed areas, if household curbside collection is not provided (and sometimes even if it is), then the community provides a drop-off location (e.g., a transfer station) for citizens to transport their own waste for disposal and recycling. This waste is aggregated and then transported from these transfer stations to the local disposal or recycling center. Sometimes drop-off at these transfer stations is free (if you are a resident of the community), and sometimes the transfer stations operate on a fee-based system, which is based upon quantity (e.g., price per bag of trash). Where curbside collection is not provided or available, illegal dumping or management "at home" rates are higher and participation in recycling programs are lower [19].

16.4.3   On-the-Go and Event Collection

On-the-go disposal and recycling is the common terminology for disposal and recycling that occurs outside the home. The purchase and consumption of much of our convenience foods with single-use packaging occur outside the home as we go about our day. In the USA, for example, an average of 2.44 lb. (1.1 kg) of waste is generated per individual at public events [20]. On-the-go solid waste disposal and recycling represent an important part of capturing more waste and recyclables, reducing litter and increasing the recycling rates in the USA and worldwide.

16.4.4   Separation of Materials

In many places in the world, materials are separated out for disposal and recycling. In most cases, the materials that can be recycled are often collected separately from the trash (e.g., different day) or need to be dropped off at a different location. *Source-separated* materials for recycling are materials that are separated by the citizen (at the source of the waste). In our homes and communities, citizens would separate out glass (sometimes also by color), newspaper, office paper, cardboard, metal (e.g., cans from canned goods), aluminum, and plastic (often into the different polymers based upon recycling number). In the early 2000s in the USA, advancement in material separation technologies at recycling facilities (MRFs) meant that citizens could now put all of their recyclables into one container mixed together, and the recyclables could be separated out at the MRF. This type of collection is called *single-stream collection.*

Single-stream collection increased rapidly, and the majority of communities in the USA and MRFs are single stream. Since citizens just put everything that is recyclable in one container (often a covered cart and not an open bin, which anecdotally encourages more recycling since your neighbors can no longer see what you put out), more people recycle and a greater mass is put in the recycle bins. However, there has also been a downside to the single-stream system. While the MRF does separate the mixed materials into the different categories, they are not as clean or pure as they would be if source separated. This impacts the commodity markets (who will take these recyclables) and pricing (what will people pay for them). Reported "contamination" rates at the USA curbside are 17% by mass [21]. See more in the mitigation section on *wishcycling*.

16.5   Waste Management Strategies

After the solid waste is collected through any of the methods described in the previous section, it has to be processed and managed for final disposition. There are different ways to process the waste at this stage. Some communities have evolved their methods according to their own needs, but other communities have adopted, though popular, ill-fitting methods of solid waste disposal that sometimes result in big investments with no return since they were never implemented ([25]. p. 82). In general, burying, burning, and similar processes have been used to make our trash seemingly disappear. In this section, we will explore these solid waste *final* disposition techniques and their potential impacts.

16.5.1   Landfill

Landfills have been a part of our solid waste management system for centuries. Unwanted material was not only disposed of on the land but used to build up the land. Since our first waste was natural materials and waste from using our natural resources, places like Boston were built on "landfill." Waste characterization has changed with time and various development stages, of which the industrial revolution and advent of plastic are worth highlighting. In the USA today, landfilling material is still the way the majority of waste is managed. The USA still has ample room for constructing landfills, and it remains the most inexpensive means of managing solid waste.

The Resource Conservation and Recovery Act (RCRA), which governs the management of solid waste in the USA, was passed in 1976. This regulation outlined the requirements for design and operation of landfill facilities. Before 1976, landfills were unregulated "dumps" of waste, where citizens drove to drop off their own waste into the pit – and vectors (animals, etc.) were allowed to interact with the waste (Fig. 16.5). In lower-income countries, landfills look similar to the landfills in the USA in the 1970s and 1980s before the regulations were universally required (phased in).

As of 1976, with the passing of RCRA, landfills were required to be lined with composite liners (geomembrane and clay), with the leachate (the liquid that percolates through the waste in the landfill) collected and removed from the landfill. The RCRA also requires landfill covers with gas collection pipes. Landfills are heavily regulated in the USA including siting (based upon hydrology and groundwater, among others), fencing and security, liner systems, filling protocols and daily cover soils, leachate collection, gas collection and post-closure, and contingency plans (Fig. 16.6).

According to RCRA, landfills should be constructed with a composite liner system (as thick as a smartphone) that lines the bottom of the entire landfill and sides. Underneath it is two feet of compacted clay that must have low permeability (i.e.,

Fig. 16.5 An unregulated dump with black bears from the USA 1970s. (Photo Credit: Paul Marenchin)

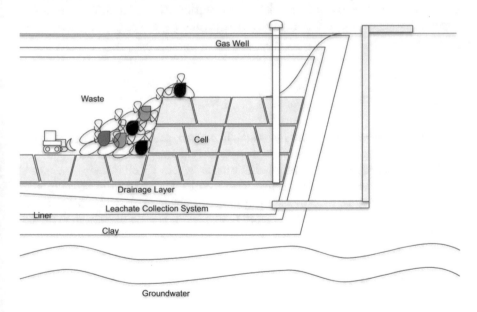

Fig. 16.6 Landfill basic components

compacted so it is a barrier later to water; it won't let water through it easily). In some cases, landfills are designed with a double liner system, so the entire liner system is constructed twice, with a secondary leachate collection system or leak detection system underneath the first liner system.

Much like how coffee is created by water percolating through coffee grounds, leachate is the liquid that drains from and percolates through solid waste (from precipitation). The bottom of the landfill is sloped to one side (typically 1–2% slope) with a drainage layer (geogrid or sand) that liquid can flow through and then be collected in perforated pipes at the lowest point. Then, the leachate can drain to a sump and be pumped out of the landfill to be partially or fully treated on-site and/or shipped off-site for treatment. Trucks drive to the working face of the landfill where they tip (place) their waste. Bulldozers move the waste while compactors compact it so that it becomes as dense as possible (then the landfill can fit as much waste as possible into its permitted airspace). Landfill gas wells are placed throughout the landfill after it has reached its fill height. Wells are installed to provide collection of landfill gas that is composed of methane (50–60%), carbon dioxide, nitrogen, and other trace gases.

The goal of typical landfill closure is to keep moisture from precipitation out and continue to collect leachate and landfill gas. Moisture is kept out by capping the top of the landfill with an impermeable layer (typically another geomembrane like the one used to make the bottom liner system) and allowing drainage off the landfill. Soil and grass might be planted on top, or sometimes the geomembrane is exposed. Solar panels can be placed on the facilities, or sometimes the sites are beneficially used for things like recreational fields. Monitoring of the landfill upon closure (landfill leachate, gas, and groundwater) is required for typically a minimum of 30 years.

16.5.2   Recycling

Recycling is one way to capture materials, like metal, paper, and plastic, for recovery and remanufacturing. Separation of materials can happen in two ways: positive sorting and negative sorting. Positive sorting is where specific items are picked manually or mechanically from a mixed waste stream (e.g., picking plastic bottles from a conveyor belt). Negative sorting is where materials that are unwanted are separated from materials that are wanted (e.g., picking plastic bags from the material on the conveyor belt).

16.5.2.1   Dirty MRF

A dirty MRF is where the entire waste stream (trash and recyclables) is delivered to a facility, and then recyclables are picked out of this mixed waste input. While dirty MRFs make it easy on citizens (there is no separation of materials required), it makes separation on the back end more challenging, since all wet waste (e.g., food

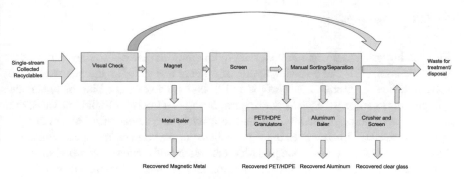

Fig. 16.7   Separation and processing steps for MRFs

waste) is mixed with things like paper, plastic, and glass, so it can be challenging to get clean recyclable streams from a dirty MRF.

16.5.2.2   Clean MRF

A clean MRF is where recyclables that are separated from trash at the source are delivered. The recyclables can be source separated into categories or can be mixed in a single stream. Recycling streams from a source-separated MRF are the most clean, followed by the recyclables from a co-mingled single-stream MRF. Separation and processing steps for MRFs include visual and manual sorting, tool-aided sorting, stacking, shredding, and granulation, among others, all-in preparation for reintroduction to the manufacturing stream (see Fig. 16.7).

In general, every community has a local approach to solid waste management (SWM) and recycling. Even within the same country, for example, in the USA, what can be recycled in Atlanta, GA may not be recyclable in Oklahoma City, OK. Many communities rely on extensive outreach and education campaigns to increase awareness about why and how to recycle. The recycling rate in the USA has not risen above 34% [7] and there is still significant contamination in the recycle streams. It is hard for communities to find messages that resonate with people and that they will remember. Moreover, the information might be dry, hard to find on the web, and challenging to communicate. Time-intensive campaigns like door-to-door education, recycling violation notices from haulers, and visits to schools are programs that have resulted in some success. However, since new residents move into and out of communities regularly, visitors often do not know local rules, and recycling programs can change. As a result, the outreach and education program is a continuous and never-ending process.

16.5.3   Composting

Composting processes organic materials in an environment where they can biodegrade. At home, this can be a simple pile of organic material like kitchen waste and leaves outside. In some cases, people build their own compost bins or pre-made compost bins can be purchased. Since composting is a microbiological process, the microbes need an optimum environment to actively and effectively biodegrade the organic material. This involves the right balance of carbon and nitrogen, moisture, and air (if aerobic). While biodegradation can occur with oxygen (aerobic) or without oxygen (anaerobic), aerobic degradation has reduced odor and occurs faster. Oxygen is added to home composting systems just by stirring or turning a compost pile. In the future, there might be widely available polymers that can biodegrade at home. The polymer closest to scale for consumer use is polyhydroxyalkanoate (PHA), which should biodegrade in home compost, although currently polylactic acid (PLA) is widely used but can only compost in industrial settings.

Industrial composting is conducted at a much larger scale. Organic materials are typically delivered to a concrete pad or flat field area. The materials like yard waste are often size reduced with shredders and mixed with other materials like food waste or sludge from a wastewater treatment plant. The composting itself can happen in windrows which are long piles of the material. The windrows are infused with oxygen through turning by on-site equipment. Industrial composting needs to meet requirements for temperature to be able to kill bacteria and pathogens. This is especially important for composting current compostable polymers like PLA that need a high enough temperature to reach hydrolysis in order to be able to biodegrade.

16.5.4   Thermal Treatment

Combustion facilities burn waste and can also potentially recover the energy. Combustion reduces the volume of solid waste typically by 90 percent. However, there is residual ash that needs to be managed after the combustion in a landfill. For a waste combustion facility to remain cost-effective, it needs a steady stream of waste to operate and create electricity. Because of the investment in air pollution control systems, which can be large and expensive, facilities taking small amounts of waste are not likely to be economically feasible nor are those located in remote areas without consistent waste material inputs. A site-specific waste flow analysis and design would be needed to determine economic feasibility. Permitting a combustion facility, which, in the USA, is conducted by each individual state, is an extensive process as it requires solid waste, air, water, and stormwater permits, as well as potential permits for land use. Combustion facilities have also historically been controversial, and public comment and input is required before permitting, construction, and operation.

Other thermal treatments exist for waste like gasification (conversion of organic waste into its basic building blocks (carbon monoxide and hydrogen) with a small amount of oxygen input in the process) and pyrolysis (thermal conversion in the absence of oxygen). Both of these processes have not been used at large scale for mixed waste and are not an easy way to make waste "disappear."

16.6 Waste Exporting and Importing

Just like consumer goods, solid waste management has become increasingly globalized. While there are international regulations on the shipment and export of hazardous waste (Basel Convention), there are no conventions on the shipment of solid waste. Plastic scrap (plastic meant for recycling) is a commodity that is exported and imported around the world, a trend that has been increasing since data became available in 1988 [22]. Before 2018, over half of the world's exported plastic was going to China [22]. China was importing plastic (and other materials) as the manufacturing hub of the world. However, this limited China's capacity for doing its own in-country recycling and became a burden as single-stream recycling became more prevalent around the word and thus the scrap became less usable and more contaminated with other unwanted materials. In 2017, China announced import restrictions for all purposes became a ban. It was estimated that 111 MMT of plastic scrap would be displaced through 2030 [22]. Then, from 2018 onward, plastic exports both increased to other countries but also decreased overall globally as alternate countries also said that they did not want to take the global scrap [23].

16.7 Mitigation Strategies

Waste mitigation depends on multiple stakeholders at the different levels of waste creation and processing. For instance, in the USA, the philosophy of "Reducing, Reusing, and Recycling" has been making a difference in waste management during the last few decades, and lately there have been new advancements in the reducing and reusing areas, besides the extensively explored recycling. In Colombia, one of the biggest examples of reducing and reusing has happened naturally for decades. Glass bottles containing beer or soft drinks are still collected and exchanged at the consumer level. That is, when a person purchases a personal size beer or soda, the said person must bring an empty bottle to buy the new one or leave a deposit and return the empty bottle after. This is the same principle being used by companies bridging the gap to wide availability (e.g., Loopstore, Algramo, etc.) all over the world, who use reusable containers to refill with products already in the market, thus, giving producers and users a way to come together in a more sustainable way. Like these, there are currently plenty of solid waste stakeholders creating methods of waste mitigation and solid waste system improvement.

16.7.1 Recycling Contamination Mitigation

There are multiple initiatives helping to solve the contamination crisis. One such company is called Can I Recycle This (CIRT). People often ask questions about what is recyclable or not in different locations across the USA. What is accepted for recycling is different depending on location and yet is often difficult for a citizen or consumer to find on a city's website, especially when it comes to plastics. People were hoping that what they put in the recycle bin was recyclable in their community and thus tended to recycle packaging and items that should not go to the MRF – this is known as *wishcycling*.

CIRT (www.cirt.tech) is a recycling technology platform that gives retailers, brands, and consumers on-demand, location-specific information about how and where to recycle materials to increase material recovery and decrease contamination in the recycling stream. Using artificial intelligence, location data, and application programming interfaces (APIs, ways to connect companies directly to the database), CIRT is fusing more technology into recycling and waste management, promoting the circular economy through trackable, actionable information. The CIRT platform collects and analyzes data of millions of products and their packaging to ultimately inform and motivate companies and their brands on packaging redesign and/or replacement. The goal is to increase recycling rates all over, starting in the USA, saving cities millions of dollars in waste management, capturing post-consumer material for reuse, motivating changes in packaging to reduce quantities, and becoming more recyclable to keep materials out of the environment and ocean.

16.7.2 Policy, Bans, and Mandates

Policy has become one of the most effective methods of waste mitigation, especially for plastics management. Increasing in popularity throughout the world, plastic bans, for instance, have helped reduce plastic littering, thus keeping it away from the natural environment ([6]. P. 117). From the phasing out of single-use plastic to plastic bags levies and bans, multiple governments have been working toward reducing the amount of post-consumer plastics in the world.

According to the United Nations (UN), there are 127 countries, out of 192 studied (66%), which have enacted some sort of plastic bag regulation ([11]. P. 10). Out of these, 83 countries have banned plastic bags from their stores (retail distribution), and 43 have instituted some sort of Extended Producer Responsibility acts (EPR). When it comes to single-use plastic, only 27 countries have enacted bans, but 63 have enacted EPR measures. It is important to mention that the majority of countries enacting this type of policy are located in the geographic south of the globe, where the majority of lower-income countries are located ([6] p. 16). There are additional regulations and mandates on different types of plastics and some specific types of solid waste in the world, but they mostly refer to hazardous materials.

16.7.3   Sustainable Product Design

Regarding plastic production, brands have started to join efforts to curtail waste through reusing and refilling initiatives [14]. Currently, some companies offer the products that people normally purchase but in reusable containers for a refundable deposit fee. When customers return their containers, they get their deposit back and the containers are put to new use.

This effort joins the goals of multiple companies who have taken a zero-waste to landfill commitment and in some cases even achieved it [12]. Some of these companies even bring their efforts to involve others by helping them achieve zero-waste goals, either by advising other companies or offering zero-waste products or packaging to end-users. So, although technology is still in a race for relevance creating products that become obsolete in very little time, manufacturers are trying to become a part of the system of sustainable materials management.

16.7.4   Circular Economy

Circularity or a circular economy strengthens the idea that it is in the best financial, environmental, social, etc., interests of all parties to create a zero-waste cycle of materials and products; taking a page from nature, creating no waste, every output becomes an input [28, 30]. The circular economy is not recycling; it concentrates on delivering products without waste in the first place, designing systems that do not promote overconsumption (like the story about the TVs at the beginning of this chapter). While this concept seems to hit all the marks to "solve our solid waste problem" – what does it really mean for our daily lives? For our cities and communities where our solid waste is managed?

One approach to applying circularity to our communities at the front line is the Circularity Assessment Protocol (CAP) born out of the Jambeck Research Group in the Circularity Informatics Lab (https://www.circularityinformatics.org/) at the University of Georgia. The CAP has been implemented successfully in 27 cities of 10 countries. The goal of this protocol is to establish the circularity status of a specific community, in order to provide the necessary data for local, regional, or national decision-making to become more circular (e.g., reduce waste and leakage). The model has seven spokes around two concentric circles that include policy, economic, and governance factors around the potential stakeholders/influencers in the community. The seven spokes include product availability (input), community attitude (community), materials and designs (product design), use and reuse in the community (use), collection infrastructure (collection), final disposition (end of cycle), and leakage of waste to the environment (leakage).

When sharing and co-owning these data analytics, the expectation is to involve and empower communities for decision-making based on their own definition of waste. Through this approach, we acknowledge and respect the circumstances of a particular cultural group in order to make appropriate decisions to manage their own solid waste as they recognize it in their community, which is a necessary first step in the right direction. After all, we need to recognize that the issue of solid waste management in the world depends on the knowledge and actions of all people, more than seven billion of them. Small actions, taken collectively make an impact, but companies and governments play larger roles in driving this system. With everyone working together on this issue, maybe we can get to a place where every "waste" is a treasure.

References

1. NRDC- Natural Resources Defense Council. (2017). Turning to the Landfill to Make Fine Art. Retrieved on August 25, 2021 from `https://www.nrdc.org/onearth/turning-landfill-make-fine-art`
2. Sanitation Foundation. (n.d.). New York Museum of Sanitation- Treasures in the Trash. Retrieved on August 25, 2021 from `https://www.sanitationfoundation.org/museum`
3. NPR: Ideas and Issues. (2012). The Landfill Harmonic: An Orchestra Built From Trash. Retrieved on August 25, 2021 from `https://www.npr.org/sections/deceptivecadence/2012/12/19/167539764/the-landfill-harmonic-an-orchestra-built-from-trash`
4. Strasser, S. (2000). Waste and want: A social history of trash. Macmillan.
5. EPA: United States Environmental Protection Agency. (2006). Final Rule: Streamlined Management Requirements for Recycling of Used Cathode Ray Tubes (CRTs) and Glass Removed from CRTs. Retrieved on August 25, 2021 from `https://www.epa.gov/hw/final-rule-streamlined-management-requirements-recycling-used--cathode-ray-tubes-crts-and-glass`
6. Kaza, S., Yao, L., Bhada-Tata, P., & Van Woerden, F. (2018). What a waste 2.0: a global snapshot of solid waste management to 2050. World Bank Publications.
7. US EPA, 2020. Advancing Sustainable Materials Management: 2018 Fact Sheet December 2020. `https://www.epa.gov/sites/default/files/2021-01/documents/2018_ff_fact_sheet_dec_2020_fnl_508.pdf`
8. United Nations, Department of Economic and Social Affairs, Population Division (2018). The World's Cities in 2018—Data Booklet (ST/ESA/ SER.A/417). `https://www.un.org/en/development/desa/population/publications/pdf/urbanization/the_worlds_cities_in_2018_data_booklet.pdf`
9. Gibbens, S. 2019. Plastic proliferates at the bottom of world's deepest ocean trench, National Geographic. `https://www.nationalgeographic.com/science/article/plastic-bag-mariana-trench-pollution-science-spd?loggedin=true`
10. Napper, I.E., Bede F.R. Davies, Heather Clifford, Sandra Elvin, Heather J. Koldewey, Paul A. Mayewski, Kimberley R. Miner, Mariusz Potocki, Aurora C. Elmore, Ananta P. Gajurel, Richard C. Thompson, Reaching New Heights in Plastic Pollution—Preliminary Findings of Microplastics on Mount Everest, One Earth, Volume 3, Issue 5, 2020, Pages 621-630.
11. Excell, C., Salcedo-La Viña, C., Worker, J., & Moses, E. (2018). Legal Limits on Single-Use Plastics and Microplastics: A Global Review of National Laws and Regulation. United Nations Environment Programme, Nairobi, Kenya.
12. ZeroWaste. (2021). Top 10 Zero Waste Companies Pushing for a Sustainable Future. Retrieved on August 25, 2021 from `https://www.zerowaste.com/blog/top-10-zero-waste-companies/`
13. Circularity Informatics Lab, The. (n.d.). Circularity Assessment Protocol (CAP). Retrieved on August 25, 2021 from `https://www.circularityinformatics.org/`

14. Pinsky, D., & Mitchell, J. (2019). Packaging Away the Planet: U.S. Grocery Retailers and the Plastic Pollution Crisis. Greenpeace, USA.
15. Tchobanoglous, G., Hilary Theisen, and S. A. Vigil, 1993. Integrated Solid Waste Management: Engineering Principles and Management Issues, McGraw-Hill.
16. Fickes, M. 2011, Off the map, Waste360, `https://www.waste360.com/route-optimization/map`.
17. Borrelle, S.B., Ringma, J., Law, K.L., Monnahan, C.C., Lebreton, L., McGivern, A., Murphy, E., Jambeck, J., Leonard, G.H., Hilleary, M.A., Eriksen, M., Possingham, H.P., De Frond, H., Gerber, L.R., Polidoro, B., Tahir, A., Bernard, M., Mallos, N., Barnes, M. and Rochman, C.M. 2020. Predicted growth in plastic waste exceeds efforts to mitigate plastic pollution. Science 369(6510), 1515.
18. Lau, W.W.Y., Shiran, Y., Bailey, R.M., Cook, E., Stuchtey, M.R., Koskella, J., Velis, C.A., Godfrey, L., Boucher, J., Murphy, M.B., Thompson, R.C., Jankowska, E., Castillo Castillo, A., Pilditch, T.D., Dixon, B., Koerselman, L., Kosior, E., Favoino, E., Gutberlet, J., Baulch, S., Atreya, M.E., Fischer, D., He, K.K., Petit, M.M., Sumaila, U.R., Neil, E., Bernhofen, M.V., Lawrence, K. and Palardy, J.E. 2020. Evaluating scenarios toward zero plastic pollution. Science, eaba9475.
19. Tunnell, KD, 2008. Illegal Dumping: Large and Small Scale Littering in Rural Kentucky. Journal of Rural Social Sciences, Vol 23, No 2. December 31, 2008.
20. Cascadia Consulting Group, 2006. Waste Disposal and Diversion Findings for Selected Industry Groups, California Integrated Waste Management Board, Executive Summary, `https://www2.calrecycle.ca.gov/WasteCharacterization/PubExtracts/34106006/ExecSummary.pdf`, Accessed August 4, 2021.
21. Recycling Partnership, 2020 State of Curbside Recycling Report, `https://recyclingpartnership.org/wp-content/uploads/dlm_uploads/2020/02/2020--State-of-Curbside-Recycling.pdf`
22. Brooks, A., Wang, S. Jambeck, J. (2018). The Chinese import ban and its impact on global plastic waste trade, Science Advances, 20 Jun 2018: Vol. 4, no. 6, DOI: https://doi.org/10.1126/sciadv.aat0131
23. Brooks, A. (2021). From the Ground up: Measurement, Review, and Evaluation of Plastic Waste Management at Varying Landscape Scales, PhD Dissertation, University of Georgia, Athens, Ga, USA.
24. Ammendolia, J, Saturno, J, Brooks, AL, Jacobs, S, Jambeck, JR (2021). An emerging source of plastic pollution: Environmental presence of plastic personal protective equipment (PPE) debris related to COVID-19 in a metropolitan city, Environmental Pollution, Volume 269, 15 January 2021, 116160.
25. Mozo-Reyes, E. (2016). *Thinking outside the chute: following a rhizome of waste and culture* (Doctoral dissertation, University of Georgia).
26. Medina-Jerez, W. (2008). Between local culture and school science: The case of provincial and urban students from eastern Colombia. Research in Science Education, 38(2), 189–212.
27. ISWA- International Solid Waste Association. (2015). ISWA Report 2015. Retrieved on September 21, 2021 from `https://www.iswa.org/annual-reports/?v=7516fd43adaa`.
28. Huysman, M (1994) Waste picking as a survival strategy for women in Indian cities. Environment and Urbanization 6(2), 155–174.
29. MozoReyes, E., Jambeck, J. R., Reeves, P., and Johnsen, K. (2016). Will they recycle? Design and implementation of eco-feedback technology to promote on-the-go recycling in a university environment. Resources, Conservation and Recycling, 114, 72–79.
30. World Economic Forum; Ellen MacArthur Foundation; McKinsey and Company (2016): The new plastics economy – Rethinking the future of plastics. http://www.ellenmacarthurfoundation.org/publications.

Dr. Jenna R. Jambeck is a Georgia Athletic Association Distinguished Professor in Environmental Engineering in the College of Engineering at the University of Georgia (UGA), Lead of the Center for Circular Materials Management in the New Materials Institute, Founder of the Circularity Informatics Lab using the Circularity Assessment Protocol at the University of Georgia and a National Geographic Fellow (2018–2021). She has been conducting research on solid waste issues for more than 25 years with related projects on marine debris since 2001. She also specializes in global waste management issues and plastic contamination.

Photo credit: Sara Hylton

Her work on plastic waste inputs into the ocean has been recognized by the global community and translated into policy discussions by the Global Ocean Commission, in testimony to US Congress, in G7 and G20 Declarations, and the United Nations Environment program. She conducts public environmental diplomacy as an International Informational Speaker for the US Department of State. This has included multiple global programs of speaking events, meetings, presentations to governmental bodies, and media outreach around the world including Chile, Philippines, Indonesia, Japan, South Africa, Vietnam, Jordan, Israel, South Korea, India, Taiwan, and China.

Jambeck has won awards for her teaching and research in the College of Engineering and the UGA Creative Research Medal, as well as a Public Service and Outreach Fellowship. In 2014 she sailed across the Atlantic Ocean with 13 other women in eXXpedition to sample land and open ocean plastic and encourage women to enter STEM disciplines. She is co-developer of the mobile app Marine Debris Tracker, a tool that continues to facilitate a growing global citizen science initiative. The app and citizen science program has documented the location of over five million litter and marine debris items removed from our environment throughout the world. Follow her work on Twitter @JambeckResearch, Instagram @JennaJambeck.

Eliana Mozo-Reyes traveled to the USA in 2008, after having graduated as an electronics engineer in her native Colombia.

After the cultural adjustment of living in a high-income country, such as the USA, she found her path in environmental engineering, where she obtained an MS degree in 2012. Eliana briefly explored engineering education, which steered her back into solid waste research, this time from a multicultural perspective.

Having dived into qualitative research, Eliana embraced the Postmodern Rhizome to study cultural differences as well as entanglements between the solid waste management styles of her two countries and obtained her doctorate in Engineering in 2016, thanks to that study.

Since then, she has become a mother to an incredibly smart little human and has maintained her research efforts through helping other researchers breach the barriers of language and culture in sustainability research.

Katherine Shayne, M.S.Env.Eng., started as an English major during her freshman year at the University of Georgia (GA). Having gotten a full scholarship through the Zell Miller program, she was ready to tackle the prelaw curriculum and go on to be an attorney. Shortly after her freshman year, her best friend was diagnosed with colon cancer, and Shayne started to reevaluate her career trajectory. She realized she could use her love of math and science to design systems that do no harm and fix toxic systems through applied science and math: enter environmental engineering. Through switching to engineering, she was able to meet Dr. Jenna Jambeck who took Shayne on a journey filled with plastics, recycling, and global materials management.

Upon graduating with her master's in environmental engineering in 2018 under the direction of Dr. Jambeck where she studied litter data from the application Marine Debris Tracker (also started by Dr. Jambeck at the University of Georgia through a grant with NOAA), she and Dr. Jambeck saw a need for more waste education through offering user-friendly technologies: enter Can I Recycle This (CIRT). Shayne is the co-founder and CEO of CIRT where she is using AI to help people and companies learn about the end-of-life of their consumer products through their Recycling Technology Platform (RTP).

Shayne is also a faculty member at UGA's College of Engineering. Her research has been focused on global materials management and marine debris for the Center for Circular Materials Management (C2M2) at UGA's New Materials Institute. She is passionate about investigating sustainable infrastructure, especially as it relates to waste management, circular materials innovation, and new materials.

She instructs engineering design courses for environmental engineers about civil infrastructure such as landfill and materials recovery facility design. Shayne worked overseeing Ellen MacArthur Foundation New Plastics Economy pilot projects for Think Beyond Plastic, working toward a plastic-free ocean with industry leaders. Katherine is also a Sustainable Ocean Alliance (SOA) Youth Leader, a OneYoungWorld ambassador, and the youngest female UGA 40 under 40 honoree. She is a Solid Waste Association of North America Young Professional and SCUBA diver.

Shayne has presented about plastic pollution to the Indonesian, Danish, and Norwegian governments, as well as at the 69th Annual Gulf and Caribbean Fisheries Institute, Students for Zero Waste National Conference, and 6IMDC. Katherine presented on innovations and entrepreneurship in a new plastics economy at Our Ocean, SOA Leadership Summit, and ACHEMA 2018.

Chapter 17
Water and Wastewater Infrastructure

Lindsey Fields Bryant

Abstract Water is a crucial human, industrial, and ecological resource, and ensuring water availability is the bedrock of our society and underpins our future. Water resources are managed by water and wastewater infrastructure, and, as such, this engineered infrastructure plays a key role in sustaining our society, environment, and economy. Wastewater and drinking water are two of the seventeen total infrastructure categories assessed and graded as part of the American Society of Civil Engineers (ASCE) Infrastructure Report Card. Though water and wastewater infrastructure are two separate categories in the Infrastructure Report Card, they are inextricably linked through the hydrologic (water) cycle and work synergistically to protect public health, the environment, and the economy. This chapter addresses both water and wastewater infrastructure, including a description of infrastructure and their purpose and general function, a brief history of the development of water and wastewater infrastructure and associated policies in the United States, and a synthesis of current challenges and emerging issues including population growth, impacts of global climate change, emerging contaminants, cybersecurity, and infrastructure funding.

Keywords Water infrastructure · Wastewater infrastructure · Emerging contaminants · Water quality · Water treatment · Wastewater treatment · Climate change impacts · Drinking water supply · Water literacy · System resilience · Water supply

L. F. Bryant (✉)
Carter & Sloope, Inc., Watkinsville, GA, USA
e-mail: Lbryant@cartersloope.com

17.1 Introduction

Wastewater and drinking water are two of the seventeen total infrastructure categories assessed and graded as part of the American Society of Civil Engineers (ASCE) Infrastructure Report Card. A thorough discussion of the methodology used for grading can be found on the Infrastructure Report Card website (http://infrastructurereportcard.org), but, briefly, the grading criteria include an assessment of capacity, condition, funding, future need, operation and maintenance, public safety, resilience, and innovation. Though water and wastewater infrastructure are two separate report card categories, they are inextricably linked through the hydrologic (water) cycle; raw water is withdrawn from various surface sources (i.e., lakes and rivers) and groundwater sources for treatment to produce potable (drinkable) water, and treated wastewater is ultimately returned to existing bodies of water (Fig. 17.1). All of these processes are governed by engineered infrastructure. This chapter will address both water and wastewater infrastructure, including a description of infrastructure and their purpose and general function, a brief history of the development of water and wastewater infrastructure and associated policies in the United States, and a synthesis of current challenges and emerging issues.

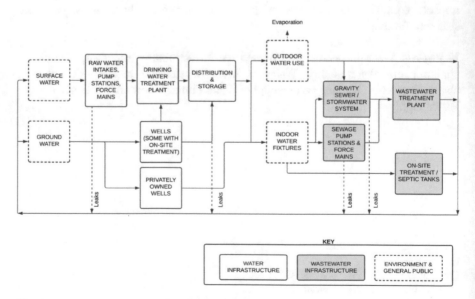

Fig. 17.1 Basic conceptual model of the full water system, highlighting water and wastewater infrastructure. (Adapted from Attari et al. [11])

17.1.1   Water and Wastewater Infrastructure

Water is a crucial human, industrial, and ecological resource, and ensuring water availability is the bedrock of our society and underpins our future [34]. Water resources are managed by water and wastewater infrastructure, and, as such, this engineered infrastructure plays a key role in sustaining our society, environment, economy, and public health. To lay the foundation for the remainder of this chapter, a brief description of water and wastewater infrastructure and their purpose and general function in the United States is presented here.

Drinking water infrastructure ensures that the US population has continuous access to safe, potable water. Water systems are responsible for the withdrawal of surface and/or groundwater, transporting raw water to treatment plants, and distribution of treated, potable water to end users. There are currently over 150,000 active drinking water systems in the United States, comprised of 2.2 million miles of underground water pipes and over 16,000 water treatment plants [6, 7]. Interestingly, a very small percentage (5.5%) of public drinking water systems serve over 90% of the population or around 272 million people [6].

Wastewater infrastructure plays a critical role in maintaining public health by ensuring the safe transport of sewage, removal of harmful constituents, and reduction of pollutant levels in wastewater that is ultimately discharged back into the environment. As defined by the ASCE, wastewater infrastructure includes all of the wastewater systems in the United States, which are comprised of collection systems (networks of sewer pipes, manholes, sewage lift stations) and wastewater treatment facilities. US wastewater infrastructure consists of approximately 14,780 wastewater treatment plants (WWTPs),[1] and over 19,700 wastewater pipe systems that are comprised of over 800,000 miles of public sewers and 500,000 miles of private lateral sewers [5, 6].

Water and wastewater infrastructure work synergistically (Fig. 17.1) to protect public health and the environment. Sewerage collection systems are responsible for isolating wastewater flows from the environment during passive (via sanitary sewer flows) and active (via lift stations and force mains) transport of wastewater from its sources (e.g., households, businesses, and industries) to treatment facilities. Wastewater treatment facilities employ various physical, biological, and chemical treatment processes that improve water quality by removing harmful toxins and reducing pollutants prior to treated effluent discharge into natural bodies of water (rivers, lakes, oceans) or onto land application sites. Minimizing the pollutants that are discharged into receiving waters ensures that human populations are safe from exposure to harmful constituents and also plays a critical role in maintaining a healthy ecosystem and environment. Similarly, water treatment plants use various levels of physical and chemical treatment to remove harmful constituents (e.g.,

---

[1] The term "wastewater treatment plant" is used in this chapter to reference treatment facilities. Treatment works that are owned by a state or municipality are often referenced as Publicly Owned Treatment Works (POTW).

viruses, emerging contaminants), organic materials (e.g., bacteria), and particulates from drinking water and to optimize taste.

Aside from the critical importance of water and wastewater infrastructure for public health, the economy and economic efficiency of a country is tightly linked to a reliable supply of clean water and adequate waste disposal [35]. The operations of countless major industries would be crippled without reliable access to clean water and wastewater services, including the healthcare industry. While thermoelectric power and irrigation are the largest freshwater consumers in the United States, other water-intensive industries include manufacturing of paint and coating, alkalis and chlorine, pesticides, synthetic dyes, adhesive, and industrial gas as well as paper-board mills, wineries, and poultry processing [8].

17.2 A Brief History

17.2.1 The Link Between Environmental Policy and Infrastructure

The importance of water and wastewater infrastructure and the protection of public health are bipartisan issues, and the ASCE report card is regularly cited and used as a reference by elected officials regardless of their political affiliation [9]. However, politics has played an important role in the development of the water and wastewater sectors because the policies implemented by the federal and local governments and enforced by various regulatory agencies largely dictate the minimum standards of functionality for water and wastewater infrastructure. Environmental rules and regulations at the federal level have historically been politically driven, with political leadership playing a critical role in framing the narrative around environmental issues and garnering public support for policy change in favor of or against environmental protections [32]. Additionally, there is a continually demonstrated relationship between politics and investment priorities in the United States and perhaps correspondingly the general state of any given infrastructure category. For example, as the United States begins to recover from the COVID-19 pandemic, the level of investment into the nation's water and wastewater infrastructure will be determined by policies and decisions at every level of the government [8]. A more in-depth discussion of infrastructure funding is presented later in Sect. 17.4.2.

While there are many intricacies and complexities of environmental policy, at its core environmental policy lies at the interface between the science of environmental issues, societal impacts of environmental degradation, and the engineering approaches to designing infrastructure to address these issues. Public policy is a course of government action in which political leaders articulate goals and strategies in formal statutes, rules, and regulations, which are in the jurisdiction of administrative agencies and/or courts to provide implementation and oversight [32]. The structure of the US government is such that the responsibilities associated with environmental policy are distributed among the federal government, the 50 states, local governments, and tribal governments [32]. Even at the federal level, responsibility for environmental policies is divided between the branches of the government, and jurisdiction over the policies spreads among various committees and subcommittees [32]. Many areas of environmental policy are dominated by governments at the state level (e.g., water management and groundwater protection). In fact, a federal policy may exist on an environmental issue, but states are often free to adopt their own environmental policies, with flexibility to adopt either stronger or more lenient mandates compared to federal guidelines. However, the US government has and will likely continue to play a role in environmental policy at the federal level because the associated risk to the environment and human health posed by environmental issues are often too great to be adequately addressed or resolved by individualized state efforts alone [32].

The Environmental Protection Agency (EPA) is the lead US agency for the protection of public health and the environment from pollution and associated threats. It was created by President Richard Nixon in 1970 through a presidential reorganization plan in response to a push from the general public for federal action against pollution and federal water supply standards [4, 38]. The reorganization plan pulled together several existing separate programs into a single new agency. Because the EPA's mission and authority do not stem from an act of Congress, the EPA is predominantly an umbrella organization that promulgates separate programs (e.g., air quality, water, and pollution prevention) that operate under separate laws and budgets [4]. Though the EPA develops regulations, provides technical assistance, gives grants, educates the general public, researches environmental issues, and sponsors partnerships, its administrator has only limited authority over programs outside of annual budget requests to Congress, and the EPA can only use tools that Congress has authorized, which are typically regulations [4, 24]. In many cases, enforcement of EPA regulations is delegated to agencies at the state level.

17.2.2   Development of Water and Wastewater Infrastructure in the United States

Efforts to dispose of human waste and to provide a supply of potable water date back to ancient societies in early civilization [44], but the development of water and wastewater infrastructure in the United States began in Philadelphia in the early nineteenth century and became widespread as the country grew in the latter half of the twentieth century [38]. In the United States, population growth and the accompanying increased water use during the Industrial Revolution led to increased concern and action for public water supply and wastewater systems, leading to the first public waterworks in America [38]. The investment of local governments into public water supply continued to grow with increasing need for water access to fight fires and other concerns, and by the beginning of the twentieth century, there were more than 3000 public water systems in the United States [38]. As recognition of disease transmittal through water (e.g., typhoid, cholera) increased in the mid-1880s, so did technological advances in water treatment methods such as slow sand filtration and rapid filtration with chemical coagulation [12, 38]. Finally, in 1912, the first water regulations were implemented under legislation passed in Congress to prevent the introduction and spread of disease [13]. Then, in 1914, the US Public Health Service Drinking Water Standards were adopted to regulate concentrations of bacteria in water, and in 1915 the United States began chlorine disinfection of drinking water [38].

Around the same time, the disposal of wastewater shifted from cesspools and septic tanks to more extensive sewerage systems throughout the United States to eliminate issues with groundwater contamination [38]. Sewer mains carried untreated wastewater to waterways, and then in the late nineteenth century, treatment of wastewater before discharging into the environment began throughout the United States to alleviate pollution of receiving waters [38]. Wastewater treatment began as land application onto farms, and this practice continued into the twentieth century until continued development and population growth prompted treatment advances such as biological treatment with trickling filters. Eventually, other types of secondary biological treatment, such as the activated sludge process that is commonly used today, and later tertiary treatment was introduced to reduce environmentally harmful levels of nitrogen and phosphorus compounds [38].

The next major strides in environmental policy relative to the water and wastewater sectors were in the 1970s, when Congress passed several laws that addressed environmental risks including pollution [4, 38]. Among these laws were the Federal Water Pollution Control Act, the Safe Drinking Water Act, and other laws that regulated pesticides, toxic substances, solid waste, and hazardous waste [4]. These laws required all point source discharges of wastewater (e.g., outfalls from municipal treatment plants, industries, etc.) to implement the "best available technology" for treatment and to obtain a discharge permit from regulatory agencies [4]. It should be noted that to this day, the EPA has no authority to regulate discharges from nonpoint sources, such as agricultural runoff. The Clean Water Act of 1972 mandated secondary treatment of wastewater be implemented in all treatment facilities in the United States. The Safe Drinking Water Act of 1974 regulates pollutants in public water supplies via a set of maximum contaminant levels that are established and regulated by the EPA. Construction of treatment plants throughout the United States was prompted by federal and state construction grant programs that incentivized treatment enhancement to meet these new regulatory requirements [38].

Since the passing of the Clean Water Act and Safe Drinking Water Act, the EPA has been regulating the country's wastewater and public drinking water by setting standards for acceptable water quality to protect public health as well as to establish protections for lands, waters, and wildlife. The EPA and state agencies continue to establish standards for contaminants in drinking water and pollutants in wastewater under these acts [8].

17.3 Status of Water and Wastewater Infrastructure in the United States

17.3.1 Recent Grades

The ASCE officially began publishing Infrastructure Report Cards in 1998, as a follow-up to a report published in 1988 by the National Council on Public Works Improvements entitled *Fragile Foundations: A Report on America's Public Works* [9]. The ASCE has adapted and improved the original approach and methodology and generated an Infrastructure Report Card every 4 years, with the most recent report card being released in early 2021.

Report card grades for both drinking water and wastewater infrastructure have been consistently poor for over two decades (Table 17.1). At the time of writing, the most recent report card (released in 2021) gave the highest grade to drinking water that it has been assigned since the 1998 *Fragile Foundations* assessment. However, this rating is still deemed "mediocre, requires attention" by the ASCE. The wastewater infrastructure grade has not improved.

17.3.2 Current State of Water and Wastewater Infrastructure

As discussed in Sect. 17.2 above, many wastewater treatment plants in the United States were originally constructed during the 1970s in response to the Clean Water Act and are now nearing the end of their lifespan, and water treatment plants that have been around since the 1970s and longer are in similar condition [6]. Much of the public sewer mains and water pipes were installed prior to World War II and are also nearing the end of their useful service life of 50–100 years [5, 6]. As collection

Table 17.1 Infrastructure Report Card grades for drinking water and wastewater infrastructure

Year	Grade	
	Drinking water	Wastewater
1988 [a]	B−	C
1998	D	D+
2001	D	D
2005	D−	D−
2009	D−	D−
2013	D	D
2017	D	D+
2021	C−	D+

[a] 1988 grade is from the congressionally chartered report titled *Fragile Foundations: A Report on America's Public Works*, written by the National Council on Public Works Improvements

systems age and deteriorate, inflow and infiltration (I&I) becomes an increasingly prevalent issue resulting in overtaxing the system and subsequent sanitary sewer overflows [6]. In water systems, leaks and breaks occur that result in losses that total over 6 billion gallons of water each day [6]. The frequency of water main breaks has increased by 27% between 2012 and 2018 alone [8].

When drinking water infrastructure fails, there are many severe consequences including disruptions in a system's supply of potable water, impediments to public safety including fire response, and damage to other types of infrastructure (e.g., roadways, etc.; [7]). Similarly, failure of wastewater infrastructure can result in sewage overflows, leaks, and spills that are hazardous to public health and the environment. The generally poor condition of pipes combined with inadequate capacity has resulted in an estimated discharge of around 900 billion gallons of untreated sewage each year [7]. According to the 2021 Infrastructure Report Card, the EPA reported that improvements were made to more than 180 of the nation's large sanitary sewer systems between 2012 and 2016, but the progress has slowed in recent years. The reason for decreased progress in recent years has not yet been conclusively determined, but we can speculate that inadequate funding has played a role.

Aside from the maintenance issues that need to be addressed with aging and failing infrastructure, there is also the matter of dealing with capacity issues. Despite the rapidly growing population, there is declining water usage in the United States due to conservation measures and increased plumbing fixture efficiencies [6]. However, ASCE [9] reports that the country's wastewater treatment systems are currently operating at an average of 81% of their design capacity, and 15% of systems have already reached or exceeded their capacity. Currently, over two million people in the United States do not have access to adequate drinking water and sanitation, which amounts to a continually unresolved humanitarian crisis [8]. Generally, the most vulnerable populations in overburdened, and socioeconomically distressed communities are disproportionately impacted by environmental consequences resulting from issues such as antiquated and inadequate water and wastewater infrastructure [25, 36]. As population levels continue to grow, capacity issues and increasing water demand will need to be addressed, and environmental justice practices will need to deepen and expand.

As infrastructure ages, the cost of operation and maintenance (O&M) is increasing, and O&M costs currently outpace available funding [6, 9]. The most recent report card identifies a $434 billion funding gap between the total cumulative investment needs between 2020 and 2029 versus what is funded in drinking water, wastewater, and stormwater infrastructure combined [6]. ASCE [9] reports the results of a recent survey that revealed that nearly half of utilities' maintenance work is reactive to system problems rather than proactive. A discussion of infrastructure funding and recent related legislation is discussed in Sect. 17.4.2 below.

Firsthand experience with municipal clients and discussions with other engineers and system owners during recent years have confirmed a labor shortage of water and wastewater operators. Treatment plant personnel report regularly working multiple shifts, being unable to take vacation time, and working large amounts of overtime to ensure treatment, collection, and distribution systems can operate as required. Aside from the concerns with burnout and employee retention, there is cause for concern for a great loss of institutional knowledge as operators retire or seek alternate careers. The recent COVID-19 pandemic has further emphasized the need for an adequate number of operators to ensure that the critical infrastructure in water and wastewater treatment plants is available at all times [31]. In the most recent infrastructure report card for drinking water, one of several recommendations to "raise the grade" was to prioritize support for finding, training, and retaining water system personnel [6].

17.3.3 Public Knowledge and Perception of Water and Wastewater Systems

The general public identifies water as a critical resource for their health and survival and understands the importance of drinking water infrastructure, but studies have shown that there are major gaps in public understanding of the water system and the role of water and wastewater infrastructure in providing access to potable water and treating wastewater [11, 34]. It is critically important for the public to understand the basic structure, function, and interconnections of water and wastewater systems so that they are able to place value on what the infrastructure does for them. This level of understanding can foster increased support from constituents on spending dollars toward infrastructure needs. It can also greatly aid conservation efforts by allowing citizens to make informed decisions about effective conservation behaviors at home and understanding increased costs for water and sewer [11].

A literature review conducted by McCarroll and Hamann [34] identified that students who are college-aged and younger lack a comprehensive understanding of water management systems, including water used in indirect ways (i.e., water used in the production of goods and services) as well as the fate of wastewater. In a study by Attari et al. [11], researchers asked experts and university students to draw their understanding of the water system, with a focus on the processes required to deliver potable water to an end user and manage that water once it leaves the home. Most students were unable to identify all components of the water system accurately and completely in their diagrams even though the majority of them believed there are risks related to water quality, quantity, and infrastructure [11]. Furthermore, one-third of students stated that they rarely, if ever, think about water quantity used in their home, 19.9% of participants did not know their local water source, 13.6% of students stated that their local water source was their house, and 58% of participants stated that they found it difficult to find information on their local water source [11]. These statistics demonstrate a disconnect between recognizing the importance of clean water and understanding the mechanism by which we access this clean water. Many of these studies provide suggestions on how to promote literacy about these topics [34], and it is crucial for these suggestions to be taken seriously as these students are our future voters and decision-makers.

While some knowledge gaps continue into adulthood, surveys suggest that the adult population tends to have a better understanding of some of the science and water systems compared to young adults and adolescents [34]. Adults are also slightly more knowledgeable about water infrastructure and related issues, and the importance of water to public health and the economy [34]. McCarroll and Hamann [34] point out that this better understanding by adults appears to be the result of life experience. However, surveys have found that most adults do not know the source of their drinking water, that they lack an understanding of the concepts of water transport, and that they do not understand how to approach conservation and protection of water resources [34]. It is important to note that people appear to be aware of the fact that they lack knowledge related to water resources and the water system, and they acknowledge that they could be doing more to conserve water but don't feel they have enough of an understanding to do so [34, 37].

Perhaps a major reason why public education and understanding of water and wastewater systems is lacking is that the related infrastructure is almost never seen by the general public. Pipes in collection and distribution systems are buried, and most people generally do not visit water and wastewater treatment plants. Despite the necessity of access to potable water and adequate wastewater treatment for society, it is easy to take an "out of sight, out of mind" approach when it comes to understanding the purpose and general functionality of infrastructure [6, 51]. Public health crises such as the highly publicized crisis in the Flint, MI, drinking water system raise public concern about drinking water safety and affordability, but this concern does not appear to translate into increased knowledge of water systems but rather fosters distrust of public drinking water [42].

McCarroll and Hamann [34] stressed the importance of translating adult water literacy to attitudes, values, and actions because their behaviors directly impact the success of implementing water resource management (e.g., conservation and pollution prevention measures) along with the funding of large infrastructure projects. Public perception also has a major impact on moving forward with conservation and resiliency initiatives such as implementing wastewater reuse on a municipal level [11]. These types of creative solutions will be a critical path forward as we face future challenges (see Sect. 17.4 below), so garnering public support to facilitate action will become increasingly important.

17.4   Looking Toward the Future

Several key recommendations on how to address current issues and raise the grade for both water and wastewater infrastructure have been generally consistent across all the Infrastructure Report Cards since the ASCE began drawing conclusions and making recommendations [5–7, 10]. At the top of the list for both water and wastewater infrastructure is the recommendation for increased federal funding for infrastructure through various grants, loans, and other programs. Other recommendations for water infrastructure include implementing asset management to evaluate risk and prioritize spending [5–7] and, in more recent years, to focus on science-based decision-making [5]. For wastewater infrastructure, recommendations for innovations such as non-potable water use and green infrastructure are made [5, 7]. Most recently, report card recommendations included taking the first steps to address an emerging contaminant (per- and polyfluoroalkyl substances or PFAS) and forthcoming impacts of climate change [6].

As we look toward the future, raising and even maintaining the grade for water and wastewater infrastructure will become increasingly difficult as a host of new challenges will arise, many of which were not anticipated when most water and wastewater infrastructure was originally designed and built [8]. These emerging issues will need to be considered in addition to recommendations already being made. Water quality and quantity will be impacted by socioeconomic changes, technological developments, and climate change impacts [53]. The more recent ASCE Infrastructure Report Cards have begun to identify and discuss pressing concerns about these future potential challenges. This section presents a discussion of the most prominent challenges that are beginning to take shape and future challenges that are anticipated.

17.4.1   Emerging Issues

Though our infrastructure needs are a country-based problem, many of the challenges that we will have to face as it relates to the function and resiliency of our water and wastewater infrastructure stem from global issues. In a recent perspective article that was written by Harmel et al. [27] to highlight the contributions of the attendees and presenters during the 2018 Global Water Security Conference for Agriculture and Natural Resources, the authors summarize the keynote address by Dr. Sonny Ramaswamy (former Director of the USDA National Institute of Food and Agriculture). In his address, Dr. Ramaswamy emphasized the need for global water security during his discussion of an impending "perfect storm" of rapid population increase, a growing middle class with its demand for more meat and produce, climate change, trade globalization (and the concurrent spread of pests, diseases, and invasive species), the disconnect between science and communication, public mistrust in science, and the challenges related to sustainability [27]. Optimized and effective water resource management is at the heart of addressing most, if not all, of these issues. A major strategy for effective water resources management will be driven by the design, resiliency, and maintenance of water and wastewater infrastructure, as water resources are managed by this infrastructure. Below is a discussion of several key issues that will likely play an increasingly important role in relation to our country's water and wastewater infrastructure in the years to come. This is not meant to be an in-depth review on each topic but rather will provide some background information on each issue, a summary of any current or anticipated approaches to dealing with these issues, and a brief discussion of challenges related to addressing each issue.

17.4.1.1   Population Growth and Limiting Water Resources

Water use has increased by more than 100% over the past decades, and continued population growth in the United States and beyond will place increased pressure on water supplies [53]. Population growth is anticipated to be one of the biggest issues for some water systems [8]. Currently, all but 20% of the American population relies on publicly owned wastewater systems, treating 62.5 billions of gallons of wastewater per day [6]. It is likely that the portion of the population dependent on publicly owned wastewater systems will increase as 86% of population growth is expected to occur in urban and suburban areas where centralized wastewater treatment dominates [6].

Despite water conservation measures that are decreasing water use and subsequent wastewater production, the continued population increase and economic growth in the United States will have a considerable impact on resource demand [6]. Not only will we have demand for per capita water use with an increased population but also the increased water demand for agriculture and water consumed during food production to support the expanding population. There will be a significant increase in water demand to meet the daily food requirements for the growing population in the next 30 years [27]. Agriculture already accounts for 70–75% of global water withdrawal, and it is thought that water demand for agriculture will increase nearly another 20% by 2050 [27, 55]. In the United States, irrigation accounts for 37% of total water use, second only to thermoelectric power which consumes 41% of total water [19]. By comparison, approximately 13% of total water use is publicly or self-supplied for domestic use, and 4.6% is classified as industrial use [19].

In addition to concerns regarding water quantity, the sustainability of water quality will also be negatively impacted by urbanization and increased volumes of wastewater from houses, industries, and agriculture [53]. Increased pollutant loading to wastewater treatment systems and to raw water sources will require enhanced treatment of both water and wastewater to meet regulatory requirements. Non-point sources of pollution, including agricultural and urban runoff, will likely play an increasingly important role in contributing to water pollution, and this will potentially have to be considered in the future from a regulatory standpoint. van Vliet et al. [53] argue that water scarcity assessments must also focus on the availability of water that is of suitable quality for each sector of water use and that water scarcity can be reduced in part by improving water quality.

Ultimately, planning for water resources must balance water needs with protecting essential ecosystem services and biodiversity to optimize long-term sustainability of natural resources [21]. Effective management of water resources will be critical to ensuring a sustainable water supply, and adequate water infrastructure that is operated optimally and efficiently will be a key component of successful water management. Scientific research efforts are focused on improving crop water use efficiency at different agricultural scales to reduce overall agricultural water usage and improve food supply [30]. In addition to implementing water conservation measures, water recycling, and reducing non-revenue water loss, many systems will also require additional raw water sources and/or storage facilities to ensure future water demands are met [8]. There are geographically water-scarce regions in the United States experiencing rapid population growth (e.g., the Southwest), and this will add additional challenges to meeting future water demand [8].

17.4.1.2   Impacts of Global Climate Change

Global climate change and its consequences have now been studied for decades and continue to be at the forefront of scientific research. Our understanding of both global climate change and its impacts have come a long way since the late 1990s, when climate predictions were speculative at best, and the potential impacts on water resources and water management were highly uncertain [33]. We now know with certainty that the global climate is warming with many unprecedented changes occurring since the 1950s [29]. There is also strong scientific evidence that more than half of the observed global warming during this time span is due to human impacts and more specifically to the documented increase in anthropogenic (i.e., human-caused) greenhouse gas emissions (e.g., carbon dioxide, methane, and nitrous oxide) driven by population increase and economic growth since the prein-dustrial era [29]. Although there is a wide range of future greenhouse gas emission projections, all emission scenarios predict an increase in global surface temperature and a high likelihood that the resulting impacts will increase in severity [29]. This, in turn, will increase risks and vulnerabilities for natural systems and human popu-lations and create new problems such as a reduction in renewable surface water and groundwater resources and increased competition for water [29].

Impacts of global climate change are spread across both natural and human sys-tems. There are many predicted impacts of climate change that will directly and indirectly affect the hydrologic cycle and water resources and thus will impact water and sewer infrastructure in a variety of ways. Many of these impacts were not con-sidered during design and construction of most existing water and wastewater infra-structure. Among these impacts are changing precipitation and melting snow and ice, an increased frequency and magnitude of extreme events (e.g., droughts, floods, heat waves, etc.), and sea level rise [29]. Direct impacts on infrastructure, such as damage due to the predicted increase in frequency and magnitude of extreme weather events (e.g., storms and storm surges, hurricanes), will require water and wastewater infrastructure to be more resilient against physical damage, especially in low-lying areas near water sources. Sea level rise and increased coastal hazard risks (e.g., storm surges, flooding, etc.) will displace people from coastal areas, and an inability to deal with extreme weather events, etc. will cause emigration out of underdeveloped/low-income countries [29] to more developed countries such as the United States, which will cause a disproportionate increase and added pressure to the water and sewerage systems to locations of immigration. There will be increased risk from declining accessible surface water (i.e., surface water flows and aquifer recharge), saltwater intrusion into freshwater, drought, and wildfires [8]. Global climate change will also cause increased pressure in competing water resource needs between agricultural, municipal, and environmental (e.g., protected federal and state waters) users, making it crucial to address not only water availability but also the inadequacies and lack of resiliency in our water infrastructure and resultant water quality [11].

Water and wastewater infrastructure must be resilient against associated potential hazards, modulate the effects of extreme events that impact water availability, and be able to return to safe and normal functionality after an extreme event occurs [46]. Engineered infrastructure is expected to have a long service life during which it must be functionally reliable, but the civil and environmental engineers who are responsible for planning and design of the infrastructure are facing uncertainty about potential climate-related impacts and their associated risks [39]. While some impacts of climate change are globally uniform, many vary across geographic regions [29]. The biggest vulnerabilities and therefore the most pressing needs to address infrastructure resiliency vary greatly across geographic locations, types of treatment systems, age, ownership status, etc. [6]. Further, there is a high level of uncertainty in the range and severity of potential impacts (e.g., How much precipitation increase will occur? How frequent will flooding be?). Engineers will need to implement a variety of improvements and factors of safety to ensure adaptation and resiliency of our water and wastewater infrastructure against the many impacts of global climate change. However, adapting engineering design to account for these impacts and their high level of uncertainty is extremely difficult and expensive [39]. There is risk of underestimating a factor of safety, which could result in infrastructure failure or overestimating a factor of safety, which could add substantial unnecessary cost to a project. Various approaches to engineering design have been proposed to begin navigating these challenges [14, 39], but there is certainly more work to be done.

A key aspect of preparedness and successful implementation of resiliency measures is going to be the ability of system owners and design engineers to anticipate the biggest risks and vulnerabilities for each specific system. Climate models are a powerful tool to assist in these efforts. Models are a critical tool for climate scientists to predict future climate variability and climate change impacts. The two major classes of climate models, Earth System Models and Global Climate Models, use atmospheric, oceanic, land surface, and sea ice data to predict climate impacts [15, 39]. Models of climate change impacts are typically done on a global scale and thus are most successful at predicting the physical impacts of global climate change over relatively large geographic areas (e.g., continents) compared to the more regional and local scales (e.g., a watershed). Even at larger scales, there are sources of uncertainty that contribute to all models, including the natural variation of climate, uncertainty in model sensitivity to anthropogenic and natural forcing, and projection of future emissions and climate drivers [28, 39].

Engineering design and planning typically occur at smaller, more localized scales than global climate models [39]. The resolution of global models is too coarse to be used effectively at regional and local scales and thus need to be downscaled to provide enough detail to effectively simulate hydrologic responses on local scales [17]. Once downscaled, climate models can be used to assess the hydrological responses of major river basins and even some subbasins including changes in basin snowpack, amount and timing of river discharge, and adaptability of dam and reservoir systems [17]. However, there is an increased level of uncertainty when global climate models are downscaled, which leads to difficulty during engineering design and planning for infrastructure resiliency [39]. There is also a financial cost associated with the computing requirements (increased data and CPU time) for downscaled models, contributing to the point of diminishing return that exists with downscaling models. Thus, engineers must address factors of safety while facing the reality of limited funding and resource availability (see Sect. 17.4.2) and strike a balance that minimizes the consequences of potential infrastructure failure [39].

Engineers and system owners must begin engaging the scientific community to gain a better understanding of future design conditions and work toward closing the "gap between climate science and engineering practice" [39]. During a recent webinar hosted by EPA on their Creating Resilient Water Utilities tool, a live virtual poll revealed that only 37% of attendees have considered climate change in their long-term planning process, 85% had not implemented any adaptation measures as a result of a climate change assessment, and 88% had not used any tools to assess climate change impacts on their water or wastewater systems [16]. The EPA is taking steps toward bridging this gap with resources such as their Climate Resilience Evaluation and Awareness Tool, which provides user-friendly, regionally modeled data to help owners discover which extreme weather events pose a threat, assess critical assets and potential solutions/actions, and perform cost-benefit analyses of risk reduction strategies (https://www.epa.gov/crwu). However, implementation of resiliency measures will lie predominantly in engineered improvements and improved design for water and wastewater infrastructure.

17.4.1.3 Emerging Contaminants

Emerging contaminants, also known as contaminants of emerging concern, include any chemical or substance that isn't regulated, has been found in detectable concentrations in natural bodies of water, and is potentially harmful to the environment and/or human health in increased concentrations [3, 47]. Even though they are typically found in concentrations that range from ng/L to µg/L, emerging contaminants are known or suspected to have negative impacts on humans and natural systems [43]. There are a wide range of emerging contaminant categories, including (but not limited to) persistent organic pollutants (e.g., polybrominated diphenyl ethers used in flame retardants, furniture foam, etc.), pharmaceuticals and hormones, personal care products, endocrine-disrupting chemicals, pesticides, dioxins, surfactants, polycyclic aromatic hydrocarbons (PAHs), alkyl phenolic compounds, nanomaterials, per- and polyfluorinated substances (PFASs), and antibiotic-resistant genes [3, 41, 47]. These contaminants can have disastrous impacts on human health, including causing cancers, have severe impacts on reproductive health, result in compromised immune systems, cause the spread of antibiotic resistance, and cause harm to aquatic organisms in receiving waters [3].

Though many emerging contaminants are not new compounds, our awareness of their prevalence in the natural environment is increasing with advancements in science and analytical technology [3]. As technological advances continue and additional harmful contaminants are exposed, more stringent regulations that require more advanced treatment are likely [8]. The EPA is responsible for protecting the public health against toxic chemicals and other pollutants in drinking water, including thousands of substances that have not been well-studied and many chemicals that are used in industrial processes [4]. There are far too many contaminants to regulate, and not all emerging contaminants have acute toxic effects to their receiving environment because they are present in such low levels [45]. There are also likely many emerging contaminants that have not yet even been detected. This makes regulatory action complicated and difficult, but the EPA has been taking a systematic approach to identifying contaminants and prioritizing regulations. A requirement of the 1996 Safe Drinking Water Act amendments is that the EPA issues a new list of up to 30 unregulated contaminants every 5 years that are to be monitored by public water systems [1]. The first list, the Unregulated Contaminant Monitoring Rule (UCMR 1), was published in 1999, and there have been a total of four UCMRs published to date. The Safe Water Drinking Act was amended again by America's Water Infrastructure Act of 2018, mandating that EPA's Unregulated Contaminant Monitoring Rule Program must require all public water systems that serve populations of 3300–10,000 people to monitor for the contaminants in each UCMR cycle, and the program will continue to include systems serving populations larger than 10,000 people [2].

Our ability to deal with emerging contaminants is critical for sustainable reuse of water [43]. Additionally, some emerging contaminants can bio-accumulate in the fatty tissues of animals and cause damage to the endocrine systems of the animals themselves as well as humans if they are not removed from wastewater prior to effluent discharge [3]. A major challenge when dealing with emerging contaminants is the difficulty in their removal from the water system once identified and regulated. Most emerging contaminants are non-biodegradable and have complex chemical composition, making existing wastewater treatment processes inadequate for complete removal [3]. While biological treatment processes such as activated sludge can remove some emerging contaminants with optimized operating conditions, they typically cannot completely remove persistent emerging contaminants [45]. There are extensive reviews in the literature of emerging contaminants and different treatment technologies that are being researched [41, 43, 45]. Several advanced treatment technologies are the subject of ongoing research, but many of these processes involve high O&M expenses for substrates and energy use [3]. Methods such as membrane filtration, activated carbon, electrochemical oxidation, and ozonation have been reported to have varying levels of effectiveness in removing emerging contaminants [3, 43, 45]. There is general agreement among researchers that use of a single treatment technology to remove emerging contaminants is likely inadequate, and a better approach might be to employ coupled treatment systems in a multiple barrier approach [43, 45]. The capital and O&M costs associated with adding more treatment technology to deal with emerging contaminants will likely widen the existing gap in wastewater infrastructure funding.

Per- and polyfluoroalkyl substances (PFAS) have been the focus of recent attention as they are the subject of ongoing legislation and litigation. There are over 5000 PFAS that have been heavily used since the 1960s in a variety of products including non-adhesive cookware and fire-extinguishing foam, and more recently they have been discovered in US-produced cosmetics [40, 54]. Their very low reactivity and virtual non-degradability (lasting thousands of years or more) has led to the popular description of "forever chemicals," and there is strong evidence of these man-made chemicals being linked to birth defects, hormone deficiencies, and cancer [23, 40]. In 2021, EPA acted by continuing implementation of their 2020 PFAS Action Plan, repurposing the Fifth Unregulated Contaminant Monitoring Rule (UCMR 5) to collect new data on PFAS in drinking water, re-issuing final regulatory decisions for perfluorooctanoic acid (PFOA) and perfluorooctanesulfonic acid (PFOS) under the Safe Drinking Water Act, and releasing an updated toxicity assessment for a group of PFAS called perfluorobutane sulfonic acid (PFBS; [23]). Most recently, the EPA established the EPA Council on PFAS, which will develop a multi-year strategy for protection against PFAS, and work with other agencies and communities to educate, exchange information, and assist with PFAS challenges [23]. Though regulation of PFAS is necessary to protect public health, there will be substantial capital costs associated with any new discharge requirements. An additional environmental concern is that even if PFAS are removed from effluent wastewater, their accumulation in treatment substrate is generating concentrated waste products that will not degrade for thousands of years.

17.4.1.4   Cybersecurity

Cybercrime is ranked by the US Federal Bureau of Investigation (FBI) as one of its top priorities. The risk of cyberattacks for theft, terrorism, and political motivations and to incite fear are a significant threat to critical infrastructure including the water and wastewater sectors in the United States [20, 26]. Such attacks pose a serious risk to public health, safety, and national security with potential outcomes including contamination of the public water supply, service outages, loss of treatment system controls and monitoring, compromised first responder efforts, and a disruption in food and fuel supply [26, 52]. In addition to these dangers, a cyberattack on water and wastewater infrastructure can also result in damage of systems components, loss of data, and theft of customer billing information [26].

The federal government has adopted policies on cybersecurity and designated oversight to ensure precautionary measures are being taken by water and wastewater systems. Individual Sector-Specific Agencies were appointed to oversee cybersecurity for different types of infrastructure to leverage their institutional knowledge and expertise regarding infrastructure in a specific sector [31]. Per the Presidential Policy Directive on Critical Infrastructure Security and Resilience under President Obama, the Sector-Specific Agency that is directly responsible for overseeing cybersecurity for water and wastewater infrastructure is the EPA [48]. The EPA issues cybersecurity mandates and is responsible for holding local water systems accountable to mandates in America's Water Infrastructure Act, in which water systems that serve more than 3300 customers must maintain risk assessments and adopt an emergency response plan for cybersecurity threats among other concerns [31]. In the event of a cybersecurity compromise, the EPA works with the FBI and the Cybersecurity and Infrastructure Security Agency to investigate the security breach [31, 48]. The EPA is suited to oversee the cybersecurity of the water and wastewater sectors because of its institutional knowledge and because this agency regulates permitting and compliance in these sectors. The directive emphasized the importance of critical infrastructure owners and operators working in concert with federal, state, local, tribal, and territorial entities as partners to strengthen cybersecurity.

More recently, the Cybersecurity and Infrastructure Agency Act of 2018 amended the Homeland Security Act of 2002 by establishing the Cybersecurity and Infrastructure Security Agency (CISA; H.R. 3359, 115th Congress). The CISA is a standalone federal agency that operates with oversight from the Department of Homeland Security to continue the National Protection and Programs Directorate (NPPD) of "leading cybersecurity and critical infrastructure security programs, operations, and associated policy." The formation of the CISA did not change responsibilities of the Sector-Specific Agencies for cybersecurity oversight. The EPA works collaboratively with the CISA and provides various resources to assist water and wastewater systems in addressing cybersecurity, and additional resources developed by organizations such as the National Institute of Standards and Technology, the American Water Works Association, and Water Information Sharing and Analysis Center are also available for use in cybersecurity preparedness [20, 26]. The National Governors Association has also been taking action and providing resources for cybersecurity [20].

In the wake of the recent cyberattack on the Colonial Pipeline, an American oil pipeline system that was hacked in May 2021, President Biden signed an Executive Order to improve cybersecurity in the United States [49]. Most notably, the Executive Order will facilitate threat information sharing between the government and the private sector by removing barriers, establishes baseline security standards for the development of software sold to the government, establishes a Cybersecurity Review Board (modeled after the National Transportation Safety Board), and creates a "standardized playbook" for federal response to cyber incidents among other enhancements to federal cybersecurity [49].

The threat of cyberattacks on water and wastewater infrastructure is growing because of a general increase in cyberattacks and also because utilities are increasing their reliance on computer technology to monitor and control treatment processes using Industrial Control System (ICS) networks and Systems Control and Data Acquisition (SCADA; [20]). Water and wastewater systems across the United States have already been facing a variety of cyberattacks such as ransomware, process manipulation, and attempts to disrupt and/or halt operations, and these attacks continue to increase [26]. While security measures can help prevent cyber threats and promote increased preparedness in response to attacks, many utilities lack the resources and capabilities to put adequate measures in place [26]. Antiquated information technology and control systems, shared and varying infrastructure, system complexity, and limited financial resources and personnel are among the challenges faced by water and wastewater systems in regard to implementing cybersecurity measures [26]. Recent areas of highest concern for security gaps in water systems are network configuration, media protection, remote access, documented policies and procedures, and staff training [20]. As cybersecurity threats continue to increase, creating a "cyber-security culture" of awareness and urgency among personnel will be an integral part of preventing cyberattacks [20].

17.4.2   Infrastructure Funding

A persistent concern dating back to the 1988 *Fragile Foundations* report and continuing throughout the ASCE report cards is the inadequacy of financial investment in our country's infrastructure to meet current operational costs and future system demands [6, 35]. In fact, a large portion of the poor report card grades for wastewater and drinking water infrastructure is attributed to the funding gap [8]. There are various federal loans, grants, and programs available to aid with funding infrastructure projects, such as EPA's Drinking Water State Revolving Fund, Clean Water State Revolving Fund, Water Infrastructure Finance and Innovation Act, and USDA Rural Development Program, which are critical for assisting communities in addressing their needs. However, the majority (66%) of capital improvement spending for water and wastewater infrastructure in the United States comes from state and local governments, and the federal government contribution to capital spending on water infrastructure has dramatically decreased (from 63% in 1977 to 9% of total capital spending in 2017) over the past 40 years [6].

As infrastructure ages, the cost of O&M is increasing and currently outpaces available funding [6, 9]. The most recent report card identifies a $434 billion funding gap between the total cumulative investment needs during the 2020–2029 time frame versus what is funded in drinking water, wastewater, and stormwater infrastructure budgets combined [6]. ASCE [9] reports the results of a recent survey that revealed that nearly half of utilities maintenance work is reactive to system problems rather than proactive. Risk and Resiliency Assessments on water systems as required by America's Water Infrastructure Act of 2018 are enabling system owners to use these assessments to justify funding for infrastructure improvements.

Though the use of water conservation appliances and fixtures has led to a decrease in water usage in the home, this decreased use has contributed to the need for rate increases. Leaks from aging infrastructure and inflation are also responsible for rate increases [6]. Across the country, both sewer and water rates have been increasing over the past 10+ years, with a 24% increase in average sewer rates between 2008 and 2016 and a 31% increase in average monthly drinking water rates between 2012 and 2018 [6]. Despite these rate increases, there is still a significant funding gap, and affordability standards will need to be considered for future rate increases. This is especially important for drinking water infrastructure, where the primary funding mechanism is user fees [6]. EPA affordability standards dictate that households should spend no more than 2% of median household income on drinking water and 4.5% on both drinking and wastewater services [6, 18]. One approach being taken at the local level to generate additional revenue is implementing innovative technologies for water reuse, energy recovery, and nutrient recycling [9]. At the state level, actions include levying local taxes, implementing restoration fees, and legislative set-asides [9].

Compounding these funding issues is the massive financial loss in water consumption and rate revenues that occurred during the COVID-19 pandemic. Negative financial impacts on both water and wastewater utilities will increase the difficulties in funding the country's multi-billion-dollar need for repairing and replacing piping, pumps, storage facilities, and treatment plants [8]. Inadequate funding will result in unreliable systems with frequent breaks and failures leaving public health and the economy at risk [8].

Both capital spending and O&M spending can be done as a proactive measure to prevent future issues or as a reaction to an issue that has already occurred. There are many obvious beneficial reasons for making every effort to spend dollars proactively rather than reactively. However, federal investment in both capital and O&M costs is chronically inadequate, and if this trend continues, there will be an annual O&M shortage of $18 billion by 2039 [8]. At the time of writing, the Biden Administration has announced support of the Bipartisan Infrastructure Framework bill, but it has yet to pass. If passed, the $1.2 trillion framework will be a monumental step toward funding infrastructure resiliency [50].

17.5 Conclusion – Striving for Resiliency

The 2021 Infrastructure Report Card notes that decision-makers are starting to shift away from focusing on only short-term issues (e.g., population growth, capacity, affordability) and are starting to include long-term concerns such as resiliency with respect to sea level rise, natural disasters, cybersecurity threats, and post-interruption recovery time. Additional references are becoming available to aid design engineers, owners, and decision-makers in assessing risk and increasing infrastructure resiliency [22]. Tools and technologies are also being developed and/or adapted for use to facilitate system resiliency in the face of more frequent extreme weather events by enabling faster response times and more streamlined operations [6].

Legislative action is also being taken to help water and wastewater systems prepare for what lies ahead. America's Water Infrastructure Act (AWIA), which was passed in 2018, requires drinking water systems serving populations greater than 3300 to develop and regularly update Risk and Resilience Assessments and Emergency Response Plans. These plans must include a risk assessment and emergency response plans for "malevolent acts" including cybersecurity threats [31].

The EPA requested approximately $7.7 million in their 2021 budget for aiding in these endeavors at the state and local levels such as training workshops and tabletop exercises [31]. Some states now require asset management plans for drinking water systems, which should help not only improve O&M efforts but also address resiliency and funding [6]. The $1.2 trillion Bipartisan Infrastructure Framework is also on the table and could have major implications for water and wastewater infrastructure resiliency if it is passed. While there is certainly more work to be done and more problems to solve, we are moving toward resiliency of our water and wastewater infrastructure.

References

1. 106th Congress (1996) The Safe Drinking Water Act as amended by the Safe Drinking Water Act of 1996
2. 115th Congress (2018) America's Water Infrastructure Act of 2018
3. Ahmed SF, Mofijur M, Nuzhat S, et al (2021) Recent developments in physical, biological, chemical, and hybrid treatment techniques for removing emerging contaminants from wastewater. Journal of Hazardous Materials 416:. https://doi.org/10.1016/j.jhazmat.2021.125912
4. Andrews RNL (2016) The Environmental Protection Agency. In: Vig N, Kraft M (eds) U.S. Environmental Policy: Achievements and New Directions, 9th edn. Thousand Oaks, CA, pp 151–171
5. ASCE, American Society of Civil Engineers (2017) 2017 Infrastructure Report Card: A Comprehensive Assessment of America's Infrastructure
6. ASCE, American Society of Civil Engineers (2021a) 2021 Infrastructure Report Card. Reston, VA
7. ASCE, American Society of Civil Engineers (2013) 2013 Report Card for America's Infrastructure
8. ASCE, American Society of Civil Engineers (2020) The Economic Benefits of Investing in Water Infrastructure: How a Failure to Act Would Affect the US Economic Recovery. Reston, VA
9. ASCE, American Society of Civil Engineers (2021b) Report Card History. In: https://infrastructurereportcard.org/making-the-grade/report-card-history/. Accessed 29 Apr 2021
10. ASCE, American Society of Civil Engineers (2009) 2009 Report Card for America's Infrastructure. Reston, Virginia
11. Attari SZ, Poinsatte-Jones K, Hinton K (2017) Perceptions of water systems. Judgment and Decision Making 12:314–327
12. AWWA (1981) Mr. Wixford and Mr. Wall and The Fair. In: American Water Works Association, Centennial Edition
13. AWWA (1999) Water Quality and Treatment—Handbook of Community Water Supplies, 5th edn. McGraw Hill Co., New York, NY
14. Ayyub BM, Walker D, Olsen J, et al (2016) Climate resilience for critical infrastructure: Needs and a new design paradigm. In: Chen SS, Ang AH-S (eds) International Symposium on Sustainability and Resiliency of Infrastructure. Taipei
15. Bader D, Covey C, Gutowski W et al (2008) Climate Models: An Assessment of Strengths and Limitations. A Report by the U.S. Climate Change Science Program and the Subcommittee on Global Change Research. Washington, D.C.
16. Baranowski C (2021) US EPA's Creating Resilient Water Utilities: Tools and resources overview. In: Building Resilience and Adapting to Climate Change Impacts for Drinking Water and Wastewater Utilities. (webinar)

17. Barnett T, Malone R, Pennell W, et al (2004) The effects of climate change on water resources in the west: Introduction and overview. Climatic Change 62:1–11
18. Black & Veatch Management Consulting L (2019) 50 Largest Cities Water & Wastewater Rate Survey: 2018-2019. Los Angeles, CA
19. Center for Sustainable Systems, University of Michigan (2020) U.S. Water Supply and Distribution Factsheet. CSS05-17
20. Clark R, Panguluri S, Nelson TD, Wyman RP (2017) Protecting drinking water utilities from cyberthreats. Journal - American Water Works Association 109:50–58
21. Clark RM (2014) Securing water and wastewater systems: Global perspectives. Water and Environment Journal 28:449–458. https://doi.org/10.1111/wej.12078
22. Committee on Adaptation to a Changing Climate (2018) Climate-Resilient Infrastructure: Adaptive Design and Risk Management. American Society of Civil Engineers
23. EPA (2021) Per- and Polyfluoroalkyl Substances (PFAS). In: https://www.epa.gov/pfas. Accessed 2 Apr 2021
24. EPA (2020) Our Mission and What We Do. In: https://www.epa.gov/aboutepa/our-mission-and-what-we-do
25. EPA (2016) EJ 2020 Action Agenda: The U.S. EPA's environmental justice strategic plan for 2016-2020. EPA-300-B-1-6004
26. Germano JH (2019) Cybersecurity Risk & Responsibility in the Water Sector. American Water Works Association
27. Harmel RD, Chaubey I, Ale S, et al (2020) Perspectives on global water security. Transactions of the ASABE 63:1–12. https://doi.org/10.13031/trans.13524
28. IPCC (2012) Managing the Risks of Extreme Events and Disasters to Advance Climate Change Adaptation. A Special Report of Working Groups I and II of the Intergovernmental Panel on Climate Change. Cambridge, UK and New York, NY
29. IPCC IP on CC (2014) Climate Change 2014 Synthesis Report. Geneva, Switzerland
30. Kang J, Hao X, Zhou H, Ding R (2021) An integrated strategy for improving water use efficiency by understanding physiological mechanisms of crops responding to water deficit: Present and prospect. Agricultural Water Management 255:107008. https://doi.org/10.1016/j.agwat.2021.107008
31. Katz J (2021) When water utilities get hacked, who should they call? FCW: The Business of Federal Technology
32. Kraft ME, Vig NJ (2016) U.S. Environmental Policy: Achievements an New Directions. In: Vig NJ, Kraft ME (eds) Environmental Policy: New Directions for the Twenty-First Century, 9th edn. Thousand Oaks, CA, pp. 2–32
33. Kundzewicz ZW, Somlyódy L (1997) Climatic Change Impact on Water Resources in a Systems Perspective. Water Resources Management 11:407–435
34. McCarroll M, Hamann H (2020) What we know about water: A water literacy review. Water (Switzerland) 12:2803-undefined. https://doi.org/10.3390/w12102803
35. National Council on Public Works Improvement (1988) Fragile Foundations: A Report on America's Public Works
36. NEJAC, National Environmental Justice Advisory Council (2016) Environmental Justice and Water Infrastructure Finance and Capacity CHARGE. Available via EPA. https://www.epa.gov/sites/production/files/2016-12/documents/nejac_environmental_justice_and_water_infrastructure_finance_and_capacity_final_charge.pdf. Accessed 3 Jun 2021.
37. Nestlé Waters (2018) Perspectives on America's Water
38. NRC (2002) History of U.S. Water and Wastewater Systems. In: Privatization of Water Services in the United States. National Academies Press, pp 29–40
39. Olsen J (2015) Adapting infrastructure and civil engineering practice to a changing climate. American Society of Civil Engineers (ASCE), Reston, VA
40. Pyles T (2020) What's in Store for PFAS Regulation in 2021? JURIST – Professional Commentary

41. Richardson SD, Kimura SY (2020) Water analysis: Emerging contaminants and current issues. Analytical Chemistry 92:473–505
42. Rodriguez G, Nagy M, Bye J et al. (2018) Public perception of Michigan water quality and affordability: A case study of Benton Harbor. Michigan Journal of Sustainability 6(1):29-50. https://doi.org/10.3998/mjs.12333712.0006.105
43. Rodriguez-Narvaez OM, Peralta-Hernandez JM, Goonetilleke A, Bandala ER (2017) Treatment technologies for emerging contaminants in water: A review. Chemical Engineering Journal 323:361–380. https://doi.org/10.1016/j.cej.2017.04.106
44. Rosen G (1993) The History of Public Health. Johns Hopkins University Press, Baltimore, MD
45. Rout PR, Zhang TC, Bhunia P, Surampalli RY (2021) Treatment technologies for emerging contaminants in wastewater treatment plants: A review. Science of the Total Environment 753:141990 https://doi.org/10.1016/j.scitotenv.2020.141990
46. Şen Z (2020) Water Structures and Climate Change Impact: a Review. Water Resources Management 34:4197–4216. https://doi.org/10.1007/s11269-020-02665-7
47. Smith GJ (2008) Aquatic Life Criteria for Contaminants of Emerging Concern Part 1: General Challenges and Recommendations
48. The White House (2013) Presidential Policy Directive-Critical Infrastructure Security and Resilience. In: Office of the Press Secretary. https://obamawhitehouse.archives.gov/the-press-office/2013/02/12/presidential-policy-directive-critical-infrastructure-security-and-resil. Accessed 5 May 2021
49. The White House (2021a) Fact Sheet: President signs Executive Order charting new course to improve the Nation's cybersecurity and protect federal government networks. The White House Statements and Releases. https://www.whitehouse.gov/briefing-room/statements-releases/2021/05/12/fact-sheet-president-signs-executive-order-charting-new-course-to-improve-the-nations-cybersecurity-and-protect-federal-government-networks/. Accessed 20 May 2021
50. The White House (2021b) Fact Sheet: President Biden announces support for the Bipartisan Infrastructure Framework. The White House Statements and Releases. https://www.whitehouse.gov/briefing-room/statements-releases/2021/06/24/fact-sheet-president-biden-announces-support-for-the-bipartisan-infrastructure-framework/. Accessed 26 Jun 2021
51. U.S. Environmental Protection Agency (2018) Drinking Water Infrastructure Needs Survey and Assessment Sixth Report to Congress
52. US EPA (2021) Incident Action Checklist-Cybersecurity. EPA 810-B-17-004
53. van Vliet MTH, Florke M, Wada Y (2017) Quality matters for water scarcity. Nature Geoscience 10:800–802
54. Whitehead H, Venier M, Wu Y et al (2021) Fluorinated compounds in North American cosmetics. Environmental Science & Technology Letters. https://doi.org/10.1021/acs.estlett.1c00240
55. WWAP (2014) The United Nations World Water Report 2014: Water and Energy. Paris, France

Dr. Lindsey Fields Bryant has taken a non-traditional career path but has held an interest in STEM for as long as she can remember. As the daughter of a math teacher, she grew up singing multiplication tables with her mother. She loved the challenge of her math and science classes throughout grade school and high school and entered the Biology program at Boston University in 2004. While in college, she focused on marine sciences, working in various laboratories and conducting research in Woods Hole, MA, and Belize, Central America. She graduated with her BA in Biology in 2008.

Her experiences in college sparked an interest in graduate research, and after completing her undergraduate work she entered the University of Rhode

Island Graduate School of Oceanography as a doctoral student. During her time in graduate school, Dr. Bryant studied the coastal ecology and biogeochemistry of Narragansett Bay, RI, and the surrounding areas. She worked extensively with local commercial fishermen and collaborators from other institutions on large-scale research projects. Her dissertation research, from which she has been published in multiple scientific research journals, focused on benthic-pelagic coupling as a function of changing organic inputs in coastal ecosystems.

After receiving her PhD in Oceanography in 2013, she moved to Athens, GA, where she began a three-year postdoctoral research position at the University of Georgia. Dr. Bryant spent her time as a postdoctoral scholar conducting research and mentoring graduate and undergraduate students. Seeking the challenge of moving out of her comfort zone, her postdoctoral research moved away from coastal ecosystems and focused instead on nutrient cycling in deep sea extreme environments such as natural oil and brine seeps and the site of the Deepwater Horizon oil spill. She participated in several multi-week research cruises in the Gulf of Mexico and the North Pacific Ocean and presented her research at various lectures and seminars.

While her budding career as an Oceanographer led to many wonderful experiences, including research cruises and even a trip to the bottom of the ocean on the HOV *Alvin* submersible, Lindsey felt a pull away from academia and toward an alternate career path. She gained exposure to engineering while working with collaborators on various research projects, and she was drawn to an engineering career where she could apply her knowledge to help communities solve tangible, real-world problems and improve their quality of life. With that in mind, she began her transition into a career in engineering.

Upon completion of her postdoctoral position, Dr. Bryant enrolled in the University of Georgia's engineering program to pursue her MS in Engineering. During her time at UGA, Dr. Bryant served as a teaching assistant and an instructor of fundamental undergraduate engineering classes and labs. Her thesis research focused on ecosystem energetics, and she constructed an ecological model of food web and energy dynamics in a local river. She also began working part-time at Carter & Sloope, Inc., a local consulting engineering firm. During her time at Carter & Sloope, she quickly recognized that her prior research experience in ecology and biogeochemistry was well-suited for work in the water and wastewater sectors. Upon receiving her MS in Engineering in 2018, she accepted a full-time position at Carter & Sloope, where she continues to work as an engineer. She currently works with communities throughout Northeast Georgia and specializes in environmental evaluations and design of water and wastewater treatment plants. At the time of writing, she is reaching eligibility to test for her licensing as a professional engineer.

As a strong advocate for women's empowerment, Dr. Bryant mentors female students in STEM as well as in Brazilian Jiu-Jitsu. In her spare time, she owns and operates a Brazilian Jiu-Jitsu School with her husband, where she teaches a women-only class. She serves as the Director of multiple Women's Self Defense Programs and has taught self-defense classes to hundreds of local women of all ages.

Part V
Improving the Quality of Life

Chapter 18
Preparing for the Electric Grid of the Future: Challenges and Opportunities

Jill S. Tietjen

Abstract The electric utility infrastructure fits within the energy category of the American Society of Civil Engineers' (ASCE) Infrastructure Report Card. That category had received a D+ for a number of years and was upgraded to a C- in 2021. Numerous challenges and opportunities face the industry as it strives to update old facilities while maintaining reliable service particularly as the demands on the electric grid change. The types of demands facing the grid include decarbonization, renewables, transmission, electric vehicles, energy storage, smart grids, nuclear power, and more.

Keywords Transmission · Distribution · Generation · Energy storage · Smart grid · Decarbonization · Electric vehicles · Solar · Wind · Renewables · Batteries · Nuclear power · Distributed energy · Microgrids

18.1 Introduction

The electric utility infrastructure has three major components – generation, transmission, and distribution. The power plants generate the electricity and are thus often referred to as the generation. The transmission lines are the wires at high voltage that deliver the generation from the power plants to the population centers. There at the substations, the voltage is lowered through transformers and then sent over distribution wires at these lower voltages to the ultimate customers. Ultimate customers fall into three broad general classes – industrial, commercial, and residential. Reliability of service is a hallmark of the electric utility industry – such that

J. S. Tietjen (✉)
Technically Speaking Inc, Greenwood Village, CO, USA

© The Author(s), under exclusive license to Springer Nature
Switzerland AG 2022
P. Layne, J. S. Tietjen (eds.), *Women in Infrastructure*, Women in Engineering and Science, https://doi.org/10.1007/978-3-030-92821-6_18

most of the public just takes the provision of electricity 24 hours a day/7 days per week for granted.

Electric utility infrastructure fits within the category of energy as part of the American Society of Civil Engineers' (ASCE) Infrastructure Report Card. In 2017, ASCE gave the energy infrastructure, including electric utilities, a grade of D+; in 2021, the grade improved to a C-. This grade as applied to the electric grid component of energy reflects that much of the generation, transmission, and distribution have gotten old, and serious steps need to be taken to update and upgrade them, particularly as the demands on the electric grid change [1, 2].

What types of demands are there? The large list includes decarbonization, renewables, nuclear, electric vehicles, distributed energy, energy storage, microgrids, and more. The challenges facing what has been called the largest machine in the world – the electric grid in the United States – are immense but so are the opportunities. All of the challenges need to be addressed, while the industry continues to provide reliable, safe, and economic electric power to customers.

18.2   How We Got Here

The electric supply in the United States at the end of 2020 totals just over 1,117,000 MW, according to the US Energy Information Administration (EIA) [3]. The major components of electric capacity are shown in Table 18.1. A number of other small categories comprise the rest of the generation fleet [4].

In 2019, the share of electricity generated by fuel was natural gas (38%), coal (23%), nuclear (20%), petroleum (1%), and renewables (17%). The renewables can be subdivided further into wind (7.3%), hydro (6.6%), solar (1.8%), biomass (1.4%), and geothermal (0.4%) [5]. For 2020, a year significantly impacted by the global pandemic, the percentages were natural gas (40.3%), nuclear (19.7%), coal (19.3%), petroleum and other gases (less than 1%), and renewables (19.8%). The renewables percentages for 2020 were wind (8.4%), hydro (7.3%), solar (2.3%), biomass (1.4%), and geothermal (0.4%) [4].

Table 18.1 Nameplate capacity of electric generating capacity by fuel type – as of the end of 2020 [4]

Type of capacity	MW	% of Total
Natural gas	482,590	43.20
Coal	218,424	19.50
Wind	117,744	10.54
Hydro	102,840	9.20
Nuclear	96,555	8.64
Solar	47,848	4.28
Geothermal	2587	0.23

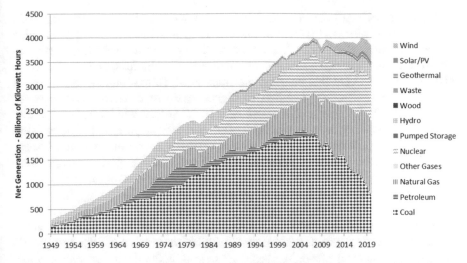

Fig. 18.1 Historical net electricity generation (electric power sector only), 1949–2020 [6, 7]

The changes in the percentage of electricity produced by each sector over history are shown in Fig. 18.1. The most striking changes of note in the last 10 years are the reduction in the percentage of electricity provided by coal and concomitantly the increase in the percentage provided by natural gas as well as the growth in wind [6, 7].

At the dawn of the electric utility industry, many of the generation facilities were hydroelectric, located remotely from the population centers. Thus as those hydroelectric facilities were built, transmission lines to deliver electricity from the remote locations to the population centers were also built. As the early coal-fired power plants were built, they needed to be near rivers or other water sources as they used steam turbines to generate electricity. Later, nuclear units also required water in order to produce electricity. Most of those facilities are also remote from population centers, and transmission was built to deliver the power they generated to the population centers. Except as required to provide reliable electric service, transmission was rarely ever built other than to deliver electricity from a power plant to a population center.

18.3   What Is the Future of Electricity Demand?

The EIA projects that electric demand in the form of electricity sales will recover from the global pandemic and then grow at a rate of under 1% per year through 2050. The use of rooftop solar photovoltaic systems and combined heat-and-power systems for commercial customers is also projected to grow over the period. The EIA does not expect significant growth resulting from electrification of the transportation sector [8]. Conversely, at least one projection reflecting significant electrification finds that the electricity demand could double by 2050 – that reflects an annual load growth of around 2.4% from 2021 through 2050 [9].

What factors can influence whether the demand for electricity through 2050 tends toward higher load growth or lower load growth? Here is a list of possible factors: the cannabis industry, the Internet of Things, electric vehicles, electrification of building heating systems, distributed energy resources, energy efficiency, and lighting. The cannabis industry, the Internet of Things, and electrification of building heating systems all influence higher levels of electricity consumption. Electric vehicles, as discussed in depth later in the article, will increase electricity consumption and the times at which electric vehicles charge have the potential to significantly impact the need for new generation resources if not controlled through policy, rates, or the charging equipment itself. Distributed energy resources, particularly if owned by industrial customers, will reduce the demand for electricity as will more efficient homes, buildings, and appliances and more efficient lighting.

The legalization of marijuana in states around the United States has led the cannabis industry to be recognized as a very energy-intensive industry. Electricity is required for lighting, moisture and heating, ventilation, and air conditioning (HVAC) since most cannabis is grown in greenhouses. As of 2020, the industry spent about as much on electricity as did the federal government with energy demand equivalent to that of about three million electric vehicles. The electricity requirements associated with home growing of four marijuana plants has been deemed equivalent to 29 refrigerators [10]. For some utilities, the cannabis industry has been their major source of electricity demand growth in recent years.

Utilities are monitoring all of these factors, so that, as electric demand grows or decreases, the best mix of resources, in compliance with local, state, and federal regulations and requirements as well as their legal obligation to serve, is in service to meet customer needs and provide safe, economic, and reliable service.

18.4   Decarbonization

As the United States – and the world – discusses and acts on decarbonization, it is helpful to understand where the greenhouse gases that are of concern regarding global warming are derived. In 2019, the US Environmental Protection Agency (EPA) reported that greenhouse gases emitted in the United States by economic sector were as shown in Fig. 18.2.

The largest emitter is seen as transportation – and thus one of the reasons for the strong interest in the electrification of automobiles and trucks. The electric power industry generates a quarter of greenhouse gas emission, thus explaining the huge focus on decarbonization for the industry. The following paragraphs describe what drives those emissions and some of the steps being taken to reduce them.

During the nineteenth century, coal was the primary fuel source that enabled the United States to transition from an agricultural society to a world economic power. In the early twentieth century, coal was used primarily as a raw material to power the nation's industrial and transportation sectors and for home heating, although

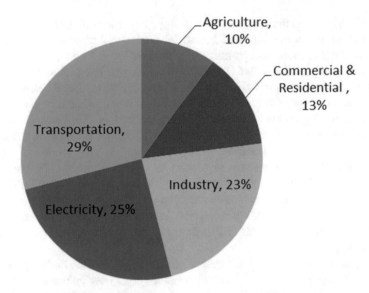

Fig. 18.2 US greenhouse gas emissions by economic sector in 2019 (US EPA) [11]

Thomas Edison used coal to fire the first electric power generation station in 1882 in New York City [12]. Most early electricity production came from hydropower.

The major expansion of the US electric utility systems occurred from the 1960s through the 1980s. During that time, coal was the primary fuel for baseload generation, and coal production nearly doubled from 1970 to 1990 [12]. The dominance of coal-fired power generation was enabled by two factors: (1) the increasing efficiency of power plants over time and (2) the abundance of local coal supply. Generating units were no larger than 150 MW from the 1930s through the mid-1950s. By 1975, however, due to technological advances, 1300-MW generating units were developed and installed – increasing the size of an individual generating unit by almost a factor of 10 as well as significantly improving energy efficiency [13]. The costs of electricity production declined as each new generating unit was installed. With coal basins located throughout the continental United States and Alaska, coal was easily accessible, available, economically priced, and readily stockpiled [14]. Coal as a fuel for electricity generation remains plentiful. The EIA estimates that at the 2014 consumption rate, known coal reserves in the United States will last for more than 250 years [15].

Why was coal and not natural gas the preferred fuel choice for electricity generating plants as demand grew? There were several reasons. The 1978 Powerplant and Industrial Fuel Use Act (FUA) prohibited the use of natural gas and oil as the primary fuel in electric utility power plants or large industrial boilers. Although these restrictions were eliminated with the repeal of the FUA in 1987 [16], the price level of natural gas, restrictions on its availability during the winter season, and its significant price volatility precluded its use for baseload generation.

The oil embargo in the early 1970s, ensuing economic conditions including rampant inflation, the Powerplant and Industrial Fuel Use Act of 1978, the growing environmental awareness as epitomized by the first Earth Day in 1970, the Public Utility Regulatory Policies Act (PURPA) of 1978, and the accident at the Three Mile Island nuclear plant in 1979 meant that the installation of new electric generating facilities no longer led to decreases in electric rates. In addition, electric consumption stopped growing at a dependable annual rate of 7%. These events in the 1970s laid the foundation for the changes in electric generation mixes that are now observed in the twenty-first century.

Developments in the economy, regulatory environment, and public attitudes are also driving the changes toward decarbonization that we are seeing today. These developments include cost decreases for solar and wind, enforcement of the New Source Performance Standards of the Clean Air Act Amendments, the enactment of Renewable Portfolio Standards (RPS) in many states, and a worldwide focus on climate change.

In the past decade, the levelized cost of electricity (the cost to build a power plant as well as the fuel and operating and maintenance costs over its lifetime) from solar photovoltaics decreased 89% (Fig. 18.3). Also, over the past decade, the levelized cost of electricity from onshore wind decreased by 70% [17].

In 2002, the Sierra Club launched its Beyond Coal campaign with the objectives of preventing new coal units from being built, having existing coal plants retire, and not mining additional coal for domestic use or for export. The campaign has been successful as coal-fired plants are retiring, and new coal-fired generation is no longer being considered in the United States [18].

Under the 1970 amendments to the Clean Air Act, the EPA had the responsibility to promulgate "new source performance standards" (NSPS) for new stationary sources and modifications to existing sources. Existing sources, including the coal-fired electric utility fleet, were assumed not to last forever – and the expectation was that when they retired they would be replaced with cleaner, new technology. Economic conditions, the passage of PURPA, and lower rates of load growth did not lead to new cleaner technology being built because the units did not retire. Instead, many utilities "life extended" their existing coal-fired units – units that had been designed for a 30-year life sometimes were in operation for 40, 50, and even 60 years. In 1999, the Department of Justice sued seven large power companies on behalf of the EPA for violations of the NSPS. Modern pollution controls were installed on many units, fines were paid, and the era of life extension was coming to an end [19].

In 1983, Iowa became the first state to enact an RPS. An RPS requires that a specific percentage of the energy that a utility sells must come from renewable energy resources. As of 2021, 30 states, the District of Columbia, and three territories have enacted RPS, while renewable energy goals have been set by seven states and one territory. From a regulatory standpoint then, states are also prodding the electric utility industry toward decarbonization [20].

Internationally, the Intergovernmental Panel on Climate Change (IPCC) was established by the United Nations and the World Meteorological Organization in

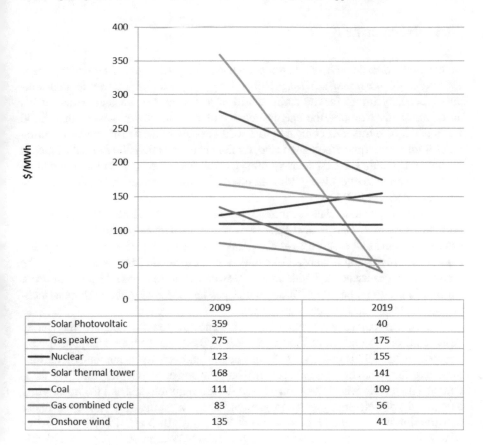

	2009	2019
——Solar Photovoltaic	359	40
——Gas peaker	275	175
——Nuclear	123	155
——Solar thermal tower	168	141
——Coal	111	109
——Gas combined cycle	83	56
——Onshore wind	135	41

Fig. 18.3 Levelized cost of electricity from new power plants [17]

1988. The IPCC "provides regular assessments of the scientific basis of climate change, its impacts and future risks, and options for adaptation and mitigation." Its efforts to protect the world's climate were recognized jointly with former Vice President Al Gore in 2007 with the Nobel Peace Prize. Reports from the IPCC have been instrumental in driving decarbonization with associated greenhouse gas emissions reductions worldwide [21, 22].

In May of 2021, a report was issued by the Berkeley Lab titled *Halfway to Zero: Progress Towards a Carbon-Free Power Sector*. It reported that in 2020, the emissions of greenhouse gases from the electric power industry were 52% lower than the 2020 value the EIA had forecast in 2005. Although some of this reduction is due to the increased penetration of solar and wind resources, most of it is due to electric utilities installing new natural-gas fired generating units and retiring older coal-fired units. This is enabling the industry as a whole to get halfway toward the target of reducing emissions to net-zero by 2035 [11].

18.5 Nuclear or Not?

In his book *Smaller, Faster, Lighter, Denser, Cheaper: How Innovation Keeps Proving the Catastrophists Wrong*, Robert Bryce argues that the way to a sustainable electricity future for the entire world is natural gas to nuclear. Much of his argument centers around the energy density of nuclear power – more than 2000 watts per square meter as compared to 1 watt per square meter for wind – meaning that for the same square footage of land, nuclear can generate 2000 times the amount of power as wind. The best solar systems exhibit energy density in the low double digits. The process of evolving from natural gas to nuclear would take place over the course of decades [23].

The natural gas to nuclear technology process addresses climate change and carbon dioxide emissions. It also addresses the approximately one billion people around the world who haven't yet electrified but desire the standard of living that electricity provides and whose electricity will be provided by some means. The many challenges associated with nuclear power will have to be addressed in order for this future to be possible. These challenges include cost, public mistrust, technology advances, and nuclear wastes [23].

The upfront capital costs associated with nuclear power plants are very large. Almost all of the nuclear units in operation in the United States today were constructed in the 1970s and 1980s. Construction stopped for a number of reasons including the escalating capital costs for nuclear power plants after the 1979 accident at Three Mile Island due to increased design requirements and the uncertainty associated with electricity demand growth. There are only two new units under construction in the United States as of mid-2021 – Vogtle 3 and 4 in Georgia. As of May 2021, the estimated capital cost for the 2200 MW of nuclear power for Vogtle 3 and 4 is $25 billion. Thus, technology advances and innovation would seem to be required to make the natural gas to nuclear future possible [24].

One such innovation well along the development path is the Small Modular Reactor (SMR) by NuScale, based in Portland, Oregon. In September 2020, the United States Nuclear Regulatory Commission approved NuScale's design and issued a final safety evaluation report. Much of the design has been underwritten by Fluor and the United States Department of Energy. Each SMR is 60 MW, and multiple SMRs can be constructed on the same site. The Utah Associated Municipal Power Systems is planning to build 12 of them to supply electricity to the US Department of Energy in eastern Idaho. In the spring of 2021, two Japanese firms invested in the NuScale venture – JGC Holdings Corporation and IHI Corporation. Also in the spring of 2021, NuScale announced that the Grant County Public Utility District had taken the first steps toward installing four SMRs at the Hanford Nuclear Reservation in Washington State [24–27].

Another entrant in the next generation nuclear power race is Bill Gates. His company TerraPower is developing Natrium Reactor plants that use liquid sodium as a cooling agent in combination with storage technology in tanks of molten salt [28].

In June of 2021, TerraPower announced plans to build Natrium reactors in Wyoming at one of Rocky Mountain Power's coal-fired power plant locations [29].

The Three Mile Island, Chernobyl and Fukushima events, and the name "nuclear power" have led the public to believe that nuclear power is not safe in spite of the fact that 96 facilities reliably, economically, and safely provide power in the United States and have for decades. Kessler suggests that rebranding is in order and probably not entirely tongue in cheek. Also, no discussion of nuclear power is complete without a consideration of what to do with the waste products, especially in the United States. Since the decision was made decades ago in the United States not to recycle the fuel, the alternative is storing it for years. However, the Yucca Mountain facility is no longer an option, leaving most spent nuclear fuel stored at the generating facility. In many people's minds, until this issue is resolved, new nuclear power technologies will not be acceptable to the American public [25, 30].

18.6   Renewables

The public, the Federal government, and the individual states are driving efforts to increase the penetration of renewable resources in the electric power industry. As described above, one mechanism for doing this is the passage of mandates or voluntary goals for renewable energy electric sales, often in the form of an RPS. There is no consistent definition for renewable energy – but the primary technologies generally include solar, wind, geothermal, hydroelectric, and biomass. Issues related to the so-called intermittency of solar and wind need to be mitigated in order for these resources to more fully contribute to the provision of electricity on the grid. Intermittency means solar energy is only generated when the sun shines, and wind energy is only generated when the wind blows.

Unintended consequences often occur in spite of regulators' and electric utilities' good intentions. One of those unintended consequences is illustrated most vividly by what is called the "duck curve."

Without any mitigating technology such as energy storage, solar energy can only be generated when the sun is actually shining. In California, Hawaii, and other areas around the world with high percentages of installed solar energy resources, this has resulted in the so-called duck curve in demand on the electric grid resulting in both potential overgeneration as well as increased ramping for generating units. In addition, the amount of solar energy that can be produced varies over the course of the day as the amount of sun insolation varies as compared to the angle of the receiving device. Thus, most solar energy is generated between the hours of 9 AM and 3 PM on days when it is not raining or snowing. As can be seen in Fig. 18.4, this results in what has come to be called the "duck curve" because of its resemblance to a duck [31, 32]. Also, this situation is not limited to the states of California and Hawaii. Duck curves are being experienced in Australia, France, Germany, India, and other areas where solar energy is being used in larger and larger amounts [33].

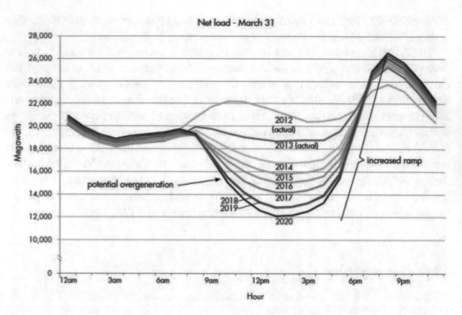

Fig. 18.4 The so-called duck curve as experienced in California [31]

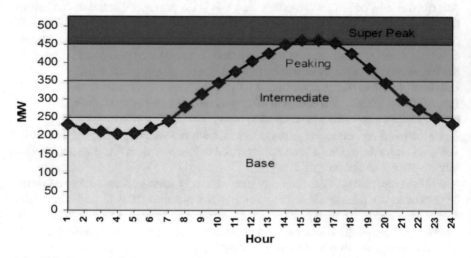

Fig. 18.5 Summer daily load curve with generation resources by type [34]

When the sun is not shining at a level that is useful for its conversion to electricity, approximately 7 PM to 9 AM, the hourly load curve during the summer months closely resembles that shown in Fig. 18.5. However, with the penetration of solar energy as now exists in California, the amount of electricity that needs to be provided by the electric utilities through the grid in the hours of 9 AM to 6 PM has declined significantly. However, the load at 7 PM is at the same level as it would have been without solar energy generation. Because of the need for generation at

7 PM and the operational imperative to not overly stress the equipment, some of it may run at a fraction of its capability during the hours that solar energy is being generated, with the resulting inefficiencies, during the hours of 9 AM to 6 PM so that it can be ramped up to meet the load at 7 PM.

This results in what is labeled "potential overgeneration" on Fig. 18.4 – electricity that is produced that is not consumed by the electric utilities' customers. Because electricity cannot be stored, that means that this overgeneration flows into the electric transmission and into other utilities' service territories – causing those utilities to decrease the generating output of their power plants to exactly match the generation to the electric demand. This is called "inadvertent flow," and the utilities that cause it must compensate those utilities that must back down their own, generally less expensive, power plants.

In addition to power plants needed to be run during the daylight hours when the sun is shining to meet the load requirements at 7 PM, other units may need to be turned on daily after 6 PM in order to meet the 7 PM load. Those units would then be shut down the following morning when the solar energy starts being generated. This so-called daily cycling increases the stresses on the mechanical equipment often leading to increased maintenance, increased maintenance costs, and decreased life spans.

Surprisingly, to many members of the public, solar and wind resources attract environmental opposition. Utility-scale solar photovoltaic projects cover much land, and the reasons for opposition to these types of projects, from many people who are renewable energy supporters and have put solar panels on their own homes, range from a desire to keep the land unspoiled to preservation of endangered species to concern about the value of their properties and their views [35]. Wind resources generate opposition based on killing wildlife (including birds and bats), noise generation, and the visual impact on the landscape [23].

18.7   Electric Vehicles

The International Energy Agency reports that 2.5% of the cars sold globally in 2019 were electric. For the United States, that percentage is around 2%. By 2025, electric vehicle sales are expected to total 10% of total automobile sales in the United States. Some projections show that all vehicles sold in the United States by 2035 will be electric. In fact, California's governor signed an executive order in 2020 banning the sale of new gas-burning cars in the state by 2035 [36–41].

Increasing levels of electric vehicle sales have significant implications for the electric utility industry and the grid. How, where, and when will people charge their vehicles? Will electric demand increase during the peak hours due to electric vehicles? Is it possible to use electric vehicle batteries as an energy resource when the vehicles are not in use for transportation?

Some of the most significant adopters of electric vehicles in the United States are expected to be transportation fleets – think FedEx, UPS, the United States Postal

Service, and amazon.com. Charging will need to occur at hub locations (depot charging) which will become new load centers for utilities – and are expected to be primarily on the distribution system. Utilities will need to think about storage, distribution system upgrades, additional monitoring equipment, advanced distribution management systems, and distributed energy resources to help manage the location and duration of these new electric demands [42].

Converting all of California's cars and trucks to electric vehicles could increase peak electricity demand in California by 25%. That won't happen in 15 years as the existing fleet of vehicles will not all be electric, but nevertheless the electric utilities have begun the needed planning for the significant increase of electric vehicles in the state. Already facing periodic rolling blackouts due to system constraints, they will need to make upgrades to the transmission and distribution system in addition to installing new generating capacity. Rules and rates will need to be put in place so that everyone is not trying to charge their vehicles at the same time [40].

How, when, and where will electric vehicles be charged? If the answer is sometimes at the office, sometimes at home, and sometimes somewhere else, this becomes what is termed in the industry a "mobile load." It requires a different design for a different system, say a refrigerator – a refrigerator does not move and therefore requires electricity only in one place. However, if a vehicle can be charged at different locations – and does – then the distribution system must be built to accommodate charging at work, at home, at school, at the mall, at the dentist's office, wherever.

So many challenges still need to be worked out with regard to electric vehicles. Who owns the public chargers? Will they be owned by private industry, electric utilities, or government? Are they fast chargers? Will people be encouraged – through rates and rules – to not charge during the peak hours of the day?

What about what has been dubbed "vehicle-to-grid" where batteries in passenger cars, buses, and other vehicles are tapped to release power to the grid? This topic has been discussed in countries including the United Kingdom and Denmark [40]. Some versions of the Ford 2022 F-150 Lightning come with a Ford Charge Station Pro, a bidirectional technology that allows electric vehicles with batteries to supply power to the home – up to 9.6 kW – and in the future, to the grid [43].

Higher penetrations of electric vehicles are without question part of the future. Public policies, rate structures, and the electric grid itself will all require modification to accommodate their use and to minimize their economic and social impact.

18.8 Transmission

As decarbonization of the grid proceeds, the electric transmission system will require significant modification. This occurs for a variety of reasons and includes (1) the location of power plants installed over many decades determined the location of the high voltage transmission lines and (2) the mismatch between the areas of high potential wind energy resources and the population centers that require the electricity. In addition, to increase the rapidity with which new transmission facilities can be built, significant modification will need to be made within the construct of the current regulatory regime, or changes to that regime will be required. The transmission system needs to respond to more severe weather related to climate change as well through hardening and other efforts. The current estimate is that the transmission system will need to be expanded by two or three times its present size in order to accommodate the coming electrification [44].

In the United States, solar intensity is highest in the Southwest as shown in Figs. 18.6 and 18.7, which demonstrate the potential for photovoltaics and concentrating solar power, respectively.

Wind speeds vary significantly across the United States as shown in Fig. 18.8 and are particularly strong in the mid-West and offshore [46].

A particular and interesting issue related to the usage of wind energy resources is the location of the wind resource itself versus the concentration of the population that is using the electricity. In Fig. 18.8, the highest wind energy speeds are shown in colors other than green. However, the population centers in the United States as shown in Fig. 18.9 almost completely coincide with the green color areas from Fig. 18.8. This means that once the wind energy is converted into electricity, it must be moved by the transmission system to the population centers.

Current regulatory systems, where involvement and permits are required at the local, state, and federal level (Table 18.2), are unlikely to be able to handle the needed expansion of the transmission system in a timely manner to support

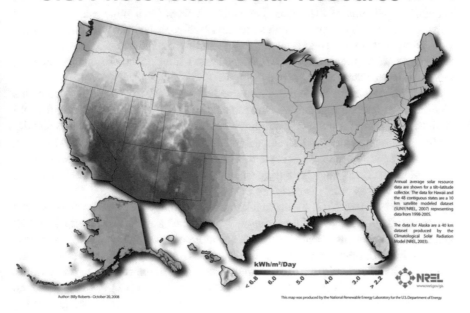

Fig. 18.6 U.S. Photovoltaic Solar Resource [45]

U.S. Concentrating Solar Resource

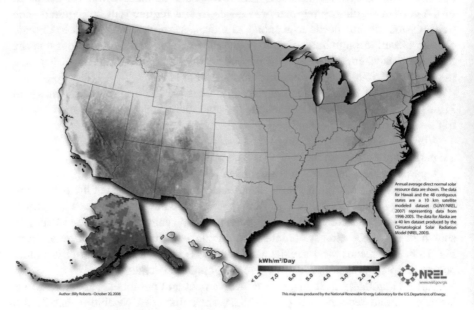

Fig. 18.7 U.S. Concentrating Solar Resource [45]

renewable buildout as well as broader electrification and the resulting growth in electricity demand. Many state siting regulations require that the facility provide benefits to residents of that state. Transmission lines from wind energy resource areas, particularly, may run through states that receive no benefit from that transmission. Options that have been floated for overcoming transmission building "paralysis" include federal authority to site transmission lines and more undergrounding along federal highway rights-of-way [44, 48].

Elements of system hardening programs being developed at utilities across the United States include vegetation management; flood hardening particularly for substations; preparing for changing weather – colder than normal – heat waves and dry spells; upgrading poles and structures with stronger materials; undergrounding; shortening span lengths; changing out wires; and investing in smart grid sensors. Some of these efforts have been triggered by experiences resulting from hurricanes and fires as well as other catastrophic events [54].

After the February 2021 Texas blackout, there is renewed talk of connecting the electric grids in the United States which is viewed as a means of diversifying risk and helping to prevent a recurrence of the economic and human costs of that blackout. There are currently three major grids in the United States that operate for the most part independently of each other – the Eastern Interconnect, the Western Interconnect, and the Electric Reliability Council of Texas (Fig. 18.10). Although the Eastern and Western Interconnect comprise the large majority of generating capacity and electricity consumption in the United States, about 950,000 MW, there

U.S. Wind Resource (80m)

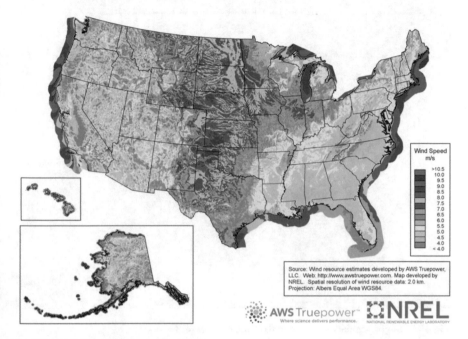

Fig. 18.8   U.S. Wind Resource (80m) [45]

are only approximately 1300 MW of connections between them – through seven high voltage direct current back-to-back converter stations installed along an imaginary line (the seam) that runs south to north from western Texas to eastern Montana. Connecting all three grids would provide an opportunity to share resources and buttress all of the systems. Preliminary studies report benefits to this arrangement [55].

18.9   Energy Storage

In order for the electric grid in the United States to integrate large amounts of solar and wind energy resources, shifting of the time as to when the energy is produced and when it is consumed will be necessary. This shifting will require energy storage mechanisms. Energy storage involves a device that accepts electric energy from the grid, converts it into an energy form that can be stored, and then converts it back to electricity, minus the efficiency losses, and returns the electricity to the grid.

Energy storage allows synchronization between the time of energy production for solar and wind energy resources and the time of system peak. In addition, energy storage devices themselves often provide operational flexibility to the system operator and the electric grid and can provide the following benefits [57]:

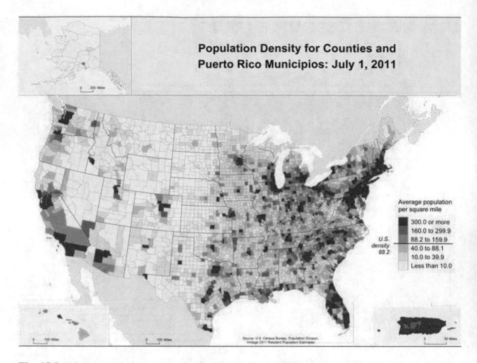

Fig. 18.9 Major population centers in the United States [47]

- Enhancement to the value of intermittent renewable energy resources on the power grid by firming their energy.
- Improvement in power quality by providing ancillary services such as voltage regulation and spinning resources.
- Ability to store low-value, excess energy when power supplies exceed demand until the energy can be economically used to meet load.
- Enhancement of the flexibility of the existing transmission grid.
- Relief of transmission congestion to defer capital expenditures on system upgrades.
- Conversion of less costly off-peak energy into higher-value on-peak power.
- Reduction of problems associated with minimum generation requirements.

Energy storage technologies of various types are being used, tested, and studied around the world. The energy storage technologies that are already commercial or have the possibility of becoming commercial in the near term include:

- Batteries: sodium-sulfur (NaS), lithium-ion, lead-acid.
- Pumped storage hydro.
- Compressed Air Energy Storage (CAES).
- Supercapacitors.
- Superconducting Magnetic Energy Storage (SMES).
- Flywheels.

Table 18.2 Agencies and regulations for transmission line siting [49–53]

Environmental Review Requirements
Clean Water Act
Endangered Species Act
Migratory Bird Treaty Act
National Environmental Policy Act
National Historic Preservation Act
Federal Agencies
Advisory Council of Historic Preservation
Bureau of Indian Affairs
Environmental Protection Agency
Federal Aviation Administration
National Park Service
US Army Corps of Engineers
US Bureau of Land Management
US Bureau of Reclamation
US Department of Agriculture
US Department of Defense
US Fish and Wildlife Service
US Forest Service
US Geological Survey
State Agencies
State Utility Regulatory Commission
Other State Energy Regulators
State Department of Wildlife
State Office of Archaeology and Historic Preservation
State Department of Transportation
State Board of Land Commissioners
State Department of Natural Resources
State Department of Agriculture
State Department of Economic and Community Development
Regional and Local Agencies
Areawide and Regional Coordination Agencies
County Offices
Planning Commissions
Historic Commissions
Municipality Offices

- Hydrogen.
- Cryogenic Energy Storage.
- Adiabatic CAES.
- Synthetic natural gas.

The commercial status of these various energy storage technologies are shown in Table 18.3.

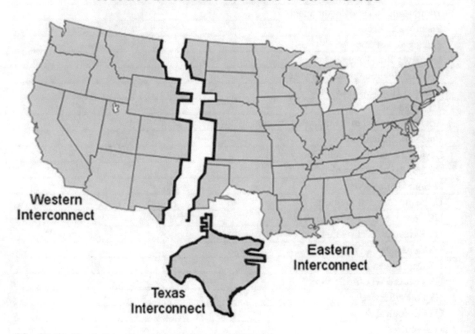

North American Electric Power Grids

Fig. 18.10 North American Electric Power Grids [56]

Table 18.3 Energy storage technology deployment status [58]

Deployed	Demonstration	Early state technologies
Pumped storage hydro	Cryogenic Energy Storage	Adiabatic CAES
Flywheels	Superconducting Magnetic Energy Storage (SMES)	Synthetic natural gas
Compressed Air Energy Storage (CAES)	Supercapacitors	
Batteries – sodium-solar (NaS), lithium-ion, lead acid	Hydrogen	
Flow batteries		
Molten salt energy storage		

One further indication of the importance of energy storage emerged in April 2021. Tesla announced a new company policy to sell its solar product only when coupled with its energy storage products. Thus, Tesla's solar product, a renewable product, also becomes a distributed energy product [59].

18.10   Distributed Energy and Microgrids

Distributed Energy Resources (DERs) are assets generally located close to the load and usually behind the meter. DERs include solar, storage, energy efficiency, and demand-side management. They can be aggregated to provide services to the grid. DERs are an important element in the modernization of the grid that will lead to bidirectional flow. Dependent on the specific technology and location, DERS can provide energy, capacity, and ancillary services for one or both of the transmission and distribution systems [60].

A microgrid is a group of loads and resources that although decentralized usually operates within the context of the entire synchronous grid. However, the microgrid can separate from the entire synchronous grid and operate autonomously when conditions so dictate. Microgrids often serve a discrete small footprint such as a college campus or hospital complex. Microgrids often incorporate DERs – usually solar, wind, combined heat and power, or generators. Many microgrids also have some form of energy storage, often batteries. In addition to being local and able to operate autonomously, microgrids have smart technology as well. The smart technology (the microgrid controller) enables the microgrid to operate autonomously when it separates from the rest of the electrical grid and controls the supply of energy (the generating resources) to match the loads [61].

In 2021, lawmakers in 20 states introduced 69 microgrid bills with an eye toward grid modernization and energy resilience. Microgrids got a boost from the Texas blackout and lawmakers are also concerned about other extreme weather events including wildfires, hurricanes, and tornadoes [62]. Microgrids and DERs are becoming a component of electric utility planning for the future of the electric grid.

18.11   Smart Grids

A definition of a smart grid was first provided by the Energy Independence and Security Act of 2007. The act enumerated ten components with the underlying theme that digital processing and two-way communication with the resulting data flow and information management are what make the grid smart. The ten components incorporate all elements of a power system including load, distribution, transmission, and generation and are associated with the use of renewables, demand-side management, energy storage, peak energy shaving, and power conditioning. The system is considered smart because of the communication technologies that enable self-healing through sensing capability with heavy monitoring and a variety of computer controls that when combined provide automatic system responses for changes in load, generation, and equipment that is out of service for whatever reason [63].

Solar and wind energy resources will benefit from the building of a smarter grid. This is because they do not rotate in synchronism with the power system and thus don't as readily provide system support in categories known as essential reliability services. A grid with extensive investment in smart grid infrastructure may have less of a need for certain essential reliability services, thus making it easier to provide for wind and solar expansion.

18.12 Conclusion

Many factors must be addressed to ensure that the US electric grid – the world's most complex machine – is prepared to face future demands. Rising from the C- that was assigned by the ASCE 2021 Infrastructure Report Card for the energy sector, of which it is component, will require thoughtful and significant efforts. Among the challenges and opportunities influencing future electric grid are decarbonization, nuclear, electric vehicles, renewable resources, transmission, smart grids, energy storage, and more. The electric utility industry has continuously risen to the challenges of the past and can be expected to rise to them in the future as well.

References

1. American Society of Civil Engineers (ASCE), America's Infrastructure Grade, https://www.infrastructurereportcard.org/americas-grades/, accessed May 4, 2020.
2. Report Card for America's Infrastructure – Energy (2021), https://infrastructurereportcard.org/cat-item/energy/, accessed May 29, 2021.
3. U.S. Energy Information Administration, *Electricity Explained: Electricity generation, capacity and sales in the U.S.*, https://www.eia.gov/energyexplained/electricity/electricity-in-the-us-generation-capacity-and-sales.php, accessed May 30, 2021.
4. U.S. Energy Information Administration, Frequently Asked Questions, What is U.S. electricity generation by energy source, https://www.eia.gov/tools/faqs/faq.php?id=427&t=3, accessed May 30, 2021.
5. *Electricity Explained: Electricity in the United States*, U.S. Energy Information Administration, https://www.eia.gov/energyexplained/electricity/electricity-in-the-us.php, accessed May 5, 2020.
6. EIA. (2016, 26 August). Monthly Energy Review (DOE/EIA-0035). Table 7.2, Electricity net generation, www.eia.gov/totalenergy/data/monthly/ (from 1949 – 2015).
7. EIA, Monthly Energy Review, Section 7, Table 7.2b, Electricity Net Generation: Electric Power Sector, https://www.eia.gov/totalenergy/data/monthly/pdf/sec7.pdf, accessed May 30, 2021.
8. Energy Information Administration. *Annual Energy Outlook 2021*. "Electricity." https://www.eia.gov/outlooks/aeo/electricity/sub-topic-01.php. February 3, 2021.
9. Walton, Robert. "Equity, security and load: FERC conference considers the challenges and potential of electrification. Utility Dive. May 3, 2021. https://www.utilitydive.com/news/equity-security-and-load-ferc-conference-considers-the-challenges-and-pot/599378/.
10. Reott, Jason. Alliance to Save Energy. "Legal Cannabis Presents Challenges For Utilities, Opportunities for Energy Efficiency," September 8, 2020. https://www.ase.org/blog/legal-cannabis-presents-challenges-utilities-opportunities-energy-efficiency.
11. Viswanath, Teri. "Achieving Biden's Climate Moonshot Requires Heavy Lifting by Electric Power Industry." https://www.cobank.com/knowledge-exchange/power-energy-and-water/

achieving-biden-s-climate-moonshot-requires-heavy-lifting-by-electric-power-industry. May 2021.
12. U.S. Department of Energy National Energy Technology Laboratory. (n.d). Key Issues & Mandates: Secure & Reliable Energy Supplies—History of U.S. coal use, energybc.ca/cache/historyofenergyuse/www.netl.doe.gov/KeyIssues/historyofcoaluse.html
13. Cassaza, J.A. (1993). The development of electric power transmission. IEEE Case Histories of Achievement in Science and Technology. New York: IEEE.
14. U.S. Energy Information Administration (EIA). (2016, 24 March). U.S. coal reserves, www.eia.gov/coal/reserves/
15. EIA. (2016, 17 June). Coal explained: How much coal is left, www.eia.gov/energyexplained/index.cfm?page=coal_reserves
16. EIA. (n.d.). Repeal of the Powerplant and Industrial Fuel Use Act (1987), www.eia.gov/oil_gas/natural_gas/analysis_publications/ngmajorleg/repeal.html
17. Roser, Max. World in Data. "Why did renewables become so cheap so fast? And what can we do to use this global opportunity for green growth?" https://ourworldindata.org/cheap-renewables-growth. December 1, 2020.
18. Sierra Club Beyond Coal Campaign, https://content.sierraclub.org/creative-archive/sites/content.sierraclub.org.creative-archive/files/pdfs/100_90-BeyondCoal_FactSheet_18_low.pdf, accessed June 5, 2021.
19. Thomas O. McGarity, When Strong Enforcement Works Better Than Weak Regulation: The EPA/DOJ New Source Review Enforcement Initiative, 72 Md. L. Rev. 1204 (2013) Available at: http://digitalcommons.law.umaryland.edu/mlr/vol72/iss4/14.
20. Shields, Laura. National Conference of State Legislatures. State Renewable Portfolio Standards and Goals. April 7, 2021. https://www.ncsl.org/research/energy/renewable-portfolio-standards.aspx
21. About the IPCC, https://www.ipcc.ch/about/, accessed June 5, 2021.
22. United Nations, https://www.un.org/en/about-us/nobel-peace-prize/ipcc-al-gore-2007, accessed June 5, 2021.
23. Bryce, Robert. *Smaller, Faster, Lighter, Denser, Cheaper: How Innovation Keeps Proving the Catastrophists Wrong*. New York, New York: Public Affairs. 2014.
24. Clarion Energy Content Directors, "Japanese manufacturer commits $20M to NuScale small nuclear reactor plan," May 28, 2021, https://www.power-eng.com/nuclear/japanese-manufacturer-commits-20m-to-nuscale-small-nuclear-reactor-plan/#gref.
25. Kessler, Andy (Inside View). "Clean Power, No Thanks to Al Gore," *The Wall Street Journal*. September 21, 2020, p. A17.
26. Ridler, Keith. Associated Press, "US gives first-ever OK for small commercial nuclear reactor," September 3, 2020.
27. Stang, John. Wash. PUD, NuScale Sign MOU to Explore Use of Small Reactors, RTO Insider, May 28, 2021. https://rtoinsider.com/rto/wash-pud-nuscale-mou-explore-small-reactors-201651/.
28. Clifford, Catherine. How Bill Gates' company TerraPower is building next-generation nuclear power. April 8, 2021. https://www.cnbc.com/2021/04/08/bill-gates-terrapower-is-building-next-generation-nuclear-power.html.
29. Gruver, Mead, "Bill Gates company to build reactor at Wyoming coal plant," AP News, https://apnews.com/article/bill-gates-wyoming-technology-environment-and-nature-government-and-politics-c647adc5f4a447eaef1c4ec28464e153, June 2, 2021.
30. Hallbert, Bruce P. and Kenneeth D. Thomas, "Sustaining the Value of the US Nuclear Power Fleet," *The Bridge*, National Academy of Engineering, Fall 2020 – Nuclear Energy Revisited. Pp. 10-16.
31. "Confronting the Duck Curve: How to Address Over-Generation of Solar Energy," Office of Energy Efficiency & Renewable Energy, U.S. Department of Energy, https://www.energy.gov/eere/articles/confronting-duck-curve-how-address-over-generation-solar-energy, accessed August 24, 2019.
32. Asano, Marc, "Hawai'i's Grid Architecture for High Renewables," *IEEE Power & Energy Magazine*, September/October 2019, pp. 40-46.

33. Wolf, Gene, "Storage Shock Absorber: Energy storage provides balance between supply and demand," *T&D World*, May 2019, pp. 78-84.
34. Author's files.
35. Carlton, Jim. Solar Power's Land Grab Hits Snag: Environmentalists. *The Wall Street Journal*. June 5-6, 2021. P. A1 and A10.
36. Hodari, David. "Wireless Charging Fuels Electric-Car Hopes." *The Wall Street Journal*. Pp. B1 and B4. January 19, 2021.
37. Elliott, Rebecca. "Utilities Wage Battle Over Charging Stations." *The Wall Street Journal*. Pp. B1 and B4. October 19, 2020.
38. A Message from CoBank. NRECA Executive News Brief. June 1, 2021. Page 1.
39. Milman, Oliver. "Advances mean all new US vehicles can be electric by 2035, study finds." The Guardian. April 15, 2021. https://www.theguardian.com/us-news/2021/apr/15/all-new-us-vehicles-electric-2035-study.
40. Gold, Russell, "For Electric Cars, California Needs a Bigger Grid," *The Wall Street Journal*. September 26-27, 2020. P. B3.
41. Office of Governor Gavin Newsom. "Governor Newsom Announces California Will Phase Out Gasoline-Power Cars & Drastically Reduce Demand for Fossil Fuel in California's Fight Against Climate Change," September 23, 2020. https://www.gov.ca.gov/2020/09/23/governor-newsom-announces-california-will-phase-out-gasoline-powered-cars-drastically-reduce-demand-for-fossil-fuel-in-californias-fight-against-climate-change/
42. Jacobs, Jonathan, "Electric Vehicle Fleets and Load Demand: Are You Ready for the Surge?" *Power* Magazine, May 2020, pp. 44-45.
43. Neil, Dan. "Ford Has Seen the Future of Home Charging – and It's Bi-Directional." The Wall Street Journal. June 5-6, 2021. P. D9.
44. Jaffe, Amy Myers. "Why We Aren't Ready for the Electrification of Everything". *The Wall Street Journal*, May 17, 2021, pp. R1 and R4-5.
45. "Renewable Energy Technical Potential," National Renewable Energy Laboratory, https://www.nrel.gov/gis/re-potential.html, accessed August 24, 2019.
46. Wind Energy Development Programmatic EIS, http://windeis.anl.gov/guide/basics/, accessed August 24, 2019.
47. "Maps," United States Census Bureau, https://www.census.gov/history/www/reference/maps/, accessed August 24, 2019.
48. Iaconangelo, David. 100% clean power? Don't wait for new technology, study says. E&E News. https://www.eenews.net/stories/1063720665. December 15, 2020.
49. Colorado Transmission Permitting. OpenEI. https://openei.org/wiki/Colorado/Transmission_Roadmap#Local_Siting_Process. accessed June 6, 2021.
50. Holtkamp, James A. and Mark A. Davidson. Holland & Hart. Transmission Siting in the Western U.S. https://www.hollandhart.com/articles/transmission_siting_white_paper_final.pdf. August 2009.
51. Baylor, Jill S. "Power By Wire – Expectations and Realities. 1988 Coal Market Strategies Conference.
52. National Coal Council. *Interstate Transmission of Electricity*. June 1986.
53. Email to author. June 10, 2021. From Chris Smith, Manager, Environmental Services, Georgia Transmission Corporation.
54. Schussler, Russell and Jill S. Tietjen, *T&D World*. "The Grid End Game." June 2017, p. 64.
55. Wolf, Gene. *T&D World* magazine. "What's A Macrogrid?", May 2021, p. 10.
56. U.S. Environmental Protection Agency, Green Power Partnership: U.S. Electricity Grid & Markets, https://www.epa.gov/greenpower/us-electricity-grid-markets, accessed June 7, 2021.
57. Tietjen, Jill S., "Hydroelectricity: Pumped Storage," From the *Encyclopedia of Environmental Management*, DOI: https://doi.org/10.1081/E-EEM-120046145, Taylor & Francis, 2012.
58. U.S. Grid Energy Storage Factsheet. University of Michigan – Center for Sustainable Systems. http://css.umich.edu/factsheets/us-grid-energy-storage-factsheet, accessed June 7, 2021.

59. Alamalhodaei, Aria, "Tesla wants to make every home a distributed power plant." https://tech-crunch.com/2021/04/26/tesla-wants-to-make-every-home-a-distributed-power-plant/.   April 26, 2021.
60. Deora, Tanuj, Lisa Frantzis and Jamie Mandel. Distributed Energy Resources 101: Required Reading for a Modern Grid. https://blog.aee.net/distributed-energy-resources-101-required-reading-for-a-modern-grid. February 13, 2017. Accessed June 7, 2021.
61. Elisa Wood. "What is a microgrid?" https://microgridknowledge.com/microgrid-defined/, March 28, 2020. accessed June 7, 2021.
62. Howland, Ethan. Microgrid Knowledge. Lawmakers in 20 States Introduce 69 Microgrid Bills. April 9, 2021. https://microgridknowledge.com/microgrid-bills-us-states/.
63. Schussler, Russell P. and Jill S. Tietjen. *T&D World*. Third-World Grid, SmartGrid or a Smart Grid? June 25, 2018. https://www.tdworld.com/grid-innovations/smart-grid/article/20971410/thirdworld-grid-smartgrid-or-a-smart-grid

Jill S. Tietjen, P.E., entered the University of Virginia in the Fall of 1972 (the third year that women were admitted as undergraduates after a suit was filed in court by women seeking admission) intending to be a mathematics major. But midway through her first semester, she found engineering and made all of the arrangements necessary to transfer. In 1976, she graduated with a BS in Applied Mathematics (minor in Electrical Engineering) (Tau Beta Pi, Virginia Alpha) and went to work in the electric utility industry.

Galvanized by the fact that no one, not even her PhD engineer father, had encouraged her to pursue an engineering education and that only after her graduation did she discover that her degree was not ABET-accredited, she joined the Society of Women Engineers (SWE) and for more than 40 years has worked to encourage young women to pursue science, technology, engineering, and mathematics (STEM) careers. In 1982, she became licensed as a professional engineer in Colorado.

Tietjen started working jigsaw puzzles at age two and has always loved to solve problems. She derives tremendous satisfaction seeing the result of her work – the electricity product that is so reliable that most Americans just take its provision for granted. Flying at night and seeing the lights below, she knows that she had a hand in this infrastructure miracle. An expert witness, she works to plan new power plants.

Her efforts to nominate women for awards began in SWE and have progressed to her acknowledgement as one of the top nominators of women in the country. Her nominees have received the National Medal of Technology and the Kate Gleason Medal; they have been inducted into the National Women's Hall of Fame and state Halls including Colorado, Maryland, and Delaware and have received university and professional society recognition. Tietjen believes that it is imperative to nominate women for awards – for the role modeling and knowledge of women's accomplishments that it provides for the youth of our country.

Tietjen received her MBA from the University of North Carolina at Charlotte. She has been the recipient of many awards including the Distinguished Service Award from SWE (of which she has been named a Fellow and is a National Past President), the Distinguished Alumna Award from the University of Virginia, and the Distinguished Alumna Award from the University of North Carolina at Charlotte. She has been inducted into the Colorado Women's Hall of Fame and the Colorado Authors' Hall of Fame. Tietjen sits on the board of Georgia Transmission Corporation and served for 11 years on the board of Merrick & Company. Her publications include the bestselling and award-winning book *Her Story: A Timeline of the Women Who Changed America* for which she received the Daughters of the American Revolution History Award Medal and *Hollywood: Her Story, An Illustrated History of Women and the Movies* which has received numerous awards.

Chapter 19
Infrastructure in a Park and Recreation Setting: The Example of the Golden Gate National Recreation Area – "The Teams behind the Partnership Brand"

Mai-Liis Bartling

Abstract The Golden Gate National Recreation Area, a large national park located in the complex and urban San Francisco Bay Area, has successfully completed many large-scale infrastructure projects. A series of case studies of relatively quick transformations of surplus, sometimes derelict, lands and facilities into national parklands makes the case for the public sector's ability to accomplish transformational work, identifying factors of setting, history, and organizational experience. Multiple goals were addressed in concert, including repair of the park's aging buildings and infrastructure, protection of natural and cultural resources, and development of visitor access. Rehabilitation of historic building complexes was achieved under public-private partnerships, including long-term leases, with reuses to benefit the public. At GGNRA, infrastructure projects included restoring function to natural systems, with the local community as a partner in carrying out the work. The complexity of projects required inclusive, multi-disciplinary professional teams, with team members who had a full array of collaborative leadership skills, a project environment in which many women thrived.

Keywords Golden Gate National Recreation Area · Urban national park · Military lands reuse · Presidio of San Francisco · Fort Baker rehabilitation · Redwood Creek restoration · Crissy Field restoration · Historic preservation in parks · Natural systems restoration · National Park Service women's voices/roles · Public private partnership in parks · National parks infrastructure

M.-L. Bartling (✉)
National Park Service (Retired), Novato, CA, USA

© The Author(s), under exclusive license to Springer Nature Switzerland AG 2022
P. Layne, J. S. Tietjen (eds.), *Women in Infrastructure*, Women in Engineering and Science, https://doi.org/10.1007/978-3-030-92821-6_19

467

19.1   Introduction

If you had lived in the San Francisco Bay Area in 1972, along with great natural beauty and the magic and gritty charm of the era, you would also have found derelict lands, unrecognized treasures and resource values, and many wonderful places that were simply off limits to you. Crissy Field was an industrial storage yard, covered with maintenance sheds, barracks, and a mostly abandoned airfield. At Fort Mason Center, there were shabby piers and empty warehouses. Lands End and Alcatraz were both neglected and unsafe.

Today, these landscapes are transformed and tied together by a single idea – a national park in an urban setting. A particular institution is at the center of the transformation – the National Park Service (NPS), specifically the Golden Gate National Recreation Area (GGNRA), working with partners and inspired by what citizens want. With sites in the counties of San Francisco, Marin, and San Mateo, the park encompasses over 300 square kilometers. In institutional terms, the transformation happened quickly. The years between 1990 and 2010 were particularly formative and included some of the most complex development projects and natural resource restoration projects in the NPS.

> This chapter makes the case that – contrary to common wisdom – the public sector can be effective and can accomplish transformative work. GGNRA is an example of this.

What accounted for the speed and for the park's effectiveness? Visionary leadership to be sure, but what actually made it work? There were many ingredients, but here the focus is on the role of dedicated professionals who were participants in this process. They had the skills and determination to transform the hopes and dreams of the public into real changes in the landscape. Much has been written about the external and political side of this park's story. Here we hear from those who carried on the work of physically building the park, sometimes despite the complex and challenging political environment they were in.

Through their unique window on events, we can understand the extraordinary collaborations that were required, some of the hurdles that were overcome, and what factors contributed to so much being accomplished, laying a foundation for what is being realized today. These through-lines of park history show how success with challenging projects seeded subsequent efforts and grew the park's capacity. In that sense, it is an origin story of how the park came to be so adept at large projects.

19.1.1   What Is "Infrastructure" in a Park Setting?

The view here is that natural systems <u>are</u> infrastructure, requiring the same detailed study and understanding of their functioning as water treatment plants, roads and highways, or electrical systems. (GGNRA had all of these as well.) In the era of rapidly changing climate, designing with the future functioning of natural systems in mind seems even more critical. Both the natural systems and cultural landscape can be understood in every way as the infrastructure that undergirds the essential activity of the park. It allows the continued thriving of habitat and wildlife – as well as the preservation of historic resources and stories – for the enjoyment of today's visitors and of future generations.

Successful park development teams are inclusive; they may include planners, architects, and engineers but also natural and cultural resource specialists, interpreters, and real estate/business specialists, among others. However, perhaps unique to this park are the fully evolved notions of "co-creating" with the public and of core and extended project teams that allow professionals from multiple organizations to work together seamlessly. The unique approach to teams was necessitated by the partnership environment and also allowed for the scale of innovative financing.

A note about the women of the GGNRA: women have always been part of the NPS story – as activists, employees, and avid users – but their voice has not always been heard within the NPS organization. As an extensive 1991 NPS workforce report showed, very few women had yet reached management levels (and even fewer minorities). NPS also lagged behind the civilian labor force in percentage of women in professional positions. In 1991, less than 20% of NPS biologists and less than 10% of NPS engineers were women [2]. GGNRA may truly have had a leadership position in the sheer numbers and sectors where women were in leadership roles, particularly on its park planning, restoration, and development teams. Here women did come to have a voice, and it became possible to work within a cohort of women. One can make the case that the number of women on park project teams at GGNRA may be part of the success story.

This chapter is grounded in my own experience and that of 19 other project professionals who were interviewed about their project experiences at GGNRA. The goal is to add to the complex story of the park.

19.2   The Park as Setting and Stage for New Ways of Working

GGNRA's particular resources and its setting in the San Francisco Bay Area laid the foundation for its working style. These factors, along with the park's founding story, set the stage for its big dreams, its partnership ethic, and ultimately its track record of success in park restoration and development projects.

GGNRA was founded on October 27, 1972, through an act signed by President Richard Nixon (Public Law 92–589), initially authorizing an area of 140 km². Congressman Phil Burton was a powerful proponent and was aided by an exceptional citizen network. This occurred at a time of national interest in urban parks, spurred by federal studies.

Historian Hal Rothman captures the context of the time and place in his "New Urban Park," the park's official history. The San Francisco Bay Area had a "history of opposing norms of American Society" [9]. There was also a vibrant community interested in preservation and in the environment. The 1950s and 1960s had brought growth to the region; 1970s environmentalism was fueled by the significant threats to what remained. The military owned many of the key properties and was in the process of actively disposing of them. Battles would be waged over the future of these places.

Local activist Amy Meyer is regarded by many as the "mother of the park." Hers is a legendary tale of citizen activism sparked by her interest in how nearby excess military lands would get used. Her book, "New Guardians of the Golden Gate," tells the story of how local citizens helped save lands in the quickly developing San Francisco Bay Area, "resolving to create a national park for the people" [8]. These activists would stay engaged with the park and insist on being consulted and involved, which had implications for how the park would do business.

The lands brought into the new national park were not pristine. Many were military lands due to long-term military presence in the Bay Area and came under NPS management not all at once, but over a period of years. They were a challenging set of resources to improve, yet they had been protected from development. GGNRA is sometimes described as a series of nine military base conversions. The park's extensive building square footage would need to be occupied with uses, setting the stage for a partnership culture. Program partners were needed to share the cost of rehabilitating buildings and spaces.

Park staff still remembers the "wild and crazy" early years of the 1970s, the adventurous spirit before all the rules were set. Visionary leadership began with Superintendent Bill Whalen, who was followed by Jack Davis, and then Brian O'Neill, who served as GGNRA superintendent for more than 20 years. Park planners under Chief Planner Doug Nadeau were youthful, including Greg Moore, who would go on to be one of the founders of the Golden Gate National Parks Association (GGNPA), which in 2003 became the Golden Gate National Parks Conservancy (or Parks Conservancy).

Partnership philosophy got started early. Bill Whalen saw it as the solution for managing the building-intensive Fort Mason Center. Brian O'Neill took partnership the furthest, making it the park's ethic and brand and forging the tightest, strongest partnership with GGNPA. Rothman, taking an historian's perspective, regarded GGNRA as an archetype of modern society: "no longer could they (land management agencies) dictate terms, but they had to negotiate, in hopes of maintaining complicated alliances that help protect the park and its budget at the national level [9]." He considered the park a blueprint for the future.

Today, the GGNRA – or the Golden Gate National Parks, as it is often called – is a vast system of parklands that serves the diverse publics of the San Francisco Bay region. GGNRA lands are seamlessly connected to the City of San Francisco and to Golden Gate Bridge District administered lands. The park's resources anchor significant economic activity in the city and region.

19.3   The Transformation of Parklands

Of necessity, this is a broad outline and simplified story of park-building, using nine case studies that exemplify the park's approach to a range of issues and opportunities (Fig. 19.1).

Projects at GGNRA were carried out within the general NPS framework, supported by centralized resources (e.g., the Denver Services Center (DSC), an internal NPS consultant providing planning, design, and construction services) and centralized processes (e.g., requiring approval by an agency-level "Development Review Board"). Park staff operated in the context of this framework, even when projects involved partners. These NPS processes provided broad sideboards, albeit with requirements to clear. Generally, projects involving partners were subject to the same quality standards, for example, rehabilitation standards and environmental

Fig. 19.1 Map of project sites

review requirements. Agency review levels were needed to maintain funding and support, but primary creativity and responsibility for solving issues was at the park level.

Projects happened within the framework of the park's 1980 General Management Plan (GMP), which provided for a kind of decentralized approach grounded in overall vision, based on land use/management zones and general objectives, rationales, and strategies developed in a participative process. The 1980 GMP was 5 years in development with over 400 workshops before its release, with the demanding public requiring a wholly new level of involvement and participation [9]. An update, begun in 2006, took almost 10 years. It reflected a better understanding of resources and an even more engaged public. Top level changes between the two plans included addressing newly added parklands in San Mateo County, a better understanding of endangered and sensitive species, recognizing the park's Biosphere designation, and recognizing demographic changes and climate change.

Projects were also carried out under a strong public-private partnership philosophy. The park's aging facilities and infrastructure were largely inherited from the Army, impacted daily by the coastal climate, and included 550 historic buildings. The park long had a strategy of relying on partners to occupy built space under agreement, assuming building maintenance but not necessarily capital costs. As noted in the park's 2001 business plan, park partners occupied 50% of the park's built space[7].

In recent years, NPS units have been required to identify the deferred maintenance values and current replacement values of their facilities and infrastructure. The large size of these numbers (Table 19.1) underscores the enormity of the park's ongoing task and the scope of creative thinking still needed.

Table 19.1 Deferred maintenance and critical replacement values[a]

	Critical systems deferred maintenance (CSDM)	Deferred maintenance (DM)	Current replacement value (CRV)
Golden Gate National Recreation Area	$147,707,211	$323,803,049	$6,186,677,595
Fort Point National Historic Site	$8,278,508	$11,846,846	$212,551,339
Muir Woods National Monument	$7,223,782	$12,100,933	$47,022,334
	Above totals are summed from the following categories: buildings, housing, campgrounds, trails, waste water systems, water systems, unpaved roads, paved roads, and all others		

aAll data was collected by NPS and is current as of 9/30/2018. The three congressionally authorized park units are administered together as the GGNRA, also known as the Golden Gate National Parks

19.3.1   The Presidio: 10 Years of Focused Attention

By any measure, the park was still young in 1989 when the closure of the Presidio of San Francisco was announced, one of 86 military facilities closed under the Base Realignment and Closure (BRAC) Act. The entire Presidio (6.0 km²) was included in the boundaries of the GGNRA when it was established in 1972, to be added to the park when it became surplus to the Army. Doug Nadeau called it the very heart of the park, and yet its closure was still unexpected.

The closure was the biggest shock yet in the park's short history, a "before and after moment" that would leave a broad impact on the park, beyond the Presidio itself.

The Presidio was established in 1776 and guarded the Golden Gate under Spanish, Mexican, and then American flags until 1994. At the time of its announced closure, it included 870 structures, 510 of which were listed as historic. With its collection of military history and cultural landscapes, it was given National Historic Landmark (NHL) status in 1962. The hilly, wooded property contained 620,000 m² of building space, 1200 units of family housing, and the last free flowing stream in San Francisco. The Presidio was a "city within a city," features of which included separate law enforcement, its own Presidio Fire Department, its own infrastructure systems, as well as quirky signage and wayfinding that were disconnected from the rest of San Francisco. Managing the Presidio would be a huge and expensive challenge for the NPS, and there was a remarkably short period of time to figure it out, as the Army was proposing to leave within 5 years. Its closure immediately raised the question of what the Presidio was to become. In what sense would it be a national park?

The announced closure soon revealed fractious divisions within the public, as they contemplated the effect of the Army's leaving and began to weigh in with their own preferences for the Presidio's future. All this was part of general debate over the "Peace Dividend." Very quickly, the closure had the attention of the city, local congresswomen, and a fiscally tough US Congress. It is hard to overstate the impact on the Bay Area of the closure announcement, which affected other Northern California Army installations in addition to the Presidio.

GGNRA began meeting with the Army's BRAC office almost immediately. Working together, park and Army counterparts developed an "umbrella agreement" with general principles for the transition from the Army to NPS, which was signed in September 1990. It was agreed to work toward a smooth transition that protected the resources and was least cost to the taxpayer, identifying eight sub-agreements that were needed to tie down the details (getting agreement on those would prove to be more difficult in the upcoming years). Among the many specific agreements reached was an Army-funded $62 million, three-year program to repair infrastructure systems. Although the Army and NPS were seemingly opposing organizations, commonalities included a sense of mission, pride in uniform, and even a shared sense that we had all won the cold war.

Even as the NPS headquarters office was doubtful of the park's ability to take on the new responsibility, the park began identifying what kinds of professional staff and expertise were required to assume operations and plan for the Presidio's future. NPS decided that development of the Presidio plan, an amendment to the park's 1980 GMP, would happen under the supervision of the NPS's DSC, taking it out of the immediate control of the local superintendent, Brian O'Neill. Managing the transition of operations would be under the park's supervision. The teams would be co-located at the Presidio (initially at a building on Crissy Field fondly called "the Beach House"). Under the supervision of Regional Director Stan Albright, an elaborate coordination system evolved between the park, the NPS regional office, and the DSC to manage the information-intensive, politically fraught planning and transition processes.

Over the next several years, the Presidio teams did the work of inventorying the Presidio's resources and assessing its condition while negotiating with the Army and engaging the broad public in imagining possible futures. Cost of future operations and rehabilitation of buildings loomed large, and there were worries that the soon-to-be-released plan would be too lofty. To broaden thinking and add business acumen, a 37-member "blue ribbon" Presidio Council was convened in June 1991 through the GGNPA. Meanwhile, in a surprise move, the Army announced it would depart by October 1994, a year earlier than planned. As the date neared, the Army and NPS held frequent joint Program, Budget, and Advisory Committee (PBAC) meetings to coordinate the transfer of operational responsibilities and funding. Tensions were high on all sides as NPS again contemplated what organizational structure was needed to navigate the next few years.

A late evening brainstorming session attended by GGNRA, GGNPA, and Grove Consultant's David Sibbet produced the idea of a Presidio Project Office (PPO) to unify a leasing and program development focus with in-progress transition planning, all outside the scope of normal park operations. The PPO would incorporate the expertise built during the planning and transition efforts. The NPS bought off on the PPO structure, which would stay in place from 1993 to 2000, but re-assigned Bob Chandler from his superintendency at the Grand Canyon to serve as general manager, reporting directly to the Washington office and bypassing even the regional director.

The Presidio's masterplan (GMPA) was released in July 1994. It envisioned the Presidio as a great urban national park, with a "swords to plowshares" role. The GMPA called for establishing an entity that would be responsible for the leasing of built spaces at the Presidio, to be given broader and more flexible leasing authorities than NPS regulations otherwise allowed. (McKinsey consultants had even modeled how governance structures for responsibilities could be divided between NPS and the new entity.) Still the NPS was surprised when the specific legislation creating the Presidio Trust passed in 1996, turning over essentially all management functions in 80% of Presidio lands and nearly all its buildings to the Trust. The Trust would operate independently of the NPS, creating two management entities at the Presidio, as NPS would retain management of the shoreline areas. The Trust was directed to manage the Presidio per the "general objectives" of the GMPA and was under a deadline to become self-sufficient.

Within just a few years, NPS had hired 275 full-time employees, built out its organization, and assumed management per a detailed transfer schedule with the Army, by any measure an enormous undertaking. Then, just as quickly, the park was required to turn around and drastically downsize to meet a 1998 deadline for handing off 5.9 km² to the Presidio Trust. GGNRA Administrative Officer Susan Hurst remembers that no one was laid off or suffered involuntary personnel actions, due to the reduction in force. NPS, committed to doing well by people, had worked hard to get special authorities for early retirement and for priority placement within the Department of Interior and the Presidio Trust.

This chapter was enormously impactful on the park and built new management capacity that could be assigned to other projects. It also provided a huge opportunity for the GGNPA's development but also tested the park and Association relationship as a new partner, the Presidio Trust, came onto the scene.

From 1989 to 1999, the NPS gave the Presidio 10 years of focused attention. When political debates raged, professional staff learned to work in the glass bowl. Yet Regional Director Stan Albright at the time acknowledged that they had "kept their eye on the ball." Rothman recounted NPS accomplishments, paraphrasing here: NPS developed the GMPA; attained and kept the support of the Bay Area public; secured considerable funding for infrastructure, building renovation, and environmental cleanup; smoothly handled a complex transition; secured a $25 million operating budget and additional revenues from interim leasing; and then successfully managed the transition to a smaller level of involvement. "All the while, it didn't bend on its core values, passing the Presidio to the Trust under those terms, as a park" [9].

19.3.2   The Case Studies

Case studies are grouped into three clusters based on their core concept.

1. Projects involving a long-term lease or agreement to finance rehabilitation of historic buildings: Thoreau Center for Sustainability at the Presidio, Rehabilitation of Cliff House and Lands End Site Improvements, and Rehabilitation of Fort Baker (Historic Core)
2. Projects involving the restoration of natural processes: Restoration of Crissy Field, Redwood Creek Restoration, and Mori Point Restoration
3. Projects exemplifying the evolution of thinking about park facilities or areas over time: Alcatraz Island Management, Fort Mason Center Pier 2 Renovation, and The Evolving Story of Muir Woods

Concurrent with any of these projects, there were always many other – equally interesting – projects underway in the park!

19.3.2.1 Thoreau Center for Sustainability at the Presidio

The Project When the Thoreau Center for Sustainability at the Presidio (now Tides Converge) opened in March 1996, it was the first large-scale (private sector) tenancy achieved under NPS management. The project combined creative private-sector financing and the federal historic rehabilitation tax credit to initially reuse four historic hospital buildings (6800 m^2), creating office and gallery space for rental by nonprofit and philanthropic organizations, preserving the buildings while moving toward a post-to-park Presidio.

The Story The four buildings were originally part of the Letterman Hospital Complex, built between 1899 and 1935, and had provided care to generations of soldiers. They were considered important examples of the Greek and Mission revival styles of their era. While the complex's floorplan had been suited to the hospital life of its day, it presented daunting issues for reuse. The Presidio GMPA had identified the buildings for education, research, and sciences, and NPS real estate/program development staff were beginning to look for applicants with programs compatible with the GMPA.

Hugely important was the 1993 passage of Public Law 103–175, which allowed the NPS to negotiate and enter a lease at fair market value for all or part of the Letterman complex for science, research, and education, with proceeds allowed to be retained and used for expenses "with respect to Presidio properties."

In 1994 (with the GMPA process wrapped up), the NPS PPO issued a Request for qualifications for leasing all or part of the Letterman complex. Tides Foundation, a public foundation dedicated to progressive social change through creative non-profit activity, responded together with Equity Community Builders, a San Francisco-based real estate firm specializing in environmentally responsible design and creative financing solutions. Their successful proposal is to develop 6800 m^2 into a nonprofit center, to be named for Henry David Thoreau. A for-profit limited partnership, Thoreau Center Partners, L.P., was created to lease the property from NPS, arrange for the real estate financing, and be responsible for the rehabilitation and ongoing management of buildings. This structure enabled the partnership to qualify for the historic rehabilitation tax credit, needed to attract equity investment.

Thoreau Center Partners, L.P., selected a San Francisco firm experienced with rehabilitations utilizing tax credits, Tanner Leddy Maytum Stacy Architects. Design began in 1994 and was completed by the summer of 1995. A municipal development model was utilized, meaning that the lease was not signed until design was approved by NPS and financing was in place; however, terms were negotiated. The lease was signed in September 1995, and work was completed by March 1996, when the center opened.

The $5.5 million project was financed by bank loans and private loans, with a balance needing to be raised from equity investors. Anticipated project revenues were not sufficient without the $5.25 million in rehabilitation expenses that qualified for the historic tax credit. The highly leveraged use of the tax credit was allowed by the long 55-year ground lease, permitted under the special leasing authority.

The National Trust for Historic Preservation regarded the project as a model for successful public/private partnership for reuse of historic structures and also praised the project's pioneering role in the use of environmentally sustainable design, materials, and construction methods, rare in historic rehabilitations. The buildings were in poor repair, needing completely new electrical, mechanical, plumbing, and life-safety systems. Through close work between all parties, the completed project met a wide range of environmental goals such as maximizing day lighting in new offices, providing natural ventilation, using energy-efficient mechanical-electrical lighting, and integrating demonstration alternative energy systems for photovoltaic. A full 73% of materials removed during construction were recycled [5].

A second phase of the project ("Tides II"), adding 7000 m², was also completed under NPS tenure, in 1998. The shared space nonprofit center continues to operate today as Tides Converge. Within the center's now 14,000 m² are housed 74 nonprofits and social organizations. Twenty years later, it has refocused on collaboration between its tenants, who enjoy the center's "welcoming space filled with natural light, a café space for casual interaction and connection" [1].

Success Factors/What's Important Here Some of the project's success has been attributed to its inspiring program (both phases of the project were quickly leased up), its setting within the Presidio as national park, as well as the key role that the historic tax credit played in its financing. Developer Tom Sargent, when asked about the project's replicability, cited basic business factors such as a financeable lease, a legally binding property description, and market demand for the space being created [5]. Sargent also underscored the commitment needed from both parties to working as a team, with trust and openness, and the importance of designating a full-time point person on the NPS side responsible for shepherding the process. Other factors that helped this project succeed include: upfront analyses of infrastructure and hazardous materials and plans for their concurrent rehabilitation in step with the project.

Steve Kasierski, NPS PPO real estate specialist and project manager, reiterated how uncommon the redevelopment model was at NPS. Thoreau Center was a milestone in the park's capacity to manage large development projects and the PPO had gained valuable experience. Required by the transaction, the NPS became a fully capable reviewing agency, able to issue permits and conduct inspections. Here, the NPS was also using the best outside business/financial expertise and had developed a fine-tuned understanding of the many roles – landlord, facilitator, regulator – that it had to inhabit. At Thoreau Center, the NPS filled in the details of how to carry out a new model for public-private partnership. A later Government Accounting Office audit of the transaction affirmed that the business deal had been appropriate with benefits to both parties [10].

From the NPS side, NPS needed private partners willing to buy into the vision and take on the risks and challenges of working with a public agency. It must be said that NPS knew it had an exceptional partner and valued the relationship. Steve

Kasierski, who had personally reached out to the Tides Foundation when the RFQ was issued (following Tides' response to an earlier Call for Interest), considered this pivotal move on his part one of his single most valuable contributions to NPS.

19.3.2.2 Rehabilitation of Cliff House and Lands End Site Improvements

The Project The rugged western corner of San Francisco, with its wild character and steep cliffs overlooking the Pacific Ocean, is known as Lands End. NPS oversaw key improvements to the area so that it could be enjoyed by its million annual visitors: rehabilitation of the iconic Cliff House restaurant; constructing of a promenade and trailhead at Merrie Way, 0.8 kilometer of accessible trail, and four scenic overlooks along the California Coastal Trail; and building of a new visitor center, the Lands End Lookout. The site improvements transformed a dense, weedy forest to a more open area with improved safety, encouraging the return of a broad array of visitors.

The Story Between 1976 and 1980, NPS acquired the Cliff House, Sutro Heights Park, and Sutro Baths, each in a different way. All are part of the "Sutro District" with its windswept views and important historic resources. By the turn of the twentieth century, the area included shops, restaurants, and carnival, with a steam train from downtown. When landslides stopped rail service after 1925, a slow decline began. The Sutro Baths closed in the 1960s, and over time the area became blighted.

NPS's 1992 masterplan for the Sutro Historical District (EDAW, Sheryl Barton) called for the District's revitalization and made a case for the value of ruins, interpreting ordinary life, and the concept of a cultural landscape. While the California State Historic Preservation Office did not sign off on the District's historic status, the public supported making the ruins safe (but not tidy) and partially restoring the Sutro Heights Park gardens.

Cliff House Rehabilitation The Cliff House itself opened in 1863, and its structure was first rebuilt in 1896. When it burned in 1907, it was replaced in 1908–1909 with a less lavish building designed by Reid Brothers, which was continually renovated into the 1970s. The 1980 GMP proposed restoration of the 1909 Cliff House. The goal was to "capture the spirit of another era." The GMP also said that if it was not structurally possible to rehabilitate the building, a new building could be built which should "assume an entirely new aspect" and "generate excitement."

The 1992 Sutro masterplan also called for the restoration of the 1909 Cliff House, despite lack of support from local preservationists who criticized the building's design for being too ordinary. Building investigations, however, were showing that a significant amount of the 1909 building was, in fact, intact. Rehabilitation would be done by NPS's long-time partner/concessioner at the Cliff House (family restauranteurs Dan and Mary Hountalas) under a new 20-year contract signed with NPS in 1998. Local firm Page & Turnbull began work on the design (lead architect Mark Hulbert).

Fig. 19.2 Cliff House construction, second phase. (Courtesy of Project Architect Carrie Strahan)

Initially, the public was not happy with the design, as it proposed keeping more of the 1909 building than they wanted, and a deadlock developed. Park staff remember the tense public meetings that were held (facilitated by Grove Consultants and architect C. David Robinson) to discuss the concerns and the public's desire for "tiers and views." Project Architect Carrie Strahan recalls, "The turning point was unveiling a physical, 3D model that allayed fears." The all-white, foam-core model translated design guidelines developed from earlier public input, allowing Hulbert's vision to come through. Mary Hountalas recounted that it had taken 4 years of "reviewing and bickering," before the design was accepted in 2002 [4]. Construction costs were estimated to be $14 million, with the concessioner covering most of it.

In addition to allowing the public access and views, the final design was for a smaller size footprint and lower profile Cliff House that maintained the dignity and integrity of the 1909 building and reused 75% of the existing structure. The construction was done in three phases: (1) replacement of the north wing with a three story "Sutro Wing," a steel building with glass on two sides, allowing viewing; (2) restoration of the 1909 building by removal of a 1950 addition and extensive seismic work (Fig. 19.2); and (3) building an atrium to serve as the entrance and adding observation decks. The construction firm Nibbi Brothers was under contract to both the NPS and concessioner, in a combined rehabilitation and new construction effort, for a seamless albeit complicated project. Further complicating the project were the harsh environment of the Cliff House, the lack of a back alley for laydown, the need to be sensitive to the natural environment, and significant materials challenges to match historic fabric.

Construction needed to be complete before the restaurant's busy season, and this tight timeline, as well as the decision to keep the business running throughout, required extraordinary collaboration between all parties to quickly resolve issues. Everyone who was critical to the project's success participated in a consultant-led partnering workshop. To this day, participants remember the BHAG (Big Hairy Audacious Goal) the team was working toward. This commitment would be needed to survive the difficult moments, which included the need to replicate the chemical make-up of the historic 1909 stucco, with its high moisture and salt content, due to the use of beach sand. Most serious was finding that the side of the foundation facing the ocean had no footings. The entire west facade needed to be shored up, reinforced, and connected to the slope – an expensive change order! The completed Cliff House (final construction cost: $18 million) opened in October 2004 and was widely celebrated.

Trail and Landscape Improvements With the Cliff House renovation completed, the park and the Parks Conservancy began to study ways to enhance visitor experience at the site and restore native vegetation. The opportunity was spotted to connect the site improvements to the California Coastal Trail, which helped attract funding to the project. Key stages were as follows:

1. Anticipating that the thinning of trees would be a sensitive issue, a full year of community outreach was done to educate about local history and provide information about the project. UC Berkeley Landscape Architect Joe McBride was enlisted to give walks and brought old aerials of an era when there had been far fewer trees, sharing information about natural and cultural history and building community connections.
2. Even with an overall plan for the area, there were details to work through. Onsite workshops – open to all interested public – were held to determine how the needs of tour buses, people, cars, and bikes could all be accommodated, while satisfying neighbors that the site would be safe and attractive. The consultant group MIG took public input and helped develop design guidelines. A charette attended by park staff, Parks Conservancy staff, and regional and DSC technical experts helped produce the detailed plan.
3. Actual construction began in 2005 with the installation of approximately fifty 18–24 m piers to stabilize the cliff drop, which would allow for construction of the trails and parking area. NPS Project Manager Steve Griswold and Parks Conservancy Project Manager John Skibbe formed the core of the seamless team. May and Associates provided a detailed implementation strategy, and full-time public information coordinators were onsite to update and respond to public concerns in real time. In 2006, the area opened to the public with a community celebration.

The remaining piece would be the visitor center at Merrie Way, with its $five million cost estimate. The community wanted the design to be almost invisible from the crest of the hill, preserving views (Fig. 19.3). The final design honored those preferences, was durable, and drew inspiration from coastal gun batteries. The Lands End Lookout opened in 2012, its construction supported by multiple funders.

Fig. 19.3 New Lands End visitor center provides views of the Pacific Ocean. (Courtesy of Parks Conservancy)

Success Factors/What's Important Here The project team identified many ingredients, starting with the holistic and area-wide approach.

Just as important was the intense listening to the public at the right time – upfront – for both the Cliff House and the trail/landscape improvements. Per John Skibbe, the challenge was to have the story come through and make the information easy to understand. As a result of the upfront work, the neighborhood came to embrace the improvements and provided volunteer labor when it came time to plant.

Another critical ingredient was the seamless-to-public team, composed of members from different organizational entities, but necessary for bringing multiple funds together, whether for the Cliff House or for the site improvements. To illustrate, the Federal Highway Administration chipped in funding to improve access to the site. NPS contracted for and paid for the parking lot, up to the curb line of the overlook. The Parks Conservancy funded trails and overlooks, as well as planning and design of the visitor center, and philanthropic dollars were instrumental in completing the improvements.

Still another ingredient is cultivating the support of the Hountalas family, who had managed the Cliff House for years (a partnership that continued until December 2020). The family's deep local roots, connection to place and history, and their ties to the community were important to the success of the project. Carrie Strahan, in her role as "owner's rep" for the NPS, emphasized the trust between parties that was needed.

19.3.2.3 Rehabilitation of Fort Baker (Historic Core)

The Project Fort Baker sits directly north of the Golden Gate Bridge and has breathtaking views of the Golden Gate. Here, NPS oversaw the rehabilitation and conversion of historic buildings into a retreat and conference center (Cavallo Point), also rehabilitating the historic parade ground that the buildings surround. The $102 million conference center project was part of a complicated site that also includes a children's museum, Coast Guard facility, marina, and sailing center. The opening of the conference center in 2008, NPS's first LEED-certified lodge, marked the ninth and final "post-to-park" conversion in GGNRA (Fig. 19.4).

The Story The Army acquired the 1.36 km^2 site in 1866 to fortify the north side of the Golden Gate and from 1901 to 1910 built the parade ground and 24 colonial-revival style buildings around it. Along with the Presidio, Fort Baker was included in the boundaries of the new GGNRA when it was established in 1972. The 1980 GMP foresaw it as a recreation spot and transportation hub for Marin parklands. With the Army base's formal closing in 1995, the site quickly fell into disrepair.

An Early Business Mindset NPS planning was kick-started by a planning grant from the Marin Community Foundation, out of concern over the buildings lying vacant. Building upon NPS and Parks Conservancy (still called the GGNPA until 2003) experience at the Presidio, the team was from the start "clear-headed about infrastructure and building condition, and how to pay for it all." A market study and economic analysis (Porat and Assoc.), in concert with the site-specific master planning effort, showed that a retreat and conference center had the best chance of succeeding and would financially allow for the rehabilitation of historic buildings. Meanwhile, the GGNRA and Parks Conservancy planners (Nancy Hornor and Cathie Barner) acted as early project ambassadors, visiting and studying other projects in park settings.

Infrastructure First Within a year of the 1995 base closure, the park had already begun infrastructure condition reports, anticipating that full disclosures would be required for the business lease transaction. Per NPS Project Engineer Melanie Wollenweber, "other projects put infrastructure last, but here we weren't playing catch-up." By 1998, the park began to push for federal appropriations to fix the infrastructure, with the Parks Conservancy organizing onsite congressional visits to make the case. With bipartisan support, Congress designated Department of Defense BRAC dollars for a $25 million 8-year program to be used only at Fort Baker. The Army did the toxics cleanup itself prior to the August 1, 2002, transfer. With BRAC knowledge developed at the Presidio and the bandwidth to pay attention to detail, Environmental Engineer Brian Ullensvang was able to negotiate a detailed protocol so that the discovery of hazardous materials would not cause delays for later construction. Infrastructure upgrade construction began in 2004.

Fig. 19.4 Rehabilitated historic buildings, now part of Cavallo Point Lodge, and parade ground. (Courtesy of Project Engineer Melanie Wollenweber)

Redevelopment Model Lease By 1999, with the addition of experienced business manager Steve Kasierski, the team started work on an RFQ/RFP solicitation, using a similar redevelopment model as had been used at the Thoreau Center in the Presidio. The NPS would sign an Exclusive Negotiating Agreement first, binding parties to good faith negotiations, followed by a detailed Lease Disposition and Development Agreement. The lease itself would not be signed until design was complete, financing was in place, and construction was "shovel ready." September 11, 2001, was set as the date to review proposals, with a public presentation by all RFP respondents to follow in the evening. With experts already convened, the evaluation panel proceeded, but the evening presentations were cancelled. Rare in the NPS world, the local park team participated in the evaluation process, contributing knowledge and community sensitivity to the selection of an experienced development team with local connections, who brought a sense of appropriate scale and public purpose.

One troubling development threatened the whole effort. While the park was committed to the "smallest economically feasible sized" project, it had been required to analyze the environmental effects of the *largest* size that might be built, and the neighboring city of Sausalito filed suit over size and traffic concerns. Project team members did months of community work through meetings with city officials, ad hoc and informal groups to build community understanding and support and calling on the respected League of Women Voters to facilitate tense public meetings. The work paid off, and when the regional permitting authority, Bay Conservation and

Development Commission (BCDC), capped the project at 225 hotel rooms, fears began to ease, and the lawsuit was able to be resolved.

Conference Center Construction The lease was signed in 2006, and conference center construction began immediately. A compact timetable was dictated by the lodge needing to open before the 2008 summer busy season. It would take the complete commitment of all parties. As at the Cliff House, all (NPS-Parks Conservancy team, private-sector developer, tenant/lessee, contractors) who were necessary for the project to succeed participated in a "partnering workshop." Team members learned to operate on the principles of "promise date" (personally committing to delivery dates they could meet) and "open kimono" (sharing all relevant information).

Throughout, the core NPS project team of 5–6 functioned as a kind of "skunkworks," able to dedicate most of their time to the project and apply creative solutions, supported by effective decision-making. Several team members brought bandwidth and expertise from their recent Presidio experience, and as Steve Kasierski recounts, "we were firing on all cylinders." The team had an NPS "project executive," with access and control over the project as well as over parkwide resources and processes, who was able to make in-the-moment decisions.

Melanie Wollenweber emphasized the complications of working around sensitive preservation issues. "The best design isn't easy, but I'm now most proud of what was difficult. Through sweat and tears – and contemplation – you ultimately get good design." The team's Parks Conservancy Project Manager John Skibbe explained the challenges of the parade ground including the many test plots to identify the appropriate scrub grass. The team's NPS Project Architect Carrie Strahan, experienced from her work on the Cliff House restoration, recalled the "small battles waged at the day-to-day inspection level." Saving historic fabric was always in mind, for example, removing paint from old tin ceiling tiles by freezing so as to preserve the tiles for reuse, even numbering them so that the ceiling patterns were also preserved. NPS Project Architect Joanne Wilkins expressed pride in the innovative solar panels that were installed on the project's new buildings (Fig. 19.5).

Both the park and the lodge developer were deeply committed to sustainability and worked together to prepare the extensive LEED documentation. The single package, prepared in partnership, included 35 historic and new buildings plus landscaping to achieve LEED Gold.

Success Factors/What's Important Here Perhaps foremost is that the NPS organization allowed the local team the autonomy to do the job. The empowered team was skilled and confident and "hit the ground running." Team synergy was high throughout, and the collaborative experience was valued! Other ingredients included:

1. Team strength from organizing around a clear business framework, including full disclosure with partner organizations and respect for timelines. The project itself was rooted in a comprehensive masterplan recognizing other onsite partners and balancing values.

Fig. 19.5 New lodge buildings blend in at Cavallo Point (L); solar installation on the new buildings (R). (Courtesy of Carrie Strahan)

2. Early and regular cultivation of members of congress and other elected officials at all levels of government and influence, including hosting site visits to make a case both for federal funding and for a site-specific leasing authority.
3. Embedding in community-based processes with ongoing work to resolve issues.

JoAnne Dunnec, private sector real estate attorney who advised NPS throughout the leasing process, underscored the use of the redevelopment model, which is still uncommon at NPS. She reflected that flexibility was needed from NPS. NPS had a regulatory interest, but also a proprietary one, and hired people who could bring this sensibility to the project. She also credits a masterplan that allowed limited new construction for economic viability. Finally, she credits the extraordinary developer, from whom transparency and community benefits were both needed and received.

The Cavallo Point Lodge opened its doors just as the 2008 recession took hold, but it has survived. Steve Kasierski reflected on the 10+ years of focused effort, "On aggregate, you have to say the project succeeded." Also, in the words of another team member, "It was the highlight of my career."

19.3.2.4 Restoration of Crissy Field

The Project NPS and the Parks Conservancy (still called the GGNPA until 2003), together with the people of the San Francisco Bay Area, restored the Crissy Field shoreline, featuring an 80,000 m^2 tidal marsh, 120,000 m^2 historic grassy airfield, 2.1 km continuous, wheelchair-accessible pathway, and environmental center, that all together forms a new focal point for the GGNRA. The area is ecologically significant and continues to inspire.

On May 6, 2001, a San Francisco Chronicle article celebrated its opening: "In this era of diminished expectations and unhappy compromises, it is something of a miracle that Crissy's restoration happened at all – and especially that it happened in San Francisco."

The Story The land underneath Crissy Field was originally a saltwater marsh, filled in for the 1915 Panama-Pacific International Exposition, and later converted to an airfield managed by the Army as part of the Presidio. The area was within the boundaries of GGNRA as established in 1972, and in 1973, a 210,000 m² shoreline area with a trail was dedicated. It was managed by the NPS, fenced off from the rest of the Army's Presidio enclave, and bordered by rubble, mud, and weeds (Fig. 19.6). In 1987, NPS completed plans for its rehabilitation, but these were put on the shelf when the Presidio closure was announced.

With the Army's departure in 1994, the whole of Crissy Field came under NPS management, with its lack of visitor amenities and degraded resource values. NPS began a detailed site design (reflecting general proposals in the just completed Presidio-wide GMPA), and with ecological restoration, historic preservation, and user recreational goals. Early hydrologic and engineering studies showed that a tidal marsh was feasible and recommended a 120,000 m² footprint as the minimum size to maintain natural tidal functioning – built smaller, the mouth to the bay would fill with moving sand, committing NPS to periodic dredging. Other site constraints (the GMPA site concept also included restoration of the historic airfield) would limit the marsh to about 80,000 m². Consultant studies suggested that the necessary "hydraulic prism" might be accomplished through steep grading, but the 80,000 m² size was regarded as a troubling limitation.

Planning staff clearly remember the tension around the "go – don't go" moment, as natural and cultural resources values were pitted against one another, and other hydrologic information began to cast doubt on marsh long-term feasibility. Also of concern, a project without a marsh would not garner the community support needed for the huge undertaking. Natural Resources Chief Terri Thomas remained a strong proponent, "who never lost faith or courage." Natural resources professionals knew that the steep sides of the marsh as designed would make it difficult for habitat to establish. (Mounds would be added to help plants take root, while allowing for onsite reuse of excavated fill.) The approved Crissy Field Plan and Environmental Assessment (Jones and Stokes), completed in 1996 2 years after the Presidio-wide GMPA, did include the marsh. Objectives for the project: the "re-establishment of an ecologically viable self-sustaining tidal marsh requiring a minimum of human intervention and providing high-quality educational and interpretive opportunities."

After 11 years of planning and 3 years of design, construction began in 1998 (Fig. 19.6). Elements included: removing 79,000 metric tons of hazardous wastes; removing 13,600 metric tons of shoreline rubble and 280,000 m² of asphalt and concrete; restoring the 80,000 m² tidal estuary marsh and 65,000 m² of dunes, with boardwalks leading to the beach; reshaping the promenade; recontouring the landscape; creating a 120,000 m² grassy area representing the historic configuration of the 1920's airfield; and adding curvilinear earthen mounds for variation (Fig. 19.7).

The project's many twists and turns required nimbleness. Two cannonballs and additional mortar shells were found and safely detonated by a bomb squad before work could go on. Thousands of artifacts of Army life were discovered and needed

Fig. 19.6 Shoreline at Crissy Field, before (L) and during (R) construction. (Courtesy of Parks Conservancy)

to be excavated, some now on display at the Presidio. Environmental permits allowed no off-haul, meaning that asphalt had to be ground and re-used on site. Rebar was recycled, and extra soil excavated for marsh restoration was used throughout the site. Native American consultation and monitoring was needed throughout the project to respond to potential discoveries. In one instance, discovery of a midden necessitated redesign of marsh configuration.

One important thread was the restoration of the 65,000 m^2 of dune habitat, requiring 105 species of shrubs, wildflowers, and marsh plants totaling over 100,000 individual plants. The park undertook a huge stewardship program to collect seed, grow seedlings, and plant them. Terri Thomas remembers, "it was like a movement." In the words of then NPS Plant Ecologist and stewardship leader Sharon Farrell, "when people perceive a need, energy is unbridled – if not us, who?" Ultimately, the "Help Crissy Field Campaign" garnered support from 3000 volunteers. The program engaged youth in particular and helped grow a new generation of stewards and scientists.

Also of note was the toxics cleanup. The area had been a landfill, and the contaminants were heavy metals and PAH's (left after combustion), some so toxic as needing incineration. Crissy Field had to be dealt with ahead of the rest of the Presidio's environmental cleanup or risk the project. The strategy became to get a single decision document approved for all of the project area. The remedy was excavation, which the Army realized would be cheaper in the context of this project, as no refill was needed. Timing was close. Brian Ullensvang remembers that construction was "right on the tails" of the cleanup.

Success Factors/What's Important Here   Project success began with NPS and its partner the Parks Conservancy committing to a result that would inspire and that included functioning natural systems. "NPS's focus was generational," in the words of one team member. The project's ambition inspired $34 million of philanthropic support and countless volunteer hours. The lead gift of $17 million from the local Evelyn and Walter Haas, Jr. Fund was up to that point the largest NPS had received anywhere in the system.

Fig. 19.7 Crissy Field grassy airfield, marsh, and promenade, after restoration. (Courtesy of Parks Conservancy)

Equally important, NPS and Parks Conservancy engaged with the various interest groups throughout – including boardsailors, dogwalkers, environmental groups, and historic and cultural groups – and worked hard to satisfy their needs and interests. Carol Prince, Parks Conservancy Deputy Director for External Affairs, emphasized that even though Crissy Field lay on the edge of the most expensive area of San Francisco, the focus stayed on people who *weren't* represented, i.e., the *potential diverse future users* from all parts of the city. The intensity of outreach required the continuous efforts of both NPS and Parks Conservancy. When the project celebrated its May 2001 official opening, 75,000 people attended, attesting to the value of co-created vision, building personal connection, and real ownership.

The core team/extended team model for combining people from different disciplines and organizations fostered the cooperation needed to complete the construction. The team met every Monday morning, bringing together Parks Conservancy Construction Supervisor Glen Angell, GGNRA Chief Planner Nancy Hornor, NPS PPO's Cicely Muldoon, Parks Conservancy's Carol Prince, and because of the importance of the project, Parks Conservancy Executive Director Greg Moore and Board Member Dave Grubb. It was an extraordinary collaboration, each partner doing what only it could do. Team members remember how NPS "went to bat" for the project with the California State Historic Preservation Office to gain needed approvals.

19.3.2.5   Redwood Creek Restoration

The Project   Following a decade of planning, NPS completed a 0.19 km^2 restoration project at the mouth of Redwood Creek at Muir Beach – a landmark park site – improving visitor access and at the same time returning the creek to a functional, self-sustaining ecosystem by realigning it and restoring the wetland system. The landscape level restoration actions, performed between 2009 and 2013, are allowing long-term natural evolution of the creek.

The Story   The Redwood Creek Watershed extends from the peaks of Mount Tamalpais to the Pacific Ocean. Within its 23 km^2 are diverse and rich ecosystems. It is one of 25 global biodiversity hotspots recognized by the Nature Conservancy and within the UNESCO designation of Golden Gate Biosphere Reserve. While most of the area is today protected by public ownership, the watershed's hydrological system had been degraded by a "century of agriculture, recreation, and development." Soil cores showed that a large open water lagoon had persisted at the site for 3000 years but in the last century had filled with sediment.

In the early 2000s, the three managing agencies (NPS, California State Parks, and Marin Municipal Water District) began engaging with each other and the public to create a vision. At the same time, NPS Project Manager Carolyn Shoulders began writing grant proposals. Project planning was launched with a first grant from the California Department of Fish and Game, out of concern over dwindling salmon. (The agency became a major funder of the construction as well.) Pivotal to the development of a concept plan was the hiring of Phil Williams, who brought his knowledge of geomorphology and his vision for water-based systems. Bill Dietrich of UC Berkeley and others at Berkeley's Stillwater Sciences contributed expertise in sediment dynamics. The project drew on the best outside technical expertise, whether it be geomorphic, sediment dynamics, engineering, or fish biology.

By 2006, the park issued a draft environmental impact statement (DEIS) that evaluated the alternatives, with the final EIS issued in 2008. Technical peer review (required by National Marine Fisheries Section 7 consultation under the Endangered Species Act) and all-day design meetings followed. The project manager remembers how the park gave lots of leeway to adapt design and engineering. Notably, already during design, project planners developed a detailed monitoring plan to provide snapshots of how the dynamic ecosystem would respond.

The central feature of the project was to relocate 439 m of the creek to the low point of the valley, reconnect it to the floodplain, and allow it to establish a more natural route. Formal construction started in 2009 and was divided into four main stages, complicated by the need for the existing channel to remain functional.

1. Early work for ecological benefits, e.g., create a new frog pond for California red-legged frogs, expand the tidal lagoon with a new backwater habitat feature, reduce the lower end of the parking lot to give flood flows more leeway, and dig out invasive grass

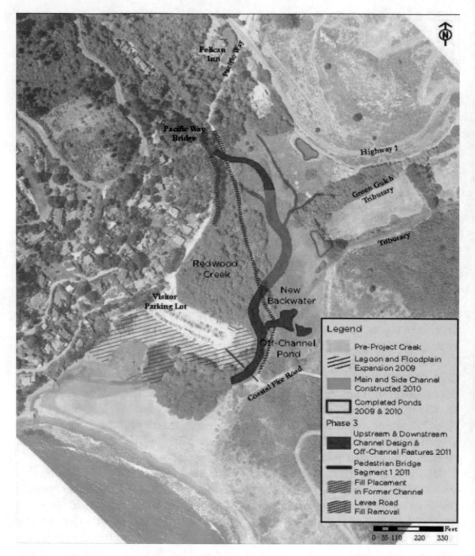

Fig. 19.8 Phase 3 plan for Redwood Creek restoration showing new channel alignment. (Courtesy of NPS)

2. Construction of half of the new channel alignment, which remained as backwater for a year and dug a second new frog pond
3. Completion of new channel realignment (since Marin County did not secure funding for a new Pacific Way Bridge, the project design needed to be altered to link to the old channel under the existing bridge) (Fig. 19.8)
4. Removal of the old parking lot from the floodplain, requiring its re-orientation, and exchange of large quantities of soil to rebuild the floodplain with appropriate riparian soil

To complicate things, the park needed to complete work during the July to November dry season each year, so that the creek could continue to function in the event of a flood. After each stage of earthwork, volunteer stewards planted a wide range of native wetland, riparian, scrub, and dune plants. The project completely converted vegetation composition, creating new riparian habitat and allowing for the reestablishment of sensitive species.

Meanwhile, for visitors, a 139 m pedestrian bridge was constructed over the floodplain to provide beach access from the parking lot, trail segments were upgraded and rerouted, and a new parking lot and picnic area were created. The trail to the beach was made wheelchair accessible, and beach access wheelchairs are now available onsite.

Of note, so many of the key players on this project were women that a Muir Beach neighbor, who stopped by the construction site, commented in surprise, "It's all women here!"

Success Factors/What's Important Here Carolyn Shoulders underscored that looking for what was needed for natural processes – and then planning visitor access – allowed the project to achieve both. She describes the scope of the project as at the "upper end of what restoration can mean."

Shoulders adds additional takeaways: "The contrast between where we started and what it is, is huge" (Fig. 19.9). What had started as the "Big Lagoon Restoration" became "Redwood Creek Restoration" when it was understood that a dredged marsh would not maintain itself, due to large sediment inputs, nor was it ever "just a fish project." Rather, it was about restoring natural processes, which in turn provide habitat for sensitive species. In fact, red-legged frogs and western pond turtles are thriving, while for a host of reasons that pertain broadly to coastal California, the coho salmon are still dwindling.

The impact of climate change? The concept design used the projected sea-level rise from the Intergovernmental Panel on Climate Change (IPCC) as of 2003, and at

Before, 2003   After, June 2016

Fig. 19.9 Redwood Creek restoration before (L) and after (R). (Credit Brian Clure, courtesy of NPS)

each stage of project design, corrections were made to reflect updated IPCC projections. Per Shoulders, "But we are now on the far end of the worst projections."

Would she have wanted to do anything differently? She might have wanted less of the floodplain used for parking, but there is a need to balance that as well. The lot is full on busy weekends! She also says, "I wish the project could have been completed 20 years earlier, because it might have helped the Coho before changing ocean conditions began to pose additional threats."

On an organizational level, the project was also significant. Firstly, the project deepened the Parks Conservancy's commitment to natural resources work. At Redwood Creek, the Parks Conservancy was hugely helpful with public engagement, managing contracts, organizing park stewards and volunteers, and finding grant funds to support the more than $14 million project. Several funders provided $one million each, attracted to the project's range of benefits, landmark location, and good planning. Secondly, what was learned at Redwood Creek directly informed GGNRA's GMP update.

19.3.2.6   Mori Point Restoration Project

The Project At Mori Point, NPS restored the natural flow of water from the hills into four newly created ponds, part of a landscape scale effort to expand wetland habitat. The project focused on trail improvements, habitat restoration, and community stewardship and on steps to assure that natural function continues. Between 2003 and 2012, a 0.13 km^2 wetland parkland was created, engaging a community of volunteers in the process.

The Story Mori Point is located on a rugged coastal promontory in Pacifica, south of San Francisco, and contains a spectacular 0.44 km^2 of coastal scrub, grassland, and riparian habitat. The habitat provides refuge for the most endangered land reptile in North America – the San Francisco garter snake – as well as the endangered red-legged frog. Past land use, including the introduction of invasive plants and the lack of established trails, had damaged the vital wetland and grassland habitats and limited visitor opportunities for enjoyment.

Mori Point was a new park addition, transferred to NPS in 2002 and one of its first forays into San Mateo County. Unusual for park areas, Mori Point has immediately adjacent neighborhoods. Within a year, NPS and Parks Conservancy began to lead onsite volunteer programs. Site improvements were followed in three key stages: (1) beginning in 2004–2005, create and expand four new pond habitats and also install box timber steps up to "the Point"; (2) improve visitor access through California Coastal Trail realignments and enhancements, including an "outdoor accessible trail" guiding visitors through sensitive habitat; and (3) create an accessible elevated trail over the wetland, enabling wildlife to move freely around the site (Figs. 19.10 and 19.11).

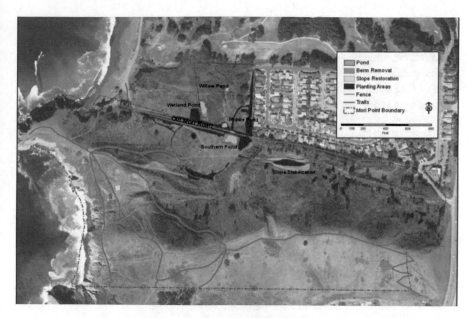

Fig. 19.10 Mori Point site plan shows key project elements and the proximity to neighbors. (Courtesy of Parks Conservancy)

Fig. 19.11 Improving wildlife habitat while improving visitor access at Mori Point. (Credit Mason Cummings, courtesy of Parks Conservancy)

A next phase of work took place between 2009 and 2012 to further improve the natural functioning of the site. This time the sequence and key activities included: (1) removing obstructing berms to restore hydrological connectivity between three of the ponds, removing fill soil and excavating a swale to redirect rainwater; (2) transferring fill soil (49 dump trucks full) to a damaged hillside, to rebuild a portion of the slope and restore natural topography and hydrology (during this process, 130 cubic meters of debris were removed from the site); and (3) re-excavating the smallest pond, Middle Pond, to increase its size and depth by 30% and lining it with natural clay to improve water retention.

Throughout, steps were taken to control weeds and erosion, and protective fencing was installed around the ponds. Over 30,000 seeds were gathered for revegetation; planting was achieved through the tremendous efforts of volunteers and interns. Water quality, endangered species, vegetation survivorship, and visitor use were monitored throughout.

Success Factors/What's Important Here The Mori Point project was San Mateo residents' early introduction to NPS. In the words of Parks Conservancy's Sharon Farrell, "we didn't want to be seen strictly as a regulator; we wanted to be seen as a proponent for care of the land, committed to listening to and engaging the community."

The Trails Forever program (a partnership between the NPS and the Parks Conservancy) brought the vision of a world class trail system running through the parks. Established in 2003, it provided a broad umbrella for accomplishments in trail enhancement, stewardship, education, engaging the community, building volunteer capacity, and raising funds. By 2008, in its fifth year, Trails Forever stretched from San Mateo to the Marin Headlands. Mori Point was selected for one of the first 10 projects under Trails Forever because of the resource values on site. As such, the project became an early demonstration of the power of Trails Forever to serve as an integrator, able to concurrently address visitor access and natural systems restoration in order to achieve both [3].

Mori Point was also a place where the NPS and Parks Conservancy staff worked on navigating changes in roles, responsibilities, and relationships. Because the park's resources were strained, the Parks Conservancy became the stronger partner in all aspects of the project, from community engagement and supporting compliance documentation to overseeing design and implementation. Along with Redwood Creek, the project pushed the Parks Conservancy further toward project management and delivery of natural resource projects as a core feature of how it would support the NPS. Earlier reticence – due to the sometimes controversial nature of natural resource work – began to give way. Today, the Parks Conservancy prioritizes climate adaptation, as well as multi-benefit and cross-boundary work.

Mori Point's success underscored the value of engaging the community as volunteers and stewards of the site. We need to ask: do we bend the definition of infrastructure to include human aspects of developing and nurturing park stewards (Fig. 19.12)?

Fig. 19.12 Nurturing future stewards while growing plants. (Courtesy of Parks Conservancy)

19.3.2.7   Alcatraz Island Management

Closed as a federal penitentiary in 1963, Alcatraz was declared excess property in 1967. Famous for escape stories, its repair needs were daunting. The question of what to do with Alcatraz and other surplus properties in the San Francisco Bay Area became part of the initial impetus for a national park to serve urban people. From 1969 to 1971, the site was occupied by Native Americans – a 17-month, now famous chapter in its history. In 1972, its neglected 89,000 m^2 became part of GGNRA, and in 1973 the site was opened to visitors. By 1977, it was the most popular visitor attraction in San Francisco, due largely to its colorful history as a penitentiary. In 1985, a self-guided audio tour (managed by the Parks Conservancy) began to generate revenues for park programs.

The island has embodied the difficulty of balancing natural resources, cultural resources, and visitor access. Damage to natural resources from the on-island premiere of the movie "The Rock" in 1996 became a turning point for concern over their protection. Due to the excess noise and outside lighting of the movie premiere during the peak of breeding season, most of the thousands of nesting birds on the island were disturbed, and a significant number of nests were lost. The importance of Alcatraz as a bird nesting site had been known to natural resource professionals, but now became widely appreciated. By the late 1990s, the park had begun preparing structural reports and submitting funding requests. The park's 2001 business plan identified a 5-year, $22 million program to "Restore Alcatraz Landmark Buildings." Over the next years, volunteers helped restore the island's historic gardens (2004); the audio tour was updated and re-routed for a "deeper visitor experience" (2006); and restoration and renovation of the main cellhouse were completed (2007). Multiple partnerships and fund sources were brought to bear on restoration of island structures, and design guidelines were developed for wayfinding, benches, trashcans, and signage.

Highlighted Here While parts of public stay focused on the escape story, NPS has moved to a holistic view of the island that has required great collaboration between NPS and partners. Also of note is the hugely important role that Alcatraz audio tour revenues play in the Parks Conservancy's ability to provide aid to the park (made all too clear during the recent pandemic-related shutdown of the island).

One cautionary tale highlights the island's complexity. The island is not on the grid but rather is powered by generators. It is the largest single contributor to the park's carbon footprint. Solar panels were installed, and when they work, it is easy to see their great potential. The original goal was to preferentially use solar, but tricky technical issues have dogged the system, and it is an area where the park is not expert. As Environmental Specialist Laura Castellini explained, "The park strives to reduce its carbon footprint, but is constrained by substandard electrical infrastructure."

19.3.2.8 Fort Mason Center Pier 2 Renovation

Historic Lower Fort Mason served as the San Francisco Port of Embarkation through the end of the Korean War. With the establishment of GGNRA, the vacant piers and warehouses (28,000 m²) were transferred to NPS. Since 1975, it has been managed by the Fort Mason Foundation (now Fort Mason Center for Arts and Culture) as a place for art, humanities, recreation, ecology, and education. The year 1980 was pivotal: the Fort Mason Foundation stopped producing its own programs and began managing the site for other groups. It was now able to pay its own operating costs, developed its own board, and launched its own capital improvement campaigns. It also began to utilize real estate financing, such as federal historic rehabilitation tax credits.

Under a new redevelopment-style lease in 2005, the Fort Mason Foundation became responsible for buildings, piersheds, and site improvements across a 40,000 m² leasehold campus. NPS retained responsibility for pier substructures. NPS architect Michelle Rios, who served as owner's rep for the shed and tenant improvements, explained that NPS managed the $12 million repairs to the pier substructure as leverage for Fort Mason Center's $20 million shed rehabilitation and as a later phase $20 million of improvements for the graduate campus of the San Francisco Art Institute.

Highlighted Here Fort Mason Center represents an adaptive use of cultural resources that speaks to community and regional needs. It has been a pioneer in the national movement to create nonprofit centers, utilizing otherwise excess properties. Over the years, NPS encouraged the Fort Mason Center's maturation as a major nonprofit real estate and arts enterprise, and in this project NPS was able to leverage its core investment into further public benefit.

19.3.2.9 The Evolving Story of Muir Woods

Redwood Creek flows through the Muir Woods National Monument, also managed by GGNRA. At Muir Woods, watershed level actions have improved the creek, trail, and endangered species habitat. Rather than a restoration, project work here was more resource enhancement and represents "an evolution of thought." Muir Woods was designated a national monument in 1908 to preserve the last remaining stands of old-growth redwood. In 1972, its 2.3 km^2 were incorporated into GGNRA. It is recognized internationally for both its magnificent trees and as a founding site of the United Nations.

Early management focused on "clearing debris and keeping the creek straight." From 1983 to 1993, critical inventory work was done by NPS staffers Mia Monroe, Terri Thomas, and Nancy Hornor, already thinking on the watershed level, with additional specialists hired later as part of the NPS Natural Resources Challenge. Over time, enhancements were done in three basic stages: (1) getting boardwalk off the ground, using reclaimed redwood, and ending "cleanup" of forest floor, all of which reverses negative soil and drainage impacts; (2) under the Trails Forever program, moving the trail further from the creek and addressing accessibility; and (3) creating a functional channel that could migrate by moving riprap and replacing bridges over the floodplain. Taking a very broad approach, NPS moved water and sewer systems, removed cars from road shoulders, reconfigured parking, initiated a shuttle system from Highway 101, and worked to quiet the soundscape.

Highlighted Here Muir Woods represents "an evolution of thought that brings us up to today," according to Sharon Farrell, who began her NPS career at Muir Woods in 1991 as a plant nursery intern, before eventually leaving NPS to become Parks Conservancy Executive Vice President, Projects, Stewardship, and Science. Muir Woods and the larger Redwood Creek Watershed became a springboard for the One Tam Collaborative, with its cross-boundary, watershed-level focus at Mt. Tamalpais that involves NPS, Parks Conservancy, and multiple other agencies and organizations.

Much has been learned at Muir Woods. Natural Resources Chief Daphne Hatch underscored Muir Woods' evolutionary impact in this way. "We worked to include an alternative in the General Management Plan update that, while not selected, helped to move the preferred alternative in the direction of environmental sustainability." She also noted how rapidly the updated GMP has been implemented at Muir Woods. "It integrated so much, restoring natural function and addressing visitor experience."

19.4 Conclusions: What Made It Work?

19.4.1 Ingredients for Park-Building

Project professionals interviewed for this chapter spoke to the value of organizational culture and their own growing project experience. They also identified these more specific ingredients.

An approach to planning that included:

- Huge aspirations and also imposed deadlines
- Federal funding responsibility within the public-private partnership
- Early connection of vision and project to business framework
- Completed and approved plans, developed with the community
- Bringing focus to an entire area holistically
- Concurrently addressing visitor experience with natural and cultural resource needs

Project management systems that included:

- Clear standards and guidelines (examples: the park's "project management handbook;" the park's Projects Division as the "standard bearer")
- Park-level project review processes that could stand up to challenges; strong internal environmental review
- Having complete information on environment, building condition, and economics
- Team empowerment with a clear decision structure
- Communication protocols that facilitated the "huge amount of coordination needed among moving parts"
- Inclusive teams that brought together partners, consultants, and park staff
- Developed notion of a core team driving the project, an extended team providing support/input, and strategic teams to address difficult issues as needed and report back to the core team

Community engagement that involved:

- Openness to study and to learning from the best outside expertise to be found
- Early and continuing public engagement, recognizing its great value to maintaining the public's long-term interest and support
- Co-creating the vision: attention to community relations as integral to the team and the project
- Building support for the project within the NPS organization itself, including at its Washington and regional levels

19.4.2 Observations and Reflections

The following are some concluding thoughts on how the GGNRA organization changed and grew in response to the challenges it faced and how, in this changing environment, women found a place.

Transformed Organization Early in its history, the National Park Service had modeled itself after the US Army. In the 1970s and 1980s, it was still a very traditional organization. Even today, superintendents are held in very high regard. However, where there was complex project work, the traditional park sense of hierarchy had to transform a bit to embrace project teams and structures. This created an environment where there were new opportunities to participate and advance for people who did not fit the traditional ranger mold, including women, if they had the skills.

An Organization that Learned During the Presidio transition's fast-paced infrastructure repair program, GGNRA learned interdisciplinary project team skills from NPS's DSC, where women were full participants on teams. GGNRA then benefitted from DSC's downsizing, which placed engineers and architects into the field. GGNRA also gained new tools and skills from engaging in Army processes during the Presidio transition years. The Presidio Council had examined new models for public-private projects and government real estate practices. At every step, GGNRA learned from its partners. There was openness to study and to seeking the best outside expertise that could be found.

Decentralized, Democratic Systems, Yet Structure The park's project management system included: a core team/extended team model, a project management handbook setting standards, functional systems for environmental and compliance review, and an organized approach to multi-year funding (including augmentation by philanthropic dollars). GGNRA made all these elements work together. There was creativity and leadership not only by the superintendent, but at all levels. Teams were empowered. They understood boundaries, while appearing seamless to the public in the complex partnership environment. Team members reflected that "structure brings freedom to be creative," and "master the structures, and then find the flexibilities."

Very Special Park Partner The Parks Conservancy and NPS relationship was at the center of many projects; each partner bringing to the project what the other could not. Within the GGNRA's public-private partnership environment were numerous other valued partners.

Team Process While time consuming and frustrating to some, patient team process allowed real constraints to be understood earlier. There were tensions between partners and between specialists. It took patience and creativity to find solutions among competing goals. Finding solutions to difficult issues was a source of pride.

How participants described it: "There was a lot of listening, a lot of reading and rehashing, just working things out together"; "tensions got worked through around the table"; "so many specialists, but you made it through again and again"; "we were all completely committed to the process and end result."

Innovation in Everything　In the words of Civil Engineer Debbie Campbell, who worked on numerous GGNRA projects at DSC and the NPS's regional office, "No one at GGNRA has been afraid of doing something differently." She noted that whether it was new ways to engage the public in planning, non-traditional financing, new approaches to design (including design-build), or operational contracts with the Parks Conservancy, the park was willing to test new ideas.

Co-created Vision, but Clear Public Agency Role as Landlord, Regulator, and Facilitator　GGNRA hired people who could understand both their regulatory and proprietary role and could function in that environment. In public-private partnership projects, GGNRA had to *want* the project to succeed. One project manager described her "passion for the owner's rep role, where the job is to hold both sides accountable, and build an environment of trust." The analog in restoration projects: NPS wanted to be seen not just as a regulator but as a "proponent for care of the land," and asked, "where is the collective opportunity?" Also important was learning to work in a glass bowl, as many projects were high stakes and had very engaged publics. GGNRA professionals needed a high-level skill set.

"Diagnosis Involves Many More Parts"　Complex problems can't be addressed in isolation; an area approach is key. By concurrently addressing resource protection and functioning with visitor access and experience, one can achieve both. The formula: tackle projects holistically, gather information ahead, include partners and stakeholders, and keep it all connected to NPS while working confidently across boundaries.

What Is the Effect of Adding Women to Teams?　Polly Kaufman's comprehensive history of women's acceptance into the NPS organization notes "that women designers brought a sense of scale and were pragmatic" and also that it has made the NPS culture "more collaborative, inclusive, less paternalistic and more open to partnerships" [6]. GGNRA's project environment valued – *made essential* – a new style of collaborative leadership, which made additional space for women. One participant said it this way, "Women may be better at connectivity and relationship building and are able to work through the difficult parts of a relationship." Also, another reflected on how important it was to "keep pushing, but in a collaborative way." Others felt women found it easier to stretch boundaries and give up control to get things done.

Supportive Human Resources　Of enormous import, ways were found to retain women who had skills. (This was not the case in all NPS units). When staff can stay, they can continue on to the next project. GGNRA benefitted from retaining staff with experience and used available authorities to the maximum (including job sharing, longer unpaid leave, intermittent or part-time status) to retain women in the workforce. Particularly in planning, natural resource management, and on project teams, there were many women at GGNRA.

Experience of a Lifetime For many, it became the experience of a lifetime, perhaps especially for women, some of whom had come from less welcoming environments and so understood how exceptional an environment they were now part of. One contrasted it with the 1990s NPS world of "good old boys" and described being re-inspired by the park's culture of creativity and innovation. This outsized contribution of women continues at GGNRA.

Speed and Replicability Projects were completed as fast as projects in the public space can be done well. For those who participated, it was a wild ride; from the outside, it took decades. From concept to implementation there were many unspoken heroes and hidden figures. Some of those I interviewed reflected on the question, "Is it replicable?" Answers included that "all of it is transferrable" but also that "many factors were aligned." The ongoing task of realizing park vision at GGNRA continues, with one effort seeding the next.

Acknowledgments Heartfelt thanks to those who shared their experiences working on projects in the Golden Gate National Parks, including providing review comments and images – Cathie Barner, Laura Castellini, JoAnne Dunnec, Henry Espinoza, Sharon Farrell, Daphne Hatch, Nancy Hornor, Susan Hurst, Stephen Kasierski, Craig Kenkel, Marti Leicester, Carol Prince, Michelle Rios, Carolyn Shoulders, John Skibbe, Carrie Strahan, Terri Thomas, Brian Ullensvang, Joanne Wilkins, and Melanie Wollenweber. Thank you to these individuals for their additional input, editing skills, loan of documents, and other support – Linnea Bartling, Debbie Campbell, Carey Feierabend, David Shiver, and Nancy Russell (National Park Service's Oral History Center). Lastly, appreciation for the extra encouragement from Marti Leicester and Susan Hurst, my colleagues in an oral history project to capture women's voices and stories from the Golden Gate National Parks.

References

1. Brotsky, C., Eisinger, S., & Vinokar-Kaplan, D. (2019). *Shared space and the new nonprofit workplace.* New York: Oxford University Press, p. 193.
2. Cripe, R. (1991). Workforce profile of women in the National Park Service. Washington, DC: National Park Service.
3. Golden Gate National Parks Conservancy (2012). Trails forever: transforming the Golden Gate National Parks video. https://www.youtube.com/watch?v=8Khh2r-d33w. Accessed 21 June 2021.
4. Hountalas, M. (2009). *The San Francisco Cliff House.* Berkeley: Ten Speed Press, p. 187.
5. Johnson, E. (1997). The Thoreau Center for Sustainability: a model public private partnership. *Preservation Information.* Washington, DC: National Trust for Historic Preservation.
6. Kaufman, P. (2006). *National parks and the woman's voice.* Albuquerque: University of New Mexico Press, p xxiii, 175.

7. Madden, P. (2001). *Golden Gate National Parks business plan*. San Francisco: National Park Service.
8. Meyer, A., & Delehanty, R. (2006). *New guardians of the Golden Gate*. Berkeley and Los Angeles: Regents of the University of California, p. 5.
9. Rothman, H. (2004). *The new urban park*. Lawrence: University Press of Kansas, p. xi, 17, 62, 197.
10. U.S. General Accounting Office (1997) Report – National Park Service: Concerns over costs relating to Army's Transfer of the Presidio of San Francisco. Washington D.C: U.S. GAO.

Mai-Liis Bartling graduated from the University of California, Santa Barbara, in 1976, with a BA in Urban Studies, an interdisciplinary major combining coursework in the emerging Environmental Studies program with her own interest in the urban environment. Her first professional position was with Heritage Conservation and Recreation Service (HCRS), researching strategies and providing technical assistance to localities in California who were facing the challenges of Proposition 13.

Some years later, she landed at the Golden Gate National Recreation Area (GGNRA), initially setting up management and information systems. The closure of the Presidio was a turning point in her career. She served as coordinator for early Presidio Planning and Transition efforts for the National Park Service, later Program Controller of the rapidly expanding Presidio Project, and then as Deputy General Manager for the NPS's Presidio Project Office. Still later, she served as Deputy Superintendent for the GGNRA. In that role, she directly oversaw the transformation of Fort Baker, as well as the park's huge program of infrastructure, building, and landscape projects, many with associated business agreements. Bartling is gratified to have been able to work in a national park on the forefront, with wonderful colleagues, and within the diverse San Francisco Bay Area. She retired in 2009, after a 31-year federal career.

Bartling is currently involved as a volunteer with the NPS's Oral History Center in a project that is dear to her heart – to record the voices and stories of the women who have worked at the Golden Gate, especially during the park's formative years, which coincide with the years of rapid growth of women in the workplace.

Bartling is of Estonian heritage, coming to the United States as a post-World War II refugee. She is active in Estonian American organizations, including serving on the boards of the Estonian American National Council and the San Francisco Estonian Society. In her free time, she likes to hike locally with her birdwatcher husband or join him on birding trips further afield. They have three adult children.

Index

Printed in the United States
by Baker & Taylor Publisher Services